D1499210

Magnetic Oxides

Part 2

Magnetic Oxides

Part 2

Volume *I*

Edited by

D. J. Craik

Reader in Chemistry,
University of Nottingham

A Wiley–Interscience Publication

JOHN WILEY & SONS

London · New York · Sydney · Toronto

Library of Congress Cataloging in Publication Data:

Craik, Derek J.
Magnetic Oxides.

'A Wiley–Interscience publication.'
1. Metallic oxides—Magnetic properties. I. Title

QC766.M4C7   546'.32   73–14378
ISBN 0 471 18354 7 (Pt. 1)
ISBN 0 471 18355 5 (Pt. 2)
ISBN 0 471 18356 3 (Set)

Printed in Great Britain by J. W. Arrowsmith Ltd.,
Winterstoke Road, Bristol

Preface

Alongside the metals, principally iron, cobalt and nickel and their alloys, oxides constitute one of the two main groups of magnetic materials. Magnetic materials are generally taken to be those which exhibit a magnetic order leading to a spontaneous magnetization at room temperature, i.e. are ferromagnetic or ferrimagnetic, though there is equal basic interest in antiferromagnetics and in materials which are only magnetically ordered at low temperatures. Above certain critical temperatures, Curie points or Néel points, the ordered materials behave as paramagnetics though not generally conforming to the simple Curie law.

With notable exceptions the magnetic metals are ferromagnetic while the magnetic oxides are ferrimagnetic and thus tend to have the weaker magnetization. The overriding contrast, however, is the relatively insulating nature of the oxides. Although a few show metallic conductivity and there is a greater difference of conductivities between common magnetic oxides than there is between the more conducting oxides and the metals, it is certainly the case that the oxide materials can have very low conductivities. The technical development of the metals for high frequency applications may be regarded as a continuous struggle against eddy current losses, involving the forming of increasingly thin laminations and insulated powers. The great significance of the development of the oxide materials in the Philips laboratories in the 1940s is thus apparent. In particular, for microwave frequencies, materials became available with very low conductivities, low dielectric loss tangents ($\sim 10^{-4}$) and low magnetic losses away from resonance which thus permitted the penetration of such radiation with negligible attenuation.

The principal classes of oxide materials are the cubic spinel ferrites developed by Philips as the 'Ferroxcubes' (mainly manganese-zinc and nickel-zinc ferrites), the hexagonal 'Ferroxdures' (barium or strontium ferrites closely related to magnetoplumbite) and 'Ferroxplanas', also developed by Phillips as permanent magnet and high frequency materials and the ferrimagnetic garnets having the cubic structure of the naturally occurring mineral or gemstone. (There is a little confusion in nomenclature: while 'ferrite' may be taken to embrace all the magnetic oxides it seems more commonly to be used as an

763

vi

abbreviation for spinel ferrite; and 'garnet' on its own may in context be taken to refer to the ferrimagnetics containing iron and rare earth or yttrium ions). While silicon-iron maintains its place as the heavy duty low-frequency material there is some overlap between the use of thin nickel–iron sheet and the high-permeability (manganese–zinc) ferrites at intermediate frequencies. At high frequencies, as in the communications field, the advantages of the ferrites become more pronounced and at the microwave level the choice ceases to exist, the garnets playing a vital role which is shared by some ferrite compositions and hexagonal oxides.

From a more general point of view, oxides are of wide natural occurrence, illustrate a great range of magnetic structures and behaviour and have un-doubtedly been more intensively studied than any other class of ionic compound. An underlying feature of both the technological and basic interest in oxides is the relatively strong exchange which occurs in these particular compounds.

At a superficial level certainly, a general description of the oxides is more straightforward than that for the metals. The mere observation that they are poor conductors of electricity suggests the relevance of a localized electron model, as opposed to the free-electron or collective-electron models appropriate to the metals, and this simple conception is justified to a considerable extent by the observed magnitudes of the ionic moments. This governs the organization of the book. Following the obviously important topic of the growth of crystals the first step is to consider how the properties of the cations are affected by their incorporations into an oxide crystal lattice (Chapter 3). This would apply to both dilute and concentrated systems. For concentrated systems in particular the next priority is clearly the exchange interactions which are the basis of magnetic order (Chapter 4) and which may give rise to a spontaneous magnetization M_s. The interactions of M_s, or of the spins, with the lattice which give rise to anisotropic behaviour and to magnetostriction are then covered in Chapters 5 and 6 and the optical and magneto-optical properties, currently of great technological as well as basic interest are dealt with in Chapter 7. Chapter 8, on electrical transport properties, completes the account of the more important basic or intrinsic properties.

The formation of domain structure is governed by both the intrinsic properties and the macroscopic structural features, and the domain structures in turn influence the observed behaviour of the materials. Thus Chapter 9 should form a link between the intrinsic and the technological properties. Microwave behaviour and properties clearly call for an extensive account (Chapter 10) as do the applications of oxides in magnetic recording (Chapter 11) since this constitutes the greatest commercial outlet for the materials and also provides for the description of some of the most remarkably particular preparative methods. The intensive study of oxides within the topic of geomagnetism (Chapter 12) represents a different type of application, workers in this field having elucidated many of the basic behavioural problems of oxides and used the results of the studies to help to explain the formation of our environment. Finally, although it was considered impossible to dwell on many particular applications, the central role of oxide materials in the ingenious and topically interesting bubble domain devices appeared to call for the special account in Chapter 13.

The introductory Chapter 1 is intended to assist those not familiar with the simpler aspects of the subject by providing a superficial survey and also, in its extensive final section, to give some background to the more general technological properties of oxide materials through a discussion of the interaction of the intrinsic properties and the microstructure. This is both one of the newest and oldest aspects of the study of magnetic oxides. Since the beginning of the technology a great amount of purely empirical effort has been devoted to achieving particular properties by the control of chemical composition and processing techniques, implicitly of intrinsic magnetic properties and microstructure. Much of the relevant information remained in the form of the recipes and routines of the manufacturers. In some recent years an increasing amount of systematic study has been devoted, by workers such as M. Paulus at CNRS, Bellevue, and A. L. Stuijts and colleagues at Philips, to the truly scientific understanding of the structure and behaviour of polycrystalline specimens of technological significance.

Sincere thanks are due to all those who gave permission for their work to be reproduced in the form of figures and tables, usually replying to requests with encouraging cordiality. Specific acknowledgement to each is given in the references accompanying the captions. We also thank the publishers of the journals in which the material originally appeared for their kind permission for its reproduction, i.e. Academic Press, New York; Academie des Sciences, Institut de France; American Ceramic Society; American Institute of Physics; Bell Laboratories; Blackwell Scientific Publications Limited; British Ceramic Society; Gauthier Villars, Paris; Institution of Electrical Engineers; Institute of Electrical and Electronics Engineers Inc.; Institute of Physics, London; Macmillan (Journals) Limited; Masson & Cie, Paris; Microfilm International Marketing Corpn., New York; North Holland Publishing Company, Amsterdam; Pergamon Press Limited; N. V. Philips, Eindhoven; Physical Society of Japan; Plenum Press, New York; Société Chimique de France; Springer-Verlag, Berlin–Heidelberg–New York; Trans-Tech Inc.; University of Tokyo Press; University Park Press.

The editor wishes to thank C. A. Bates for checking his own chapter and P. V. Cooper for checking other parts of the manuscripts most carefully, and M. Paulus for providing invaluable notes on microstructure.

More general acknowledgements are felt to be due to all those who have contributed to the study of magnetic oxides, particularly in the pioneering stage. Most immediately recalled are the celebrated explanation of ferrimagnetism by Néel and the early work of J. L. Snoek and his successors at the Philips Laboratories. Classical experimental results by C. Guillaud remain as vital demonstrations of ferrimagnetic principles and it is with special feeling that one recalls the demonstration of the compensation phenomena by E. W. Gorter.

Contributors

J. W. ALLEN   *Lincoln Laboratory, Massachusetts Institute of Technology, Cambridge, U.S.A.*

G. BATE   *International Business Machines Corporation, Boulder, U.S.A.*

C. A. BATES   *Department of Physics, University of Nottingham, England.*

R. R. BIRSS   *Department of Pure and Applied Physics, University of Salford, England.*

A. H. BOBECK   *Bell Laboratories, Murray Hill, U.S.A.*

D. J. CRAIK   *Department of Chemistry, University of Nottingham, England.*

K. M. CREER   *Department of Geophysics and Planetary Physics, University of Newcastle-upon-Tyne, England.*

I. G. HEDLEY   *Department of Geophysics and Planetary Physics, University of Newcastle-upon-Tyne, England.*

E. D. ISAAC   *Department of Pure and Applied Physics, University of Salford, England.*

S. KRUPIČKA   *Institute of Solid State Physics, Czechoslovak Academy of Sciences, Praha, Czechoslovakia.*

P. M. LEVY — Department of Physics, New York University, U.S.A.

H. MAKRAM — Centre National de la Recherche Scientifique, Laboratoires de Bellevue, Meudon, France.

W. O'REILLY — Department of Geophysics and Planetary Physics, University of Newcastle-upon-Tyne, England.

R. PARKER — Portsmouth Polytechnic, England

C. E. PATTON — Department of Physics, Colorado State University, U.S.A.

M. ROSENBERG — Institute of Physics, Academy of the Socialist Republic of Romania, Romania.

K. W. H. STEVENS — Department of Physics, University of Nottingham, England.

C. TĂNĂSOIU — Institute of Physics, Academy of the Socialist Republic of Romania, Romania.

M. VICHR — Centre National de la Recherche Scientifique, Laboratoires de Bellevue, Meudon, France.

K. ZÁVĚTA — Institute of Solid State Physics, Czechoslovak Academy of Sciences, Praha, Czechoslovakia.

Contents

2. Crystal Structures and Fabrication of Crystals

H. MAKRAM AND M. VICHR

3. Crystal Field Theory

K. W. H. STEVENS AND C. A. BATES

4. Exchange

P. M. LEVY

7. Optical Properties of Magnetic Oxides

J. W. ALLEN

9. Magnetic Domains

M. ROSENBERG AND C. TĂNĂSOIU

10. Microwave Resonance and Relaxation

C. E. PATTON

11. Magnetic Oxides in Geomagnetism

K. M. CREER, I. G. HEDLEY AND W. O'REILLY

12. Oxides for Magnetic Recording

G. BATE

13. Magnetic Bubbles

A. H. BOBECK

xxi

9 *Magnetic domains*

M. ROSENBERG and C. TĂNĂSOIU

9.1 INTRODUCTION

There are many reasons why an investigation of domain structure is of interest. In many cases it is possible to estimate the wall energy density γ_w, a material constant which characterizes such magnetic properties as permeability, coercive force and critical dimensions of single domain particles. Also it is possible to estimate the exchange constant which is fundamental for characterizing the interactions responsible for magnetic ordering.

The behaviour of magnetic materials in constant fields is a direct consequence of the domain structure modification under the influence of external magnetic fields while the behaviour in alternating fields is basically a problem of Bloch wall dynamics.

A special interest in domain structure investigation arose in the last few years from the possibility of using bubble domains as memory and logic devices or for the magnetic modulation of light.

The main difficulty in domain structure investigation is connected with the necessity to obtain sound single crystals where the observations are more easily interpreted. But at the same time the results obtained on single crystals cannot be entirely applied to polycrystals, which is by far the most frequent state of oxidic materials for applications. The difficulties in observing domain structure in polycrystalline materials partly explains the small number of works devoted to this topic.

The present chapter contains a survey of some fundamental, and the most recent, work done in the field of the theory of the domain structure, the most usual methods of domain observation as well as the main results obtained in the investigation of the domain structure of magnetic oxides.

For a detailed presentation of the theory and methods of observation the reader may have recourse to some excellent works published in the last fifteen years (Kittel and Galt, 1956; Craik and Tebble, 1961; Dillon, 1963; Craik and Tebble, 1965; Carey and Isaac, 1966).

9.2 SURVEY OF THE DOMAIN THEORY

9.2.1 Energies involved

The theory of domain structure must equally apply to all ferromagnetic and ferrimagnetic materials regardless of the type of chemical bond which characterizes them. The domain structure in a perfect crystal of a certain shape and dimensions depends only on the values of material constants such as saturation magnetization M_s, anisotropy constants K, magnetostrictive constants λ and exchange constant A.

The theory is a phenomenological one. It is based upon the assumption that either magnetic dipolar and spin–orbital interactions or magnetic quadrupolar and higher moment interactions cause a small perturbing effect upon the parallel or antiparallel orientation of the spins throughout the whole crystal owing to the main exchange interaction. It therefore seems legitimate to approximate the direction angles of the spins with continuous functions of position and to define a vector magnetization M_s whose direction angles vary continuously with position. Thus $M_s = M_s v(r)$ is a function of the position r, where v, the unit vector along M_s, is a continuous function of the position r. The domain theory deals with the situation in which changes through appreciable angles may occur on a scale that is small in comparison with the regions in which M_s is constant, which may be considered as the magnetic domains of the given structure.

The expressions of the energy densities corresponding to the main interactions occurring in the calculations of the domain structure parameters are:

(a) *The excess exchange energy density* w_E, considering only nearest-neighbour interactions between atoms with identical spins S is

$$w_E = A(\nabla v)^2 \tag{9.1}$$

where A is the exchange constant, described in Section 1.4.

In most of the ferrimagnetic oxides A depends on several exchange integrals. For barium ferrite, for instance, van Loef and van Groenou (1964) took into account the five kinds of magnetic sublattices a, b, c, d, e, in which the magnetic lattice of this material may be divided and gave

$$A = 14J_{ac}S_aS_c + 8J_{bc}S_bS_c + 12J_{ad}A_aS_d + 8J_{de}S_dS_e$$

where J_{ij} are the exchange integrals for the interaction between the sublattices i and j, and S_i the spin at the sites of the i sublattice.

(b) *The anisotropy energy density* w_A, which for cubic crystals is

$$w_A = K_1(v_x^2v_y^2 + v_y^2v_z^2 + v_x^2v_z^2) + K_2v_x^2v_y^2v_z^2 \tag{9.2}$$

where the cubic axes are chosen as coordinate axes.

For hexagonal crystals

$$w_A = K_1 \sin^2\theta + K_2 \sin^4\theta + K_3' \sin^6\theta + K_3 \sin^6\theta \cos 6(\varphi - \psi) \tag{9.3}$$

where the hexagonal axis is chosen as the polar Oz axis, θ and φ being the polar coordinates which determine the position of the unit vector $\mathbf{v}$ and ψ a reference angle in the hexagonal plane (cf. Chapter 5).

(c) *The magnetoelastic energy density* w_{ME} caused by a tensile stress σ_{ij} is for cubic crystals

$$w_{ME} = -\tfrac{3}{2}\lambda_{100}(\sigma_{11}v_x^2 + \sigma_{22}v_y^2 + \sigma_{33}v_z^2)$$
$$- 3\lambda_{111}(\sigma_{12}v_xv_y + \sigma_{23}v_yv_z + \sigma_{13}v_xv_z) \tag{9.4}$$

where λ_{100} and λ_{111} are the magnetostrictive constants. For a uniform stress $\sigma_{ij} = \sigma g_i g_j$, where g_i are the direction cosines of the stress components relative to the crystal axes, one gets

$$w_{ME} = -\tfrac{3}{2}[\lambda_{100}(g_1^2v_x^2 + g_2^2v_y^2 + g_3^2v_z^2)$$
$$+ 2\lambda_{111}(g_1g_2v_xv_y + g_2g_3v_yv_z + g_1g_3v_xv_z)] \tag{9.5}$$

In the case of an isotropic magnetostriction $\lambda_{100} = \lambda_{111} = \lambda$ and

$$w_{ME} = \tfrac{3}{2}\lambda\sigma \sin^2 \theta \tag{9.6}$$

where

$$\cos \theta = g_1v_x + g_2v_y + g_3v_z.$$

(d) *The magnetostatic energy density* w_M which has the expression

$$w_M = -\tfrac{1}{2}\mathbf{H_D} \cdot \mathbf{M_s} \tag{9.7}$$

where $\mathbf{H_D}$, the demagnetizing field is

$$\mathbf{H_D} = \nabla\left[\int_v (\operatorname{div} \mathbf{M_s})r^{-1}\, dv - \int_s (M_sv_n)r^{-1}\, ds \right] \tag{9.8}$$

The first integral is taken over the whole volume of the crystal and the second one over its surfaces. M_sv_n is the normal component of the magnetization at the crystal surface.

By introducing the volume density of magnetic poles and the surface density σ as $\rho_M = -\operatorname{div} \mathbf{M_s}$ and $\sigma_M = M_sv_n$ respectively and the scalar magnetic potential of the magnetic poles, φ_M, the magnetostatic energy density may be written as

$$w_M = \tfrac{1}{2}\rho_M\varphi_M \tag{9.9}$$

for the volume magnetic poles, and as

$$w_M = \tfrac{1}{2}\sigma_M\varphi_M' \tag{9.10}$$

for the surface magnetic poles, where φ_M satisfies the Poisson equation inside v,

$$\nabla^2\varphi_M = 4\pi M_s\nabla\mathbf{v} \tag{9.11}$$

and φ_M' the Laplace equation outside v,

$$\nabla^2\varphi_M' = 0 \tag{9.12}$$

with the boundary conditions on S

$$\varphi_M - \varphi_M' = 0 \tag{9.13}$$

and

$$(\partial\varphi_M/\partial n) - (\partial\varphi_M'/\partial n) = 4\pi M_s v_n \tag{9.14}$$

(e) *The magnetic energy density w_H in the external field*, given by

$$w_H = -\mathbf{H}.\mathbf{M}_s \tag{9.15}$$

The domain structure with all its characteristics should result from the function $\mathbf{v}(\mathbf{r})$ throughout the crystal in thermodynamic equilibrium. $\mathbf{v}(\mathbf{r})$ may be obtained by solving the equations for the equilibrium state with the condition that the free energy variation of the crystal equals zero for arbitrary variations $\delta\mathbf{v}$ of $\mathbf{v}$. This variation is $\delta\mathbf{v} = \delta\boldsymbol{\theta} \times \mathbf{v}$ where $\delta\boldsymbol{\theta}$ is a small vector rotation. Thus, ensuring the satisfaction of the constraint $\mathbf{v}^2 = 1$, one obtains for equilibrium the torque equation (Brown, 1957, 1963)

$$\mathbf{v} \times [A.\nabla^2\mathbf{v} - (\delta w_A/\delta\mathbf{v}) + M_s(\mathbf{H}_0 + \mathbf{H})] \tag{9.16}$$

with the above-mentioned boundary conditions for the scalar magnetic potential φ_M. But, for a crystal with all three dimensions finite, the difficulty of a rigorous calculation of the domain structure based on Equation (9.16) has so far proved insuperable. For this reason domain theory is developed in successive approximations, based upon the postulation of the existence of the domains and of the regions between them, through which the magnetization gradually changes from the orientation in a domain to that in the neighbouring one.

These transition regions, first suggested by Bloch (1932), are the domain, or Bloch, walls when the separated domains are magnetized at 180° to each other. The problem was separated in two parts, according to Landau and Lifshitz's well known paper (1935). Firstly, the excess energy of a domain wall γ_w and its width δ were calculated, assuming $\delta \ll l$, where l is the transverse dimension of a domain. In the second stage, the geometric parameters of the domain structures, based upon plausible models, were calculated.

9.2.2 Domain walls

In bulk crystals the most usual type of wall, in the absence of an external magnetic field, may be approximated by the Landau and Lifshitz (L–L) model (1935) on the assumption that the equilibrium spin distribution within the wall is such that the associated magnetostatic energy is zero. Such a situation can be achieved if on passing across the wall the component of $\mathbf{M}_s$ along the normal to the wall surface is constant or zero at each point and if the perturbing effect due to the finite dimensions of the wall is neglected. In this case it is possible to solve Equation (9.16).

9.2.2.1 Uniaxial crystals

9.2.2.1.1 Crystals with preferred axis. A rigorous treatment is possible for a one-dimensional domain structure with Bloch walls in a thick boundless plate. Brown's equation has been solved by Shirobokov (1939, 1945) and Forlani and Minnaja (1969) under the following assumptions:

(a) $\mathbf{v}$ depends upon a single coordinate, say x, and lies in the (yz) plane.

(b) w_A can be written as $K(1 - v_z^2)$, where $K > 0$, i.e. the easy magnetization axis is in the z direction.

Introducing a non-dimensional coordinate $X = x(K/A)^{\frac{1}{2}}$ the first integral of Brown's equation becomes

$$dv_z/dX = \pm[(1 - v_z^2)(b - v_z^2)]^{\frac{1}{2}} \tag{9.17}$$

where b is an integration constant. The solution is

$$X = b^{-\frac{1}{2}}F(v_z, b^{-\frac{1}{2}}) \tag{9.18}$$

where F is the incomplete elliptic integral of the first kind. It is seen from Equation (9.18) that v_z is a periodic function of the coordinate, and the period L is given by

$$L = 4b^{-\frac{1}{2}}F(\pi/2, b^{-\frac{1}{2}}) \tag{9.19}$$

In Figure 9.1 the solution is plotted for some significant cases, i.e. in zero field and for several values of the reduced field $h = HM_s/2K$. The mean energy density (per square centimetre of the surface of the Bloch wall and per centimetre in the direction normal to the wall) can be computed, giving

$$E = K[1 - b + 2E(\pi/2, b^{-\frac{1}{2}})/F(\pi/2, b^{-\frac{1}{2}})] \tag{9.20}$$

where $E(\pi/2, b^{-\frac{1}{2}})$ is the complete elliptic integral of the second kind.

The last expression can be easily computed in the limit $L \to \infty$, which corresponds to the L–L approximation (1935) of the density of the Bloch wall energy γ_w, assuming that $\delta/l \ll 1$.

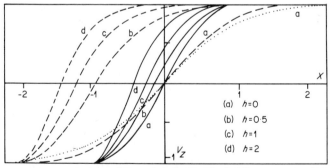

Figure 9.1 Solutions of Equation (9.18). The full lines represent M versus X for $L = 4$ and $h = 0, 0.5, 1$ and 2. The dashed lines represent the corresponding cases for $L = 8$; the dotted line represents the 'classical behaviour' (Forlani and Minnaja, 1969).

For $L \to \infty$ (9.20) reduces to

$$E \simeq 8K/L \qquad (9.21)$$

Taking into account that for every centimetre length of the normal to the wall there are l^{-1} domains and that $L(K/A)^{\frac{1}{2}} = 2l$ the density of the Bloch wall energy is

$$\gamma_w = E . l = 4(KA)^{\frac{1}{2}} \qquad (9.22)$$

In the limits of this approximation one obtains the wall width δ defined as the distance over which $\mathbf{v}$ varies through 180°. If $\mathbf{M}_s$ is parallel to (Oy) and the wall to (yOz) plane, then

$$\delta = (A/K)^{\frac{1}{2}} \qquad (9.23)$$

and

$$\gamma_w = 4K\delta \qquad (9.24)$$

From (9.24) it can be seen that γ_w may be estimated by direct measurement of δ. So far, owing to experimental difficulties, it has not been possible to get reliable values for γ_w in such a way.

The variation in spin direction within a 180° domain wall can have the sense of both right- or left-handed helix. Thus a Bloch wall can be divided into parts differing by the sense of rotation of the spins. Bean (1962) introduced the term 'Bloch line' to designate the line of contact between two sections of wall of opposite sense. The Bloch line has both anisotropy and exchange energy densities much higher than the rest of the wall.

Taking into account the second term in (9.3), Kaczer (1964) and Träuble et al. (1965) derived for the density of the Bloch wall energy the expression

$$\gamma_w = 2(AK)^{\frac{1}{2}}\{1 + [(K_1 + K_2)/K_1 K_2] \text{ arc } \sin[K_2/(K_1 + K_2)]^{\frac{1}{2}}\} \qquad (9.25)$$

9.2.2.1.2 Crystals with preferred plane. In this case the domain walls must lie in the hexagonal plane and because of the hexagonal symmetry of the anisotropy in this plane, one expects three kinds of wall: 180°, 120° and 60° walls. Taking into account the fourth term in (9.3) Kaczer (1962) obtained

$$\gamma_{180°} = 4(AK_3)^{\frac{1}{2}} \qquad (9.26)$$

$$\gamma_{120°} = (8/3)(AK_3)^{\frac{1}{2}} \qquad (9.27)$$

$$\gamma_{60°} = (4/3)(AK_3)^{\frac{1}{2}} \qquad (9.28)$$

9.2.2.2 Cubic crystals

In crystals with cubic symmetry the possible walls are: 180° and 90° if the easy directions coincide with the [100] crystallographic directions and 109° 28′ and 70° 32′ when the easy axis lies along the [111] crystallographic directions.

These wall energies and widths were calculated by Lilley (1950) in the L–L approximation. Lilley used the older results by Landau–Lifshitz (1935),

Shirobokov (1939, 1945), Néel (1944a, 1944b) and Lifshitz (1944). In the case of 90°, 109° and 71° walls the condition div $M_s = 0$ across the wall can be fulfilled under the assumption that M_{sn} (the normal component of $\mathbf{M}_s$ on the wall plane) should have the same value not only on each side of the wall but also within the wall.

It is convenient to specify $\mathbf{v}$ by the polar coordinates θ and φ, where θ is the angle between $\mathbf{M}_s$ and the normal to the wall, and φ the angle between the projection of $\mathbf{M}_s$ on the wall plane and some selected zero direction in this plane. The initial and final directions of $\mathbf{M}_s$ in adjacent domains may be denoted by θ, φ_1 and θ, φ_2, constancy of M_{sn} being equivalent to the constancy of θ.

In the L–L approximation, if M_{sn} is constant across the wall, then in each point of the wall the equilibrium exchange energy density w_E equals the aniso-tropy energy density w_A adjusted by subtracting appropriate constant terms $(w_A)_{min}$ to make it zero for $\mathbf{M}_s$ along (θ, φ_1) and (θ, φ_2). Thus

$$w_E = w_A - (w_A)_{min}$$

or in the chosen conditions

$$A \sin^2 \theta (d\varphi/dx) = K_1 f_A(\theta, \varphi) \tag{9.29}$$

where $K_1 f_A(\theta, \varphi) = w_A - (w_A)_{min}$ and w_A (reduced to the first term in (9.2)) is expressed in polar coordinates. The wall energy density is thus given by

$$\gamma_w = \int_{-\infty}^{+\infty} [w_E + w_A - (w_A)_{min}] \, dx \tag{9.30}$$

Substituting dx from (9.29), the integral transforms to

$$\gamma_w = 2(AK)^{\frac{1}{2}} \sin \theta \int_{\varphi_1}^{\varphi_2} [f_A(\theta, \varphi)]^{\frac{1}{2}} \, d\varphi \tag{9.31}$$

The only possible walls fulfilling the condition $M_{sn} = $ constant are those for which the normal is a $[11l]$ direction, l being the third Miller index. By choosing $[\bar{1}10]$ and $[\bar{l}l2]$ as directions of the Oy and Oz axes in the plane of the wall, the expression of the anisotropy energy density becomes

$$
\begin{aligned}
w_A = {}& [K_1/4(l^2 + 2)^2][4(2l^2 + 1) + 4(l^2 + 1)(l^2 - 4) \sin^2 \theta \\
& - (3l^2 + 4)(l^2 - 4) \sin^4 \theta \\
& - 4(l^2 - 1)\{6 + (l^2 - 4) \sin^2 \theta\} \sin^2 \theta \sin^2 \varphi \\
& + 4(l^2 + 3)(l^2 - 1) \sin^4 \theta \sin^4 \varphi \\
& - 2^{\frac{3}{2}}l \cos \theta \sin \theta \sin \varphi \{2(l^2 - 1) - (5l^2 + 4) \sin^2 \theta \\
& + 2(l^2 + 5) \sin^2 \theta \sin^2 \varphi\}]
\end{aligned}
\tag{9.32}
$$

and for $K_1 > 0$, $(w_A)_{min} = 0$, while for $K_1 < 0$, $(w_A)_{min} = K_1/3$.

Denoting by $\gamma_0 = (AK)^{\frac{1}{2}}$ a unit for the wall energy density Equation (9.31) gives

$$\gamma_w/\gamma_0 = 2 \sin \theta \int_{\varphi_1}^{\varphi_2} [f_A(\theta, \varphi)]^{\frac{1}{2}} \, d\varphi \qquad (9.33)$$

Conventionally one defines a wall width δ, expressed in units of $\delta_0 = (AK)^{\frac{1}{2}}$. The ratios γ_w/γ_0 and δ/δ_0 computed by Lilley for the main types of walls in cubic crystals are presented in Table 9.1.

For 180° domain walls the excess magnetoelastic energy of the wall must be taken into account (Lifshitz, 1944; Lilley, 1950). For crystals with $K_1 < 0$

Table 9.1 Reduced values of domain wall energies and widths in cubic crystals (from Lilley, 1950)

Type of domain wall	Directions of the normal to the wall	Easy directions initial / final	γ/γ_0	δ/δ_0
70° 32′	[001]	[$\bar{1}11$] / [$\bar{1}\bar{1}1$]	0·5443	3·8476
109° 23′	[001]	[$\bar{1}11$] / [$1\bar{1}1$]	1·0887	∞
70° 32′	[110]	[$\bar{1}1\bar{1}$] / [$\bar{1}11$]	0·4611	4·2642
109° 23′	[110]	[$\bar{1}11$] / [$1\bar{1}1$]	1·3680	3·3093
109°23′	[111]	[$\bar{1}11$] / [$1\bar{1}1$]	1·2903	3·8476
90°	[001]	[100] / [010]	1·0000	3·1416
90°	[110]	[100] / [010]	1·7274	3·9738
90°	[111]	[100] / [010]	1·1852	3·1416

the corrected values of γ_{180° are

(a)
$$\gamma_w/\gamma_0 = 3^{-\frac{3}{2}}[(8 + \tau)^{\frac{1}{2}} + \{(9 + \tau)/9\} \sin^{-1}(9 + \tau)^{-\frac{1}{2}}$$
$$+ (2^{\frac{3}{2}}\tau/9) \sinh^{-1}(8/\tau)^{\frac{1}{2}}] \tag{9.34}$$

for a wall normal to [110], and

(b)
$$\gamma_w/\gamma_0 = [(1 + \tau)/3]^{\frac{1}{2}}[(\sigma + 1)^{\frac{1}{2}} + \sigma^{-\frac{1}{2}} \sinh^{-1}(\sigma^{\frac{1}{2}})] \tag{9.35}$$

for a wall normal to [$\bar{1}\bar{1}2$].

In (9.34) and (9.35) the notations

$$\tau = (54C_{44}\lambda^2_{111})/|K_1|; \qquad \sigma = 7/(1 + \tau)$$

were used, where C_{44} is one of the elastic coefficients (C_{11}, C_{12}, C_{44}) and λ_{111} the magnetostriction coefficient.

For crystals with $K_1 > 0$ the corrected values for γ_{180° and δ_{180° are

(a)
$$\gamma_w/\gamma_0 = 2[(1 + \tau')^{\frac{1}{2}} + \tau' \sinh^{-1}(\tau'^{-\frac{1}{2}})]$$
$$\delta/\delta_0 = 2(1 + \tau')^{-\frac{1}{2}}\{\sinh^{-1}[(1 - \tau')/\tau']^{\frac{1}{2}}$$
$$+ 2(1 + \tau')^{-\frac{1}{2}} \sin^{-1}[(1 + \tau')/2]^{\frac{1}{2}}\} \tag{9.36}$$

for a wall normal to [001] and

(b)
$$\gamma_w/\gamma_0 = 2[(1 + \tau')^{\frac{1}{2}} + \{(1 + 4\tau')/3^{\frac{1}{2}} . 2\} \sinh^{-1}\{3/(1 + 4\tau')^{\frac{1}{2}}]$$
$$\delta/\delta_0 = 2(1 + \tau')^{-\frac{1}{2}}[\sinh^{-1}\{(2 - 4\tau')/(1 + 4\tau')\}^{\frac{1}{2}}$$
$$+ \{3^{\frac{1}{2}}/(1 + \tau')\} \sin^{-1}\{2(1 + \tau')/3\}] \tag{9.37}$$

for a wall normal to [110].

Here

$$\tau' = [9(C_{11} - C_{12})/2\lambda^2_{100}]/2K_1$$

9.2.3 Theoretical domain structures

The present theory explains at least in principle, the variety of domain structures observed in ordered magnetic materials. In simpler cases it gives some models for calculation which allow a comparison with the experimental data.

The method consists in calculating the free energy of various possible structures. For a given crystal the most probable domain structure is that which corresponds to the minimum free energy. Generally, two extremal cases are treated: flux closed and open structures. There are also some models for intermediate structures.

9.2.3.1 Uniaxial crystals

9.2.3.1.1 Closed flux domain configurations. It is possible to conceive a domain structure for a uniaxial single crystal of thickness c along the easy axis with limiting surfaces perpendicular to it, which will have no magnetic poles as shown in Figure 9.2 (Landau and Lifshitz, 1935). The 'flux circuit' is entirely closed within the crystal by triangular prismatic domains on the upper and lower surfaces of the crystal. The absence of poles results from the continuity of the normal component of $\mathbf{M}_s$ across the prism sides.

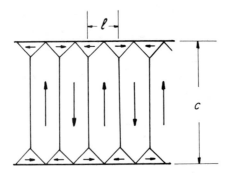

Figure 9.2 Landau–Lifshitz model with domains closing the magnetic flux inside the crystal (Landau and Lifshitz, 1935).

The total energy per unit area of the crystal surface is the sum of the anisotropy energy of the closure domains and the wall energy (the energy of the zigzag walls is neglected).

$$E = K_1 l/2 + \gamma_w c/l \qquad (9.38)$$

By minimizing E with respect to l the equilibrium values of the domain width l_0 and the energy E_0, are

$$l_0 = (2\gamma_w c/K)^{\frac{1}{2}} \qquad (9.39)$$

and

$$E_0 = (2\gamma_w Kc)^{\frac{1}{2}} \qquad (9.40)$$

If the crystal surfaces make the angles θ_1 and θ_2 with the easy axis, the domain width l_0 will be

$$l_0 = [4\gamma_w c/K(\sin\theta_1 + \sin\theta_2)]^{\frac{1}{2}} \qquad (9.41)$$

Values of γ_w can be easily obtained by measuring the domain spacing l_0 for a given c.

With increasing c, the anisotropy energy of the closure domains becomes more and more important, as does the magnetoelastic energy (which was neglected in (9.38)) and finally the simple L–L structure ceases to be stable. For c greater than a certain value c_0 the structure of Figure 9.3 becomes energetically more favourable (Lifshitz, 1944).

Denoting by ξ a geometric parameter to characterize the degree of branching shown and assuming that the angle between the walls of the spikes domains

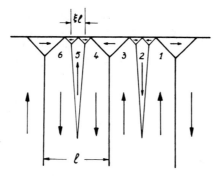

Figure 9.3 Lifshitz model with both closure and spike domains (Lifshitz, 1944).

and the easy axis is very small, Lifshitz found two conditions for the energy minimum, namely

$$(K/4M_s)(\mu l/\pi\gamma_w)^{\frac{1}{2}} = (1 - \xi)^{\frac{1}{2}}/(1 - 3\xi) \tag{9.42}$$

and

$$c/c_0 = [(1 - \xi)^2/3(1 - 3\xi)][-1 + 4\xi + 6\xi^2 + 4(1 - 3\xi)/(1 - \xi)^{\frac{1}{2}}] \tag{9.43}$$

where

$$c_0 = 27\pi^2\gamma_w M_s/\mu^2 K^2; \qquad \mu = 1 + 2\pi M_s^2/K \tag{9.44}$$

The domain width corresponding to the critical thickness c_0 is

$$l_c = (16\pi\gamma_w M_s^2)/(\mu K^2) \tag{9.45}$$

From (9.42) and (9.43) it is obvious that for the structure with branching domains the dependence $l(c)$ must differ from that given by (9.39).

9.2.3.1.2 Open flux domain configurations. The calculation of energy and geometrical parameters may be easier for simple periodical structure with magnetization lying everywhere parallel to the easy axis. The energies involved are: magnetostatic energy, due to the magnetic free poles on the surface of the crystal, and Bloch wall energy.

Kittel (1949) gave the procedure for calculating the magnetostatic contribution for periodic surface distributions of magnetic poles with a periodic surface density $\sigma(x, y)$ having the periods $2\pi L_x$ and $2\pi L_y$. The function σ may be expanded in a double Fourier series

$$\sigma(x, y) = \sum_{-\infty}^{+\infty} \sum_{-\infty}^{+\infty} C_{mn} \exp[i(mx/L_x + ny/L_y)] \tag{9.46}$$

Denoting $\xi = x/L_x$ and $\eta = y/L_y$, the Fourier coefficients become

$$C_{mn} = (4\pi^2)^{-1} \int_0^{2\pi} \int_0^{2\pi} \sigma(\xi, \eta) \exp[-i(m\xi + n\eta)] \, d\xi \, d\eta \tag{9.47}$$

The potential $\varphi(x, y)$ can also be expressed in a double Fourier series since it has the same periodicity as the magnetic charge density.

The Fourier coefficients of the potential are expressed in terms of C_{mn} by using the Laplace equation and the boundary conditions at the surfaces $z = 0$ and $z = c$.

Owing to the symmetry of the problem, the corresponding trial function is

$$\varphi(\xi, \eta) = \sum_{-\infty}^{+\infty} \sum_{-\infty}^{+\infty} \varphi_{mn} \exp[i(m\xi + n\eta)] \exp(-P_{mn}z) \tag{9.48}$$

where, from Laplace's $\nabla^2 \varphi = 0$, one obtains

$$P_{mn} = [(m/L_x)^2 + (n/L_y)^2]^{\frac{1}{2}} \tag{9.49}$$

The continuity of the induction $\mathbf{B} = \mathbf{H} + 4\pi\mathbf{M}_s$ at the surface gives

$$(-\partial\varphi/\partial z)|_{z=0} = \mp 2\pi\sigma(x, y) \tag{9.50}$$

and therefore

$$\varphi(\xi, \eta) = 2\pi \sum_{-\infty}^{+\infty} \sum_{-\infty}^{+\infty} \frac{C_{mn}}{P_{mn}} \exp[i(m\xi + n\eta)] \exp(-P_{mn}z) \tag{9.51}$$

By substituting (9.46) and (9.51) into the expression for the energy

$$E_M = (1/2) \int \int \varphi\sigma \, dx \, dy \tag{9.52}$$

where integration is performed over the surface of the basic cell in the planes $z = 0$ and $z = c$ one obtains (Ignatchenko and Zaharov, 1964)

$$E_M = 2\pi C_{00}^2 c + 2\pi \sum_{-\infty}^{+\infty} \sum_{-\infty}^{+\infty}{}' C_{mn}C_{-m-n}P_{mn}^{-1}[1 - \exp(-P_{mn}c)] \tag{9.53}$$

where the prime index stands for the absence of the term with $m = n = 0$.

For a unidimensional distribution of parallel magnetized strips of alternating polarity of period $l_1 + l_2 = 2\pi L_x$ and $P_{mn} = m/L_x$, and $L_y = \infty$, the coefficients are

$$C_0 = M_s(l_1 - l_2)/(l_1 + l_2)$$

and

$$C_m = (-2M_s/\pi m) \exp[(-im\pi l_1)/(l_1 + l_2)] \sin(m\pi l_1)$$

and since $(l_1 - l_2)/(l_1 + l_2) = M/M_s$ the corresponding magnetostatic energy is

$$E_M = 2\pi M_s^2 c(M/M_s) + 8M_s^2(l_1 + l_2)\pi^{-2} \sum_{m=1}^{\infty} m^{-3} \sin[(m\pi/2)(1 + M/M_s)]$$

$$\times [1 - \exp(-2\pi mc/(l_1 + l))] \tag{9.54}$$

This equation was derived in a somewhat different manner by Kooy and Enz (1960).

In the case of alternating strips of opposite polarity but equal width l, Equation (9.54) transforms in

$$E_M = 16M_s^2\pi^{-2}l \sum_{m\,\text{odd}}^{\infty} m^{-3}[1 - \exp(-m\pi c/l)] \tag{9.55}$$

which is the result obtained by Malek and Kambersky (1958).

In many cases of practical interest $c \gg l$ and therefore $\exp(-P_{mn}c) \ll 1$ and for the equilibrium states (zero magnetic field) the opposite polarized surfaces must be equal, so that Equation (9.53) reduces to

$$E_M = 2\pi \sum_{-\infty}^{+\infty}{}' \sum_{-\infty}^{+\infty}{}' C_{mn}C_{-m-n}P_{mn}^{-1} \tag{9.56}$$

Under such an assumption Equation (9.55) reduces to

$$E_M = 16M_s^2\pi^{-2}l \sum_{m\,\text{odd}}^{\infty} m^{-3} \tag{9.57}$$

where

$$\sum_{\substack{m=1\\m\,\text{odd}}}^{\infty} m^{-3} = 1\cdot0518\ldots$$

so that

$$E_M = 1\cdot71M_s^2l \tag{9.58}$$

which is the well known result obtained by Kittel (1949) in the first attempt to calculate the magnetostatic energies of periodical distributions.

This result permits the calculation of the equilibrium domain width l_0 and energy E_0 for a strip or Kittel-like domain structure. The total energy of the domain structure per square centimetre of the xy plane is

$$E = 1\cdot71M_s^2l + \gamma_w c/l \tag{9.59}$$

and from $\partial E/\partial l = 0$ one obtains the equilibrium values

$$l_0 = [(\gamma_w c)/(1\cdot71M_s^2)]^{\frac{1}{2}} \tag{9.60}$$

and

$$E_0 = 2(1\cdot71)^{\frac{1}{2}}M_s\gamma_w^{\frac{1}{2}}c^{\frac{1}{2}} \tag{9.61}$$

In connection with the increased interest in cylindrical (bubble) domains, several calculations of the magnetostatic energies for a periodic distribution of circles of charge density $+M_s$ in a background of $-M_s$ for two-dimensional cases have been made (Craik and McIntyre, 1969; Charap and Nemchik, 1969; Craik and Cooper, 1970; Craik et al., 1971; Druyvesteyn and Dorleyn, 1971). For a square array of circular cylinders of radius a and spacing $2b$, the energy is

$$E_M = M_s^2\left\{2\pi c\left[\frac{\pi}{2}\left(\frac{a^2}{b^2} - 1\right)^2 + \frac{2a^2}{b}F\right]\right\} \tag{9.62}$$

496

where

$$F = \sum_{-\infty}^{+\infty}{}' \sum_{-\infty}^{\infty}{}^+ \frac{J_1^2[(a\pi/b)(m^2 + n^2)^{\frac{1}{2}}]\{1 - \exp[(-c\pi/b)(m^2 + n^2)^{\frac{1}{2}}]\}}{(m^2 + n^2)^{\frac{3}{2}}}$$

and J_1 is the first order Bessel function.

For a hexagonal array of cylinders

$$E_M = M_s^2\{2\pi[(a^2\pi/b^2 3^{\frac{1}{2}}) - 1] + 2a^2[H + 2G/3\pi]/bc\} \tag{9.63}$$

where

$$H = \sum_{\substack{-\infty \\ m\,\text{even}}}^{+\infty}{}' \{J_1^2[a\pi m/b][1 - \exp(-c\pi m/b)]/3m^3$$

$$+ 3^{\frac{1}{2}}J_1^2[a\pi m/b3^{\frac{1}{2}}][1 - \exp(c\pi m/b3^{\frac{1}{2}})]\}$$

and

$$G = \sum_{\substack{-\infty \\ (m+n)\,\text{even}}}^{+\infty}{}' \sum_{-\infty}^{+\infty}{}' \frac{J_1^2[(a\pi/b)(m^2 + n^2/3)^{\frac{1}{2}}]\{1 - \exp[(-c\pi/b)(m^2 + n^2/3)^{\frac{1}{2}}]\}}{(m^2 + n^2/3)^{\frac{3}{2}}}$$

For the demagnetized case and $c/b > 1$ it has been shown (Craik and Cooper, 1970; Craik et al., 1971) that these energies reduce to

(square array)

$$E_M = 1{\cdot}506bM_s^2/c \quad \text{(in erg/cm}^3\text{)} \tag{9.64}$$

(hexagonal array)

$$E_M = 1{\cdot}345bM_s^2/c \quad \text{(in erg/cm}^3\text{)} \tag{9.65}$$

Structures of undulating walls were analysed by Goodenough (1956) using the periodic element shown in Figure 9.4 with two geometrical parameters α and β.

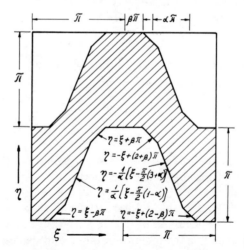

Figure 9.4 Surface pole configuration used in calculations of energy associated with rick rack domain patterns (Goodenough, 1956).

If $L_x = L_y$ and denoting $l = \pi L_x$ one obtains

$$P_{mn} = (\pi/l)(m^2 + n^2)^{\frac{1}{2}} \tag{9.66}$$

and the Fourier coefficients would be functions of α and β. The energy is then

$$E_M = lM_s^2 f(\alpha, \beta) \tag{9.67}$$

where $f(\alpha, \beta)$ depends on α and β for a given structure.

If $L_x/L_y = \gamma$ and $l = \gamma\pi L_x = L_y$ one obtains

$$P_{mn} = (\pi/l)(\gamma^2 m^2 + n^2)^{\frac{1}{2}} \tag{9.68}$$

and the function f depends on γ also.

The wall energy depends on the shape of the wall at the surface. If $S(\alpha, \beta, \gamma)$ is the ratio of the wall length to the unit surface, then the total energy can be written as

$$E = f(\alpha, \beta, \gamma)M_s^2 l + S(\alpha, \beta, \gamma)\gamma_w c/\gamma p^2 \tag{9.69}$$

The equilibrium values of the energy E_0 and of the geometrical parameters of the domain structures shown in Figure 9.5 were calculated by several authors (Kittel, 1949; Goodenough, 1956; Kaczer and Gemperle, 1961; Kozlowski and Zietek, 1966; Druyvesteyn and Dorleyn, 1971) and the main results are presented in Table 9.2.

Table 9.2 Reduced values of the equilibrium energy and geometric parameters of the domain patterns drawn in Figure 9.5

Model	$E_0/M_s\gamma_w^{\frac{1}{2}}c^{\frac{1}{2}}$	$l_0/M_s^{-1}\gamma_w^{\frac{1}{2}}c^{\frac{1}{2}}$	$a_0/M_s^{-1}\gamma_w^{\frac{1}{2}}c^{\frac{1}{2}}$	$b_0/M_s^{-1}\gamma_w^{\frac{1}{2}}c^{\frac{1}{2}}$
a	2·97	1·40		
b	2·84	1·90		
c	2·61	0·76		
d $\begin{cases}\alpha = 30° \\ \alpha = 60°\end{cases}$	3·59		2·03	0·07
	2·59		1·79	2·15
e $\frac{a}{b} = \sqrt{3}$	2·56		3·91	2·26

For the computation of E_0, approximations in the Fourier expanded series of the magnetostatic energies were used. The differences between the values of the energies of the structures C, D ($\alpha = 60°$) and E are insignificant and therefore on the basis of the above results no unique conclusion about the most favourable structure could be drawn. However, taking into account the influence of magnetic fields and using more degrees of freedom for the investigated configuration (Druyvesteyn and Dorleyn, 1971) the strip domain structure seems to be energetically the most favourable.

So far, we have assumed that the domain magnetizations are fixed in the direction of easy magnetization. Actually as shown by Lifshitz (1944), Néel (1944c), Shockley (1948) and Williams *et al.* (1949) the demagnetizing field

498

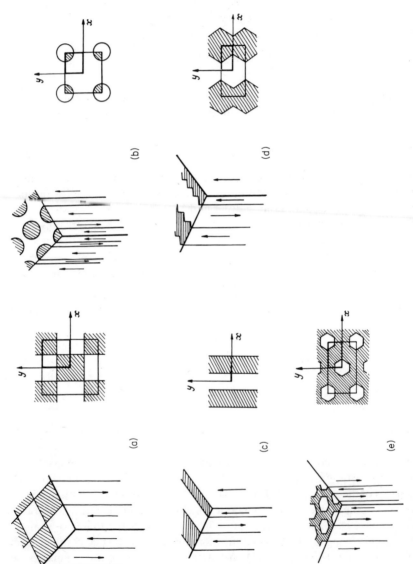

Figure 9.5 Various possible domain structures in uniaxial crystals: (a) check board patterns; (b) cylindrical domains; (c) simple strip domains; (d) simple rick rack patterns; (e) honeycomb domain structure.

exerts a torque on the domain magnetization rotating it more or less out of the easy direction, thus changing the energy of the free poles. The small deviations of the magnetization can be characterized by the μ^* effect, i.e. by introducing a permeability $\mu = 1 + 2\pi M_s^2/K$. The μ^* effect is important in materials with lower values of K and for the cases where the easy axis is not normal to the surface but parallel to the demagnetizing field. The magnetostatic energy density is multiplied by $2/(1 + \mu^*)$ at the surface of the crystal and by $1/\mu^*$ when poles are on both sides of the wall.

Experimental observations show that strip domain structures are not unique; structures as shown in Figure 9.6 are at least as favourable. Goodenough (1956)

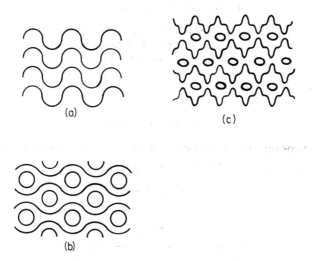

Figure 9.6 Possible domain patterns on the basal plane of uniaxial crystals: (a) simple sinusoidal patterns; (b) sinusoidal patterns with circular spike domains; (c) complex sinusoidal patterns with ellipsoidal spike domains.

explained this situation by assuming that there are deviations of the Bloch walls from the easy axis. In such a case an extra magnetostatic energy must be considered. Favourable structures may occur if the walls undulate at the surface and the amplitude of the undulation decreases with the depth of penetration (Figure 9.7). Possible superficial configurations of this kind are presented in Figure 9.6. They are the more favourable as c increases because the spike domains then have a length/diameter ratio sufficiently large to give a small demagnetizing factor and consequently a low magnetostatic energy. Up to the present there is no satisfactory theoretical treatment of these structures.

Kaczer (1964) adapted Lifshitz's (1944) calculations to the open structures in Figure 9.8 and found

$$c_0 = 16\pi^2\gamma_w/1{\cdot}7^3 M_s^2\mu^2 \tag{9.70}$$

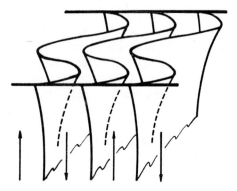

Figure 9.7 Schematic illustration of the decrease of amplitude of undulations with the depth of penetration into the crystal.

and

$$l_c = 4\pi\gamma_w/1 \cdot 7^2 M_s^2 \mu \tag{9.71}$$

as critical dimensions for which the Kittel-like structure is no longer favourable. For such a structure the dependence $l(c)$ is given by

$$l_0 = [(3/8M_s)(\gamma_w\mu/\pi)^{\frac{1}{2}}]^{\frac{2}{3}}c^{\frac{2}{3}} \tag{9.72}$$

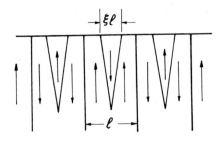

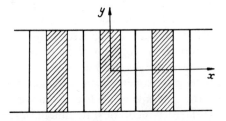

Figure 9.8 Lifshitz's modified model adapted by Kaczer for calculating the energy of complex patterns in magneto-plumbite.

Szymczak (1966) succeeded in calculating the dependence of l and c on the geometric parameters of a simple domain structure with wavy walls. Assuming the wall shape shown in Figure 9.9, the following expressions for the critical

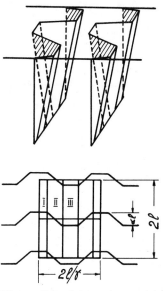

Figure 9.9 Schematic model of a wave domain structure (Sczymczak, 1966).

values of c_0, l_c and γ_c at which the transition from a Kittel-like structure to a wavy one occurs were obtained

$$c_0 = 1 \cdot 7 M_s^2 l_c / \gamma_w; \qquad l_c = 3 \cdot 94 \gamma_w / \mu M_s^2; \qquad \gamma_c = 1 \cdot 25 \qquad (9.73)$$

By using the experimental $l(c)$ dependence for magnetoplumbite it is possible to calculate the dependence $l(c)$ for any material with wavy structure from the relation

$$l = 0 \cdot 46 \mu^{0 \cdot 2} \gamma_w^{0 \cdot 4} M_s^{0 \cdot 3} c^{0 \cdot 6} \text{ [cm]} \qquad (9.74)$$

The wavy structure model gives also the dependence of the parameters α and γ on l as shown in Figure 9.10 for various values of the parameter $R = \gamma_w / \mu M_s^2$.

Using Equation (9.74) it is possible, in principle, to determine wall density energy for structures with undulating walls. The value of γ_w can also be estimated simply by the dependence of α and γ on l without the necessity of knowing the thickness c.

9.2.3.1.3 Partially closed structures. An intermediate structure between L–L and Kittel, as shown in Figure 9.11 could also exist. Introducing the parameter

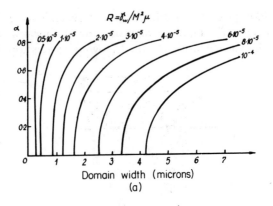

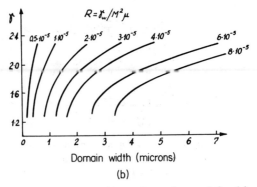

Figure 9.10 Dependence of α and γ, as defined in Figure 9.9, on domain width for various values of R (Sczymczak, 1966).

$\alpha = d/l$ two external structures are obtained for $\alpha = 1$ and 0. The total energy for such a structure has been calculated by Szymczak (1966) as

$$E = Kl\alpha^2/2 + \gamma_w c/l + 16M_s^2 lf(\alpha)/\pi^2 \tag{9.75}$$

where

$$f(\alpha) = \sum_{n=0}^{\infty} \left[\cos^2(2n+1)\frac{\pi\alpha}{2} \right] \Big/ [(2n+1)^3] \tag{9.76}$$

The equilibrium conditions

$$(\partial E/\partial \alpha) = (\partial E/\partial l) = 0 \tag{9.77}$$

(taking into account the μ^* effect) give

$$l_0 = \{[\gamma_w c]/[K\alpha^2/2 + 32M_s^2 f(\alpha)/\pi(1 + \mu^{\frac{1}{2}})]\}^{\frac{1}{2}} \tag{9.78}$$

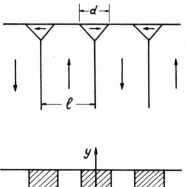

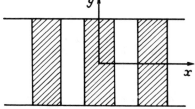

Figure 9.11 Simple model for partially closed domain configuration.

and

$$E_0 = 2\{\gamma_w[K\alpha^2/2 + 32M_s^2 f(\alpha)/\pi(1 + \mu^{\frac{1}{2}})]\}^{\frac{1}{2}} \cdot c^{\frac{1}{2}} \tag{9.79}$$

where α can be determined from equation

$$K\alpha + 32M_s^2 f'(\alpha)/\pi^2(1 + \mu^{\frac{1}{2}}) = 0 \tag{9.80}$$

From (9.59), Equation (9.62) becomes

$$\pi/8(\mu^{\frac{1}{2}} - 1) = \sum_{n=0}^{\infty} [\sin(2n + 1)\pi\alpha]/[(2n + 1)^2\pi\alpha] \tag{9.81}$$

and this allows an analysis of the stability conditions for L–L and K models with respect to the intermediate structure. Thus the K model is given by $\alpha \to 0$, i.e.

$$\pi/8(\mu^{\frac{1}{2}} - 1) = \sum_{n=0}^{\infty} (2n + 1)^{-1} = \infty$$

which gives $\mu = 1$ or $M_s^2/K = 0$, i.e. M_s finite and $K \to \infty$.

L–L model is given by $\alpha = 1$ which finally reduces to $K/M_s^2 = 0$, i.e. $K = 0$.

In this way L–L and K models have asymptotic values as extremal cases for $K \ll M_s^2$ and $K \gg M_s^2$, the stable structures being intermediate.

9.2.3.1.4 Crystals with preferred plane. Two models for infinite cylinders of radius a (Figure 9.12), were analysed by Kaczer (1962). In case (a) the flux is only partially closed inside the cylinder with div $\mathbf{M}_s \neq 0$ within the walls.

504

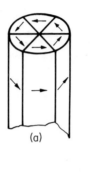

(a)

(b)

Figure 9.12 Possible domain configurations for non-vanishing hexagonal anisotropy (Kaczer, 1962).

In case (b) there is a non-uniform helical magnetization and the cylinder is divided into disc-shaped domains in which the magnetization rotates by discrete angles $\varphi = 2\pi/i$ ($i = 2, 3, 6$). Kaczer has shown that for large enough values of the radius a, the structure (b) is energetically more favourable, with $i = 3$. The equilibrium domain width is then given by

$$l_0 = [(\gamma_0 \pi a)/(1 \cdot 244 M_s^2)]^{\frac{1}{2}} \tag{9.82}$$

where $\gamma_0 = 8(AK_3)^{\frac{1}{2}}$ and the equilibrium energy

$$E_0 = 2a(1 \cdot 244 \pi a \gamma_0 M_s^2)^{\frac{1}{2}} \tag{9.83}$$

9.2.3.2 Cubic crystals

Owing to the fact that in cubic crystals there are six or eight easy directions of magnetization the domain structures are more complicated than in the uniaxial case. Theoretical models proposed up to the present qualitatively explain the patterns observed, but generally quantitative estimates do not agree well with experiment. However, there are situations in which the details of the observed structure may also be quantitatively explained.

The simpler models in Figure 9.13(a) and (b), for crystals in which the easy directions are [100] or those in Figure 9.13(c) and (d) for crystals with easy directions [111], were analysed and sometimes the agreement between experimental and theoretical data was quite satisfactory (Néel, 1944b, 1944c; Kocinski, 1958; Schwink and Spreen, 1965). The structures shown in Figure 9.13(b), (c) and (d) are obtained after a magnetic field is applied, along [110], to saturate the main domains by displacement of all 180° walls.

To estimate the total energy of the structure in Figure 9.13(a), the magnetoelastic energy associated with the closure domains magnetized along [010] direction must be included. For equilibrium in the demagnetized state the free energy expressions in (9.39) and (9.40) may be used, when K is replaced by the magnetoelastic energy density.

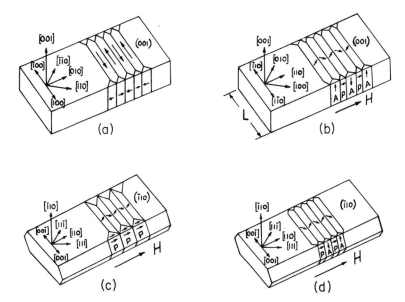

Figure 9.13 Simple possible domain configurations in cubic crystals on various crystallographic planes (Néel, 1944b, 1944c).

In the case of Figure 9.13(b), the magnetostatic and magnetoelastic energies are to be considered, while in the cases of (c) and (d) only the anisotropy and magnetoelastic energies associated with A domains and magnetostatic and magnetoelastic energies associated with P domains respectively (Kocinski, 1958; Schwink and Spreen, 1965).

The real structures are generally more complicated, even for crystals cut as in Figure 9.13. This is due to the angles between the easy axes and the crystal surfaces, giving rise to superficial structures whose details may hardly be forecast. Therefore the general method adopted for domain structure investigations in such cases consists in first observing the real structure and then fitting a theoretical model to it.

9.2.4 Domain structure and magnetization processes

Under the influence of an external magnetic field, the domain structure changes both by domain wall displacements and rotation of magnetization. Experimental observations are made on single crystals by applying a field along various directions and considering the connections between the parameters of the domain structure and the field strength in various stages of the magnetization process. The theoretical models usually apply to ideal crystals. The influence of imperfections is considered as a critical field for wall displacement H_{0w} or eventually as a nucleation field H_n for reverse domains. The origin of these fields, connected with the interaction of the wall with the crystal defects, will not be dealt with here.

506

Experimental observations show that the magnetization of the crystal does not take place only by wall displacement or spin rotation but also by sensible changes of the total wall area, starting with certain values of the applied field, which strongly modify the shape and dimensions of the reverse domains. We shall present in this section only the models which fit fairly well the experimental results on single crystals.

9.2.4.1 Uniaxial crystals

Kooy and Enz (1960) showed that for crystals thin enough to have a Kittel-like domain structure a theoretical treatment is possible. The total energy per square centimetre of an infinite plate, as shown in Figure 9.14, is the sum of

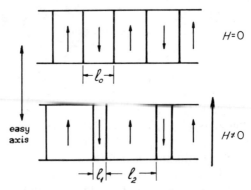

Figure 9.14 Partially magnetized simple strip domain structure.

the wall energy $E_w = 2\gamma_w c/(l_1 + l_2)$, the energy of the magnetic field $E_H = -HM_s c(M/M_s)$ and the magnetostatic energy E_M given by Equation (9.54) which corrected for the μ^* effect has the form

$$E_M = 2\pi M_s^2 (M/M_s)^2 + 16 M_s^2 c\mu^{\frac{1}{2}}/\pi^2 \alpha \sum_{n=1}^{\infty} n^{-3} \sin^2[(n\pi/2)(1 + M/M_s)]$$

$$\times \frac{\sinh(n\pi\alpha)}{\sinh(n\pi\alpha) + \mu^{\frac{1}{2}} \cosh(n\pi\alpha)} \quad (9.84)$$

where

$$\alpha = (c\mu^{\frac{1}{2}})/(l_1 + l_2) \quad (9.85)$$

For not too small values of α one has to replace $(\mu^{\frac{1}{2}} - 1)\cosh(n\pi\alpha)$ by $2^{-1}(\mu^{\frac{1}{2}} - 1)\exp(n\pi\alpha)$. The expression given for E_M can be interpreted as follows: the first term represents the demagnetizing energy of a uniformly magnetized plate with magnetization M and the second term is the correction due to the deviation of the magnetization from uniformity as a result of the domain structure.

Since both l_1 and l_2 change with the applied field, the equilibrium values of l_1 and l_2 for a given value of H can be found by equating to zero the derivatives $[\partial E/\partial(M/M_s)]$ and $(\partial E/\partial\alpha)$ of the total energy $E = E_W + E_H + E_M$.

The system of the two simultaneous equations obtained was solved numerically for the case $\mu = 1$ and for different values of the reduced plate thickness $\tau = L/(\gamma_w/16M_s^2)$. Theoretical magnetization curves for various τ are given in Figure 9.15. It may be seen that saturation generally occurs in fields smaller

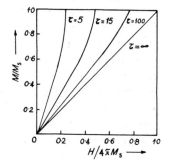

Figure 9.15 Theoretical magnetization curves in the case of the simple structure represented in Figure 9.14, for various values of $\tau = c/(\gamma_w/16M_s^2)$ (Kooy and Enz, 1960).

than $4\pi M_s$, H_s decreasing with c. The width l_2 of the reversal domains decreases only slowly with magnetization while l_1 increases very rapidly near saturation. The period $(l_1 + l_2)$ is almost constant at lower fields but increases abruptly near saturation. In this case, the sum $(l_1 + l_2)$ cannot quantitatively describe the magnetization process. The increase of the sum $(l_1 + l_2)$ shows only that starting with a certain value of the field (assuming the walls are still plane parallel) the magnetization must take place by disappearance of the reversed domains. It is also possible that at a certain field value, the reverse domains become cylindrical. Cooper and Craik (1973) survey results for both strip and cylindrical domains (particularly by Druyvesteyn and Dorleyn (1971)) and give useful approximations. In zero field the strip domain structure is energetically the most favourable, whereas for some values of the applied field, depending on both the ratio $c/(\gamma_w/4\pi M_s^2)$ and the ratio of the distances between the nearest neighbours in the x and y directions, the structure with cylindrical domains becomes more favourable.

De Jonge et al. (1971) have shown that ring domains (hollow bubbles) can exist in uniaxial platelets within a narrow region of applied field values. This region falls within that for stable bubbles. The lowest field at which a ring is stable coincides with that at which the periodic strip domain becomes unstable. Within a region $\Delta H/4\pi M_s = 0.0012$ the size of the ring changes dramatically. Thus by decreasing the reduced field $H/4\pi M_s$ from 0.2130 to 0.2125 the inner radius increases by 200 %, whereas the width of the ring remains almost the same. This field variation is about two orders of magnitude smaller than that needed to change the bubble radius by the same amount. The calculations also show that once the maximum field for a stable ring has been exceeded, the ring collapses into a bubble by a jump of the inner radius from a finite value to zero.

To summarize, the following stages of the magnetization processes are to be expected in uniaxial crystals thin enough to show strip domain structures:

(a) plane parallel displacement of the walls;

(b) changes in the wall area of the reversed domains without sensible changes of their width l_2;

(c) disappearance of some strip domains;

(d) bubble domain formation;

(e) cylindrical domain collapse at a critical minimum radius.

For bulk crystals with more complicated domain structure there are no theoretical calculations regarding the magnetization processes. Qualitatively the domain structure modifications under the influence of the magnetic field may follow the stages (a) to (e) with some particular differences. Shur *et al.* (1964) assumed that in bulk uniaxial crystals, near saturation, a curling mechanism might be possible at the surface of the crystal. Mitzek (1966) has shown that such a mechanism is possible only if in the superficial region the anisotropy constant is negative and has an order of magnitude equal to the geometric average of the exchange and anisotropy constants inside the crystal, but this condition seems to be unlikely. More recent experimental results (Craik, 1967) showed that even in bulk crystals the last stage of magnetization does not occur by curling but by disappearance of cylindrical reverse domains of finite radius.

Stability conditions for single cylindrical (bubble) domains were analysed by Bobeck (1967) and Thiele (1969) in connection with their application for new magnetic storage and logic devices. This is dealt with in Chapter 13.

9.2.4.2 *Cubic crystals*

Both older (Néel, 1944) and newer (Kocinski, 1958; Schwink and Spreen, 1965) calculations of the magnetization curve for structures such as those in Figure 9.13 gave only partial results. The models used give a dependence of l of the form

$$l = [\gamma_w c)/E]^{\frac{1}{2}} \tag{9.86}$$

where γ_w and the energy per unit volume of the closure domains, E, depends on the applied field. For the model in Figure 9.13(c) with only closure domains of the P type, Schwink and Spreen (1965) obtained

$$E_p = (-1/16)K_1 \sin\theta(1 - \cos\theta)(9\cos^2\theta + 6\cos\theta - 1) + [(9/8)C_3\lambda_{111}^2$$

$$+ (27/32)(C_2\lambda_{100}^2 - C_3\lambda_{111}^2)\sin^2\theta][\sin^2\theta/(1 - \cos\theta)] \tag{9.87}$$

where C_2 and C_3 are the elastic moduli of the crystal and θ is the angle between $\mathbf{M}_s$ in the main domains and the field. The wall energy density is

$$\gamma_w = (KA)^{\frac{1}{2}} \sin^2\theta(6 - 10\sin^2\theta)^{\frac{1}{2}}$$

$$+ \{(6 - 7\sin^2\theta)\arcsin[3\sin^2\theta/(6 - 7\sin^2\theta)]\}/(3^{\frac{1}{2}}\sin\theta) \tag{9.88}$$

Expressions for the energies of the other structures in Figure 9.13 were also obtained by Schwink and Spreen (1965).

9.2.4.3 Hysteresis and nucleation fields

According to Brown's paradox (Brown, 1945), in perfect crystals the only magnetization process to be expected is rotation which for fields parallel to the easy axis, takes place at a field value equal to the anisotropy field. But in real crystals remagnetization occurs at much lower fields than the anisotropy field. This is due to imperfections in the crystals. All the theories attempt to explain the finite values of the permeability and the coercivity on this basis taking into consideration the interaction of the Bloch walls with the crystal defects, stressed regions and non-magnetic inclusions.

In the case of uniaxial crystals, magnetized along the c axis, if one can define a mean critical field which renders the wall displacement more difficult, the magnetization curve should have the shape in Figure 9.16(a), or that in 9.16(b) if there is a distribution of critical fields.

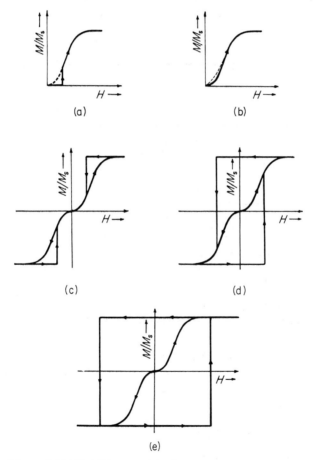

Figure 9.16 Possible magnetization curves and hysteresis loops for various values of the nucleation field, for uniaxial crystals.

510

If the nucleation field H_n is positive or negative the hysteresis loops have the shape in Figure 9.16(c) and (d) respectively. It is also possible for H_n to be negative but much greater than the saturation field, giving an ideally rectangular hysteresis loop as in 9.16(e). There are also situations where the nucleation field depends on the magnetization field. In such cases the hysteresis loop will have each form of Figure 9.16(c), (d) and (e), depending on the value of the applied field.

9.2.5 Few domain crystals

9.2.5.1 Uniaxial crystals

The preceding calculations assumed the crystal to be infinite in two dimensions, and do not apply to crystals with a small number of domains. The total energy can be calculated for simpler structures like that in Figure 9.17. The

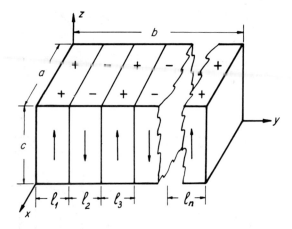

Figure 9.17 Simple structure in a rectangular block of a small uniaxial single crystal with few domains.

mutual potential energy of two surfaces with constant charge densities σ_1 on surface (1) and σ_2 on surface (2), separated by a distance c, may be written as

$$E_m = \sigma_2 \iint V(x_2, y_2, c)\, dx_2\, dy_2 \qquad (9.89)$$

where

$$V(x_2, y_2, c) = \sigma_1 \iint [(x_2 - x_1)^2 + (y_2 - y_1)^2 + c^2]^{-\frac{1}{2}}\, dx_1\, dy_1 \quad (9.90)$$

is the potential created at the surface 2 by the charge distribution σ_1.

Rhodes and Rowlands (1954) have shown that the magnetostatic energy of the domain structure in Figure 9.17 for which $\sigma_i = \pm M_s$, may be obtained as a sum of type (9.89) integrals, as

$$E_M = 2 \sum_{i=1}^{n} E_{M_i} + \sum_{i=1}^{n} E_{M_{ii}} + 2 \sum_{i<i'}^{n} \sum^{n} E_{M_{ii'}} + \sum_{i<j}^{n} \sum^{n} E_{M_{ij}} \qquad (9.91)$$

where E_{M_i} is the self energy of the surface distribution on the sheet i, $E_{M_{ii}}$ the mutual energy of the opposite sheets with different polarities, $E_{M_{ii'}}$, the mutual energy between the sheets i and i' on the same surface of the crystal and $E_{M_{ij}}$ the mutual energy of the sheet i from one side of the crystal and of the sheet j from the opposite one.

By using the parameters $p = b/a$ and $q = c/a$, the above four types of energy may be expressed in terms of only one function $F(p, q)$

$$E_{M_i} = a^2 M_s^2 F(p, 0)$$

$$E_{M_{ii}} = a^3 M_s^2 2F(p, q)$$

and

$$E_{M_{ij}} = a^3 M_s^2 [F(p_i + p_j + r, q) + F(r, q) - F(p_i + r, q) - F(p_j + r, q)] \qquad (9.92)$$

where $p_i = l_i/a, q = c/a, r = l_{ij}/a$.

The total energy is obtained by adding the wall energy of the $(n - 1)$ walls

$$E_w = (n - 1)ac\gamma_w \qquad (9.93)$$

and the contribution of the external field

$$E_H = ac \left(\sum_{i=2k+1} l_{oi} + \sum_{i=2k} l_{ej} \right) M_s H \qquad (9.94)$$

where l_{oi} is the width of odd ith domain and l_{ej} the width of even jth domain.

The equilibrium state is obtained by minimizing the total energy with respect to n and l_0. In the demagnetized state, $E_H = 0$, and

$$E_m/2a^3 M_s^2 = 4 \sum_{k=1}^{n-1} (-1)^{k+1}(n - k)\{F(k/n, 0) - F(k/n, q)\}$$

$$+ (-1)^{n+1}\{F(1, 0) - F(1, q)\} = \mathscr{F}(n, c/a) \qquad (9.95)$$

so that it reduces to a derivative with respect to n. For a block of dimensions $a \times a \times c$, the variation of the domain spacing with crystal size may be calculated as by Rosenberg (1968) and Craik and McIntyre (1967, 1969). Craik and McIntyre introduced the parameter γ_w/M_s^2 as the unit of length. In this case $a = a'\gamma_w/M_s^2$ and $c = c'\gamma_w/M_s^2$.

The crystal will have n domains if

$$a'_{n+1,n} > a' > a'_{n,n-1}$$

where

$$a'_{n,n-1} = (c'/a')/[\mathscr{F}(c'/a', n - 1) - \mathscr{F}(c'/a', n)] \qquad (9.96)$$

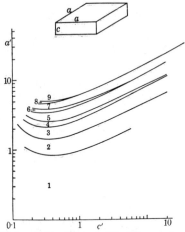

Figure 9.18 The equilibrium number of 180° domains as a function of the size and shape of a rectangular crystal with square cross section normal to the easy axis (Craik and McIntyre, 1967, 1969).

Figure 9.18 shows the equilibrium number of 180° domains as a function of the sizes c' and a' of the rectangular crystal (Craik and McIntyre, 1969). It can be seen that structures with an even number of domains do not exist for large values of c' and a'; an odd number of domains gives non-zero remanence, with a reduced interaction energy between opposite faces.

Magnetization curves for a cubic shaped crystal have been calculated by Craik and McIntyre by minimizing E with respect to n and l_0 (Figure 9.19). It is to be noticed that the magnetization curves exhibit a negative curvature

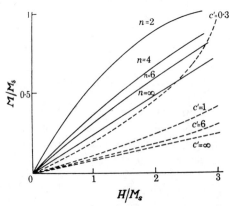

Figure 9.19 Calculated magnetization curves for a uniaxial crystal of cubic shape with the number of domains n indicated, showing the dependence of the anhysteretic susceptibility on n and thus the size (Craik and McIntyre, 1967, 1969).

$(\partial^2 M/\partial H^2) < 0$ and the anhysteretic susceptibility decreases with increasing particle size (cf. the broken lines for infinite sheets).

It is possible to estimate γ_w from the dependence of the number of domains on a and c, by using Equation (9.96). Thus, knowing $a'_{n,n-1}$ and measuring experimentally the parameter $a_{n,n-1}$ we obtain γ_w from the relation

$$a'_{n,n-1} = a_{n,n-1} M_s^2/\gamma_w$$

If the number of domains changes by one at a certain temperature, then the above and (9.96) gives γ_w at that temperature.

A finite configuration for which the magnetostatic energy can be analytically derived (Craik, 1970) is an annular charge distribution relating to a cylindrical domain in a cylindrical crystal as in Figure 9.20.

$$E_M = (8/3)\pi a^3 M_s^2 F(k) \tag{9.97}$$

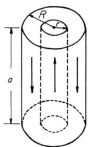

Figure 9.20 A cylindrical crystal containing a single cylindrical domain with an annular charge distribution for which the magnetostatic energy can be derived analytically (Craik, 1970).

where $F(k)$, the reduced energy density, is plotted in Figure 9.21 as a function of $k = r/R$.*

The magnetostatic energy is minimum close to $k = 1/2^{\frac{1}{3}}$, the value for which the volume of the inner cylinder equals the volume of the remainder, i.e. the

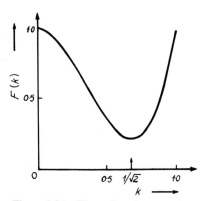

Figure 9.21 The reduced energy density calculated for the charge configuration of Figure 9.20 (Craik, 1970).

* Expressions for the energies of interaction between the surfaces are given by Craik, D. J. (1974) *J. Phys. D*, **7**, 1566.

514

crystal is demagnetized. But taking account of the finite value of the Bloch wall energy leads to the striking result that at equilibrium the volumes of the inner and outer cylinders cannot be equal and the magnetization within the central cylinder should always be less than that of the remainder to some extent. In the presence of a magnetic field the same stability relationship holds in terms of magnetic fields as in the case of the reverse cylindrical domains in an infinite plate (Bobeck, 1967). It follows that the effect of the applied fields should depend markedly upon whether they are parallel or antiparallel to the magnetization of the central cylinder. In the former case for some values of H a point is reached in which the inner cylinder becomes unstable and collapses, giving a discontinuity in the magnetization curve. In the case of H parallel to the magnetization direction of the inner cylinder, the anhysteretic susceptibility should thus be lower than for fields oppositely oriented.

9.2.5.2 Cubic crystals

When flux closure domain structures are favourable the simple few domain configurations in a crystal of volume $a \times b \times c$, with $a < b < c$ must be of the L L type and the equilibrium energy and domain width are given by Equations (9.39) and (9.40).

If the crystal is a whisker and has $K_1 > 0$ very simple configurations may be obtained as in Figure 9.22(a) and (b). For configuration (b), the magnetization process in its initial stage may be reversible and if the field is applied along the long axis of the whisker, flux closed configurations are formed, as in Figure

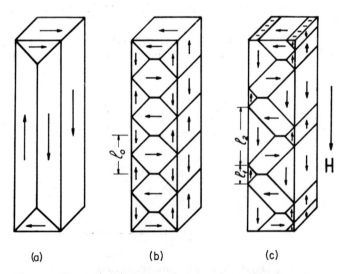

(a)   (b)   (c)

Figure 9.22 (a) and (b) Simple domain configurations in a whisker cubic crystal with positive anisotropy. (c) The initial stage of the magnetization process of the configuration in (b) (Dillon, 1963).

9.22(c). If we neglect the demagnetizing energy due to the free poles at the ends of the whisker, simple calculations show that the energy is

$$E/ab = [(l_1^2/2l_0 - l_1)K_\sigma]/a - 2^{\frac{3}{2}}\gamma_{90°}l_0/a$$

$$+ (1/2l_0 - 1/2a)\gamma_{180°} - HM_s(l_1 - l_0)/a \qquad (9.98)$$

with $l_1 + l_2 = 2l_0$ and l_0 being the equilibrium width in the absence of the field. From the equilibrium condition it results

$$l_1 = l_0 + l_0 HM_s/K_\sigma$$

which gives

$$M = l_0 M_s^2 H/K_\sigma \qquad (9.99)$$

i.e. a linear dependence of M on H with the susceptibility $\chi = l_0 M_s^2/K_\sigma$.

9.2.6 Single domain crystals

Below a critical size the reduction of the magnetostatic energy by a non-uniform magnetization configuration ceases to outweigh the increase of the exchange energy due to the non-uniformities in the parallel orientation of the spins. Thus it is possible to define a size for a single domain behaviour, under which the magnetization is expected to remain essentially uniform under all circumstances, i.e. using a criterion introduced by Kondorski (1952) when the coercive force of the crystal reaches the maximum value characteristic for a coherent rotation. Kondorski calculated, by a variational method, the radius below which a prolate ellipsoid still remained single domain even in the reverse field for uniform rotation, $H = (N_b - N_a)M_s + 2K_1/M_s$, as

$$b_c = 3·06A^{\frac{1}{2}}N_b^{-\frac{1}{2}}M_s^{-1} \qquad (9.100)$$

The same expression differing slightly in the value of the numerical factor, results from the micromagnetic calculations by Brown (1957), Frei, Shtrikman and Treves (1957) and Aharoni (1959)

$$b_c = (2\pi k)^{\frac{1}{2}}A^{\frac{1}{2}}N_b^{-\frac{1}{2}}M_s^{-1} \qquad (9.101)$$

where k is a numerical factor which equals 1·38 for a sphere and 1·08 for an infinite cylinder.

In the case of uniaxial crystals with high anisotropy field the critical size may be found by comparing the energies of a cube or a sphere with one and two domains, respectively.

For a rectangular crystal of dimensions $a \times b \times c$, with c axis along the easy axis, the two energies can be calculated using Rhodes and Rowland's (1954) expression as done by Craik and McIntyre (1967, 1969b). The results of the calculations expressed in the unit of length γ_w/M_s^2 are shown in Figure 9.23 for a range of different values of b/a. The critical size for a cube is $a' = 0·95$ which may be compared with the critical diameter for a spherical particle with $d = 1·33$, obtained using the expression for the magnetostatic energy of a

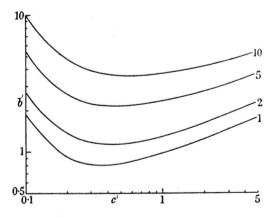

Figure 9.23 Critical size of an isolated particle as a function of the shape of the crystal. The numbers on the curves represent b/a (Craik and McIntyre, 1967).

sphere divided into two domains given by Néel (1944a). It was found that the wall is always parallel to the shorter side, which is a in Figure 9.23 (see caption).

It is more difficult to find the critical size in the meaning proposed at the beginning of this section, i.e. a single domain behaviour under all circumstances. Craik and McIntyre (1967, 1969b) showed that a cubic single domain particle in zero field may have a higher energy in a reverse field than if the particle were divided into two unequal domains by a domain wall. The lower the dimension a of the particle below the critical size for the single domain behaviour in zero field, the higher would be the value of the field needed for the nucleation of the reverse domain.

9.3 OBSERVATION OF DOMAIN STRUCTURES

This section briefly describes the most common techniques for observing domain structures in magnetic oxides. A systematic collection and detailed description of various methods currently in use is given in several excellent works (Dillon, 1963; Craik and Tebble, 1965; Carey and Issac, 1966).

9.3.1 Colloid technique

The colloid or 'powder pattern' technique is the most widely used method both for its simplicity and for the wide range of materials to which it may be applied. Ordinarily it consists of viewing, with a microscope, the sample surface coated with a colloidal suspension of magnetite.

The mechanism of pattern formation is based upon the interaction between the single domain particles from the suspension and the stray field above the specimen surface created by the boundary between domains (Kittel, 1949).

The distribution and magnitude of the stray field was studied in some detail by Craik both theoretically (Craik, 1966) and experimentally (Craik, 1967) on the basal plane of a barium ferrite single crystal. Bergman (1956) estimates that the particle diameter must be of the order of magnitude of 100 Å, in good agreement with the experimental observations by Craik (1956) and Garrood (1962).

Elmore's original colloid recipe (Elmore, 1937, 1938) is as follows: 2 g of $FeCl_2 \cdot 4H_2O$ and 5·4 g of $FeCl_3 \cdot 6H_2O$ are dissolved in 300 cm^3 of water at 70 °C, to which a solution of 5 g NaOH in 50 cm^3 of water is added with constant stirring. The heavy precipitate of magnetite formed is filtered and carefully washed with distilled water in order to remove excess salt and NaOH. Peptization is then obtained with a 0·01 N solution of HCl and the precipitate is added to a 0·5 % solution of sodium oleate which is stirred vigorously for a while and then boiled for some 20 minutes for a better dispersion. Colloids based on ethyl cellulose (Gustard, 1967) or 1-chloro-2-methyl propane (Bates, 1964) as dispersion medium, permit observations down to -30 °C and -90 °C respectively. For higher temperatures paraffin oil is generally used as dispersion medium (Andrä, 1956, 1959; Rosenberg et al., 1970).

Craik and Griffiths (1958) developed a new colloid which could be allowed to dry on the sample surface without aggregation of the particles. The film thus formed is stripped away and examined in a metallographic or electron microscope. This method has the advantages of the higher resolution of the electron microscope and of studying the patterns on an irregular surface of the specimen. Moreover it is not necessary to have the sample on the microscope stage while the colloid is drying. But its main disadvantage is that it can be used only to record static domain patterns. A colloid also used in a similar way was developed by Schwartze (1957) who used magnetite particles suspended in lacquer, and Andrä and Schwabe (1955) described patterns using dry iron oxide powder.

9.3.2 Magneto-optics effects

In these methods the magnetic domains are made visible by the direct inter-action between the plane polarized light and M_s due to the effects described in Chapter 7.

9.3.2.1 Faraday method

A beam of plane polarized light, passing through a magnetic material undergoes a rotation of its plane of polarization due to the component of the specimen magnetization lying along the direction of propagation. The angle of rotation depends upon the magnitude of sample magnetization and its sense of rotation is determined by the magnetization direction. In the domains with M_s in a plane perpendicular to the light direction, the contrast between domains is achieved under certain circumstances by magnetic birefringence (Dillon, 1958). It is also possible to achieve the necessary contrast in such cases by inclining the sample with respect to the light direction (Boersch and Lambeck, 1961).

The basic experimental requirement is a standard polarizing microscope. Though this technique works quite well for a wide range of materials and for practically all temperature ranges, it is limited by the sample thickness: a maximum of 10^{-3} cm for ferrites, 10^{-2} cm for garnets and 10^{-1} cm for ortho-ferrites so that the main effort in using this technique is directed toward preparation of suitable thin specimens. Infrared polarizing microscopes, however, allow observations in garnet specimens up to 5 mm thick (Enoch and Lambert, 1970: see Figure 9.24). The main element used in this technique is an infrared image converter tube. The contrast between adjacent domains in which the magnetization lies in the plane of the specimen is greatly enhanced by the insertion of a quarterwave plate between the specimen and the analyser (Enoch and Lambert, 1970).

9.3.2.2 Kerr method

The magnitude of the Kerr rotation (see 7.1.5) is much smaller than that due to the Faraday effect in transmission. The polar effect produces the greatest rotation and is generally used on materials such as uniaxial ferrimagnetic oxides (Fowler et al., 1963). The longitudinal as well as transverse effects produce a rotation less by a factor of 5 than the polar one, and therefore the images suffer both from a low brightness and a lack of contrast, but efforts have been made to improve this technique (Fowler and Fryer, 1954a, 1954b; Prutton, 1959).

The great advantages of this technique consist in its applicability to bulk samples of any material over a wide range of temperature and field.

The magneto-optic methods are able to follow very rapid wall motion and the only limit on the speed of changes in domain structure is the intensity of illumination available. Along this line Dreschel (1961) used an electronic flash tube as a light source, thus achieving exposures of the order of 1 millisecond while Conger and Moore (1963) used a stroboscopic technique which makes it possible to follow in slow motion the magnetization reversal processes which occur in times of the order of 1 microsecond.

9.3.3 Lorentz electron microscopy

9.3.3.1 Transmission microscopy

Domain observation by transmission electron microscopy is based on the Lorentz force experienced by a beam of electrons passing through a specimen in which M_s has several orientations (Hale et al., 1959). A focused image will not show any magnetic domains while out-of-focus images reveal features of the magnetization structure (e.g. Grundy and Tebble, 1968).

The specimens must be in the form of thin sections not more than about 3000 Å thick. Magnetic oxides cannot be electropolished but some success has been achieved by chemically etching specimens which have been previously thinned mechanically by grinding or cleaving. Hot phosphoric acid has been found suitable as the polishing solution (Cockayne and Robertson, 1964).

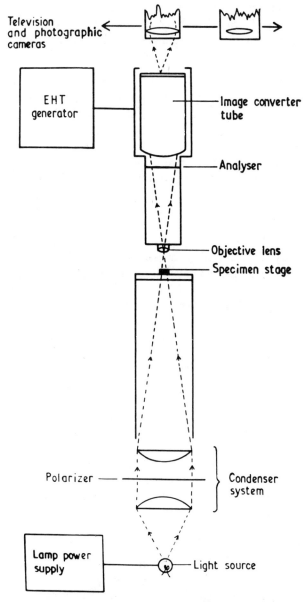

Figure 9.24 Schematic diagram of infrared polarizing
microscope (Enoch and Lambert, 1970).

Grundy (1965) succeeded in observing domain structure on thin sections
(1000–2000 Å) of magnetoplumbite crystals prepared by etching specimens of
200 µm for about 15 minutes. Magnetic domains on r.f.-sputtered polycrystalline
Mg–Mn–Zn ferrite films, observed by Lorentz microscopy have revealed
structures similar to strip domains (Lo *et al.*, 1969).

This technique is ideally suited for investigating the interaction between the magnetization and structural imperfections with high resolution. The specimen can also be subjected to strain, temperature changes or magnetic field (Silcox, 1963).

9.3.3.2 Reflection microscopy

The magnetic domains may be visualized by this technique because there is a contrast over a particular surface area depending upon the variations in the normal component of the stray magnetic field above the specimen. Spivak *et al.* (1955) and, independently, Mayer (1957) first applied this technique to the investigation of the domain structure, the latter on barium ferrite single crystals. Its main advantage lies in the possibility of obtaining direct micrographs of the surface field without harming the specimen. Great care must be taken in preparing very smooth and flat surfaces, and it has not been extensively used.

9.3.4 X-ray technique

Domains may be observed as a result of diffraction contrast, arising from magnetostrictive strains, in X-ray diffraction topographs (Lang, 1959; Polcarova and Lang, 1962; Polcarova and Kaczer, 1967; Roessler, 1967). An increase of the X-ray reflecting power due to heterogeneous elastic strain has also been observed.

Using this technique with a double crystal X-ray diffractometer Merz (1960) has observed ferromagnetic domains in Co–Zn ferrite ($\lambda = -355 . 10^{-6}$). Magnetic domains have also been observed by X-ray topography on YIG ($\lambda = -2\cdot4 . 10^{-6}$) and TbIG by Patel, Jackson and Dillon (1968) and Basterfield and Prescot (1968) respectively.

The main advantage is that domains may be observed at the same time as dislocations and other crystal defects, as well as the interaction between domain structure and crystal defects.

Particularly suitable for investigating antiferromagnetic domains is the Berg–Barrett method, which consists in recording on a photographic plate the X-ray micrographs of a crystal surface which exhibits microscopic defects and misorientation of the crystal lattice.

9.4 DOMAIN STRUCTURES IN OXIDIC MATERIALS

During the last decade much effort has been devoted, both experimentally and theoretically, to domain structures in magnetic oxides, aided by the availability of these materials in the form of sound single crystals. Also the facilities provided by computer calculations have led to a deeper understanding of the phenomena.

Though it is often possible in the very simplest cases to infer the general type of structure which should exist in a certain specimen, direct observation is the only way to obtain an insight into the detailed domain structure. This explains in part the very considerable number of experimental works in comparison with the theoretical ones.

9.4.1 Uniaxial ferrimagnetic compounds: single crystals

Owing to their relatively low crystal symmetry, high values of K_1 referred to the single easy axis are to be expected in these materials, giving basically simple arrays of 180° domains within the body of the specimen. The differences that may exist are found at the basal (0001) surfaces.

Usually the domain structures are dependent on the intrinsic magnetic properties (such as M_s and K) and the geometry of the specimen. The structure is also dependent on the magnetic history of the specimen, as well as external factors such as temperature, stress and the magnitude and direction of applied fields.

9.4.1.1 Compounds with M type structure

The M compounds are oxides of the type $A^{2+}0 \cdot (6 - x)Fe_2O_3 \cdot xB_2^{3+}O_3$ where A = Pb, Ba and Sr (or a combination of them) and B = Al, Cr, Ga (or a combination of them). Ferrimagnetic compounds are obtained for $0 \leqslant x < 3$ while for $x > 3$ paramagnetic compounds are formed. The magnetic properties and consequently the domain structure are strongly dependent on x. In the following we shall briefly describe the domain structure for all these compounds.

9.4.1.1.1 Compounds with $x = 0$ and A = Pb, Ba, Sr. Since Pearson's 1957 paper the domain structure of these compounds has been the subject of many experimental and theoretical investigations (Kaczer, 1962; Kooy and Enz, 1960; Kaczer and Gemperle, 1961; Rosenberg et al., 1963; Kooy, 1958). They have similar magnetic structures and almost the same values of the M_s and K; their domain structures are also similar. On any plane of the specimen parallel to the easy axis the domain structure in the demagnetized state consists of an array of 180° walls parallel to the c axis (see Figure 9.25) (Rosenberg and Tănăsoiu, 1963a).

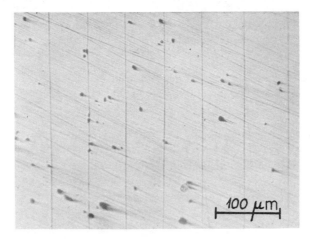

Figure 9.25 Domain patterns on an axial plane of a barium ferrite single crystal.

On basal planes the equilibrium domain patterns are strongly dependent on the thickness of the sample c, along the easy axis. For all these materials four main types of patterns were revealed as a function of c (Kaczer and Gemperle, 1960; Rosenberg and Tănăsoiu, 1963a, 1963b; Rosenberg et al., 1966). For $c < 10\,\mu m$ the patterns consist of strip domains separated by walls parallel to the easy axis as in Figure 9.26(a) and in accordance with Equation (9.60)

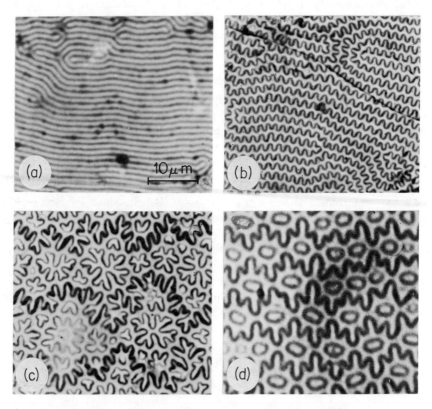

Figure 9.26 Domain patterns on the basal plane of barium ferrite crystals with thicknesses: (a) $c = 8\,\mu m$; (b) $c = 25\,\mu m$; (c) $c = 80\,\mu m$; (d) $c = 750\,\mu m$.

the domain wall spacing l_0 is proportional to $c^{\frac{1}{2}}$ (Kaczer and Gemperle, 1960; Kandaurova, 1967; Rosenberg et al., 1966). In this case γ can be easily calculated from Equation (9.60) by measuring l_0 as a function of c. Measurements for all three compositions have indicated that γ_w is around 5 erg/cm². In Table 9.3 the values of γ_w as reported by various authors are given.

For specimens with $10\,\mu m < c < 50\,\mu m$ the patterns on the basal plane consist of undulatory walls as in Figure 9.26(b). Both the amplitude A and the period P of the undulation increase with increasing c, as in Figure 9.27 for magnetoplumbite (Kaczer, 1966). P shows a smooth increase while A changes

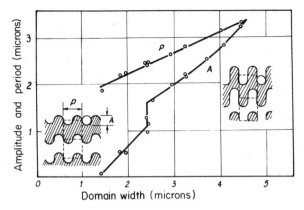

Figure 9.27 Dependence of amplitude A and period P of undulations on domain width of a magnetoplumbite crystal (Kaczer, 1964).

discontinuously at a certain value of l when $A = P/2$ and then the increase of amplitude ceases at $A = P$ corresponding to a thickness $c \simeq 4 \cdot 5 \, \mu m$. The increase of A decreases the magnetostatic energy. The jumps in A are due to a complicated mechanism for decreasing the magnetostatic energy with increasing c (Kaczer, 1966). Similar measurements on barium ferrite and magnetoplumbite were made by Szymczak (1968) but no discontinuity in amplitude increase was observed, as illustrated in Figure 9.28. The full lines represent the theoretical curve determined by numerical minimization of the total energy for the wavy model shown in Figure 9.9, with respect to α and γ. The experimental findings agree fairly well with the simple model proposed (Szymczak, 1968).

For $50 \, \mu m < c < 100 \, \mu m$ the complex undulation of the walls is accompanied by the occurrence of spike domains of circular shape as in Figure 9.26(c). The fourth model, for specimens with $c > 100 \, \mu m$, consists of patterns with both complex undulating walls and multiple spike domains (Figure 9.26(d)). The undulation of the walls at the specimen surface as well as the occurrence

Table 9.3 Wall energy density for M type hexagonal ferrites

Compound	$\gamma_w \, (\text{erg/cm}^2)$	Reference
$PbO \cdot 6Fe_2O_3$	5	Isaac, 1959
	4·8	Kaczer and Gemperle, 1960
	4·7	Rosenberg and Tănăsoiu, 1963a
$BaO \cdot 6Fe_2O_3$	2·8	Kooy and Enz, 1960
	9	Goto, 1966
$SrO \cdot 6Fe_2O_3$	4	Rosenberg et al., 1966
	8	Goto, 1966

524

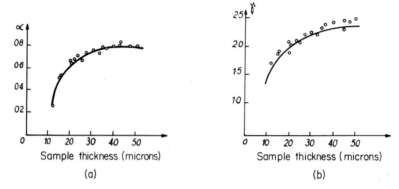

Figure 9.28 Dependence of the amplitude α and period γ on crystal thickness for a barium ferrite crystal (Szymczak, 1968).

of the spike domains decrease the magnetostatic energy arising from the free magnetic charges, by decreasing the domain width at the surface (Gemperle, 1964). In some cases these complex patterns are not accompanied by spike domains though the sample is quite thick (Figure 9.29).

Such patterns are only superficial and as proposed by Goodenough (1956) the amplitude of the undulation decreases with the depth of penetration into the crystal as shown in Figure 9.7.

For the last three models it has been experimentally found that l_0 is proportional to $c^{\frac{2}{3}}$ for all these materials (Kaczer and Gemperle, 1960; Kandaurova, 1967; Goto, 1966; Rosenberg et al., 1966). Theoretically it was not possible to predict such a behaviour, Kaczer (1964) found an approximate relationship

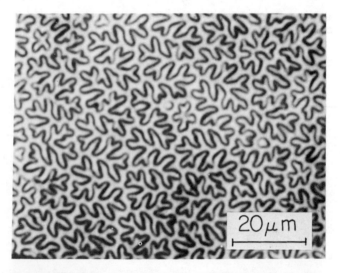

Figure 9.29 Domain patterns without spike domains on the basal plane of a strontium ferrite crystal of thickness $c = 410$ μm.

of this type between l_0 and c (see Equation (9.72)) for magnetoplumbite but the theoretical curve, calculated with this formula, lies about 20% below the experimental points. Using Kaczer's equation to estimate γ_w by measuring mean domain width from such complex patterns, values of about 10 erg/cm^2 are obtained. This is an indication of the very approximate model used.

More appropriate values are obtained both for $l(c)$ dependence and l_c and c_0 using the calculations developed by Szymczak (1968). In Figure 9.30 the

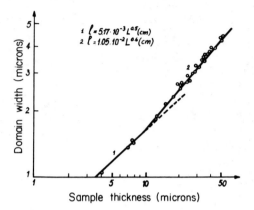

Figure 9.30 The domain width l as a function of the sample thickness c for barium ferrite crystals (Szymczak, 1968).

theoretical curve $l(c)$ was calculated by relation (9.74). Using Equation (9.73) the critical values for c and l are: $c_0 = 8.5\,\mu m$ and $l_c = 1.55\,\mu m$ and they confirm the experimental findings. Similar results are found for magneto-plumbite.

Experimental observations indicate that the magnetization process with fields parallel to the easy axis takes place in the following manner (Kooy and Enz, 1960; Rosenberg et al., 1964, 1965): for crystals thin enough to have a strip domain structure, in low fields the displacement of the walls occurs in such a way that the period $(l_1 + l_2)$ remains approximately constant while l_1 and l_2 separately, increases or decreases as the field is directed parallel or antiparallel to the inner magnetization vector of the domain. In this stage the pattern remains essentially the same. In higher fields, approaching saturation, l_1 increases very rapidly with increasing field whereas l_2 decreases only slightly, giving rise to a rapid increase of $(l_1 + l_2)$ which finally abruptly goes toward infinity for some critical field which is usually well below the field $4\pi M_s$ usually considered necessary for saturation. The agreement of the calculations and experimental data in this respect is quite excellent (Kooy and Enz, 1960). Reversed domains do not disappear by reducing their width to zero but by diminishing their lengths so that the total wall surface drastically decreases. The strip domains contract towards a cylindrical form with a diameter of the

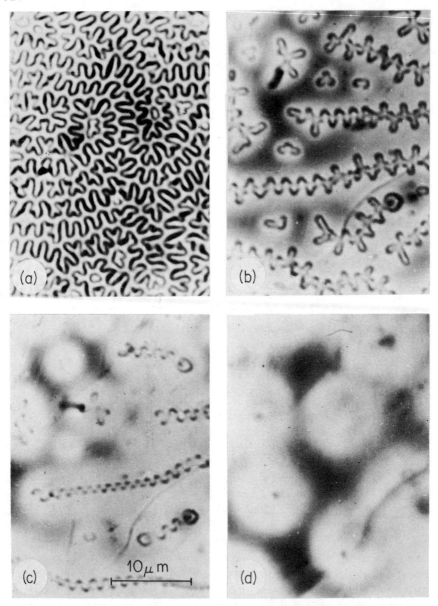

Figure 9.31 Modification of the domain patterns on the basal plane of barium ferrite by an external field parallel to the easy axis (colloid technique): (a) $H = 0$; (b) $H = 2000$ Oe; (c) $H = 3000$ Oe; (d) $H = 3450$ Oe.

same order as the minimum width of the strip domain. Such cylindrical domains are stable within a small range of fields. Their diameter decreases with increasing field until a critical field is reached at which they suddenly collapse. Kooy and Enz (1960) were the first to observe such cylindrical domains in thin platelets of barium ferrite ($c \sim 3\,\mu m$) and they found the stability conditions of such domains and calculated their collapse radii. More recently it has been found that cylindrical domains are characteristic for all uniaxial materials (Gianola et al., 1969) especially in orthoferrites and garnet films where they can be manipulated and used as logic elements in magnetic memory devices (see Chapter 13).

Magnetization processes in bulk uniaxial crystals are similar to those in thin platelets (Rosenberg et al., 1964, 1965) but there are some particular aspects in the field region near saturation when reverse domains disappear.

Optical microscopy reveals patterns of continuous dark networks enclosing rather circular light regions as shown in Figure 9.31(d) for barium ferrite single crystals. As the fields are increased toward complete saturation the light areas, remarkably, grow in size instead of diminishing. Accepting these areas as representing the intersection of reverse domains, which increase in volume in positively increasing field, it has been proposed (Shur et al., 1964) that they become progressively transformed into flat discs in which the direction of spins make a small angle with the crystal surface. The converse of this process was then supposed to correspond to a nucleation of domains by a curling mechanism on the reduction of initially saturating field. Craik (1967) showed by the powder pattern replica technique, using electron microscopy, that this anomaly is only apparent and is associated with the method of observation. The results of his observations, which are reproducible, are shown in Figure 9.32 for the

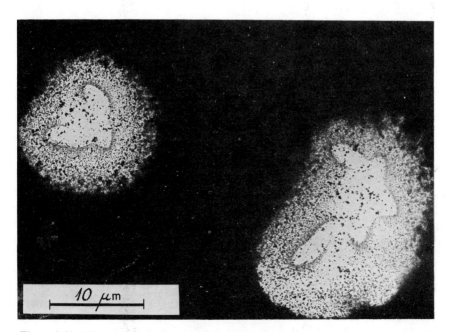

Figure 9.32 Fine details of the magnetization process near saturation illustrated by the electron microscope replica technique (Craik).

basal plane of a cube-shaped crystal of magnetoplumbite with a field of 2000 Oe applied parallel to the easy axis. It can be seen that within the light areas there are reverse domains with a smaller diameter which decrease with increasing field. Thus the general representation of the approach to saturation is as follows. The bulk domain structure is first replaced by arrays of reverse domains, of complex cross-section at the surface and then on further increasing the field there is a reversible decrease in size and the outlines approximate to circles. The smaller domains reach a critical diameter of about $5 \cdot 10^{-4}$ cm and then collapse and disappear irreversibly, corresponding to extremely small Barkhausen jumps. When a saturating field is reduced reverse domains nucleate and grow spontaneously to the critical diameter and grow further as the field decreases.

9.4.1.1.2 Compounds with $x \neq 0$ and $B = Al$. From this series the domain structures of strontium ferrite aluminates of composition $SrO \cdot (6-x)Fe_2O_3 \cdot x \cdot Al_2O_3$ with $0 \leqslant x \leqslant 2 \cdot 4$ were investigated by the colloid technique (Rosenberg et al., 1966a, 1967, 1968, 1970; Florescu and Rosenberg, 1970; Carey and Tănăsoiu, 1969). Some results on the compound $PbO \cdot 4 \cdot 5Fe_2O_3 \cdot 1 \cdot 5Al_2O_3$ were also reported by Sherwood et al. (1959). These compounds tend to form very simple structure as the Al content increases. This is governed by the decrease of M_s and the increase of the anisotropy field H_A.

From the ratio of the energies corresponding to the two extremal configurations of the uniaxial crystals, i.e. Landau–Lifshitz (E_L) and Kittel (E_K) like structures given by

$$E_L/E_K = (0 \cdot 147 H_A/M_s)^{\frac{1}{2}}$$

it can be seen that it is possible in principle to achieve a continuous approach to a Kittel domain structure if H_A/M_s increases. This is the case for crystals of this compositional series. Measurements of M_s and H_A (Bitetto, 1964) have shown a drastic decrease of M_s from about 380 G for $x = 0$ to 25 G for the composition with $x = 2 \cdot 4$. At the same time H_A increases from 20 kOe to approximately 80 kOe for the same compositions.

Figure 9.33 shows typical domain structures for various compositions for rather thick samples. Measurements of $l(c)$ for the compositions with $x = 0$; 1; $1 \cdot 5$; $1 \cdot 7$; $1 \cdot 8$; $1 \cdot 9$; 2 have also shown a rather gradual change from $l \sim c^{\frac{2}{3}}$ to a Kittel law $l \sim c^{\frac{1}{2}}$. Thus starting with $x = 1 \cdot 8$ Kittel's law is valid up to $c = 1000$ μm. The changes in structure are also accompanied by a drastic increase of the upper limits for the appearance of curved walls or reverse spike domains for each composition as x increases.

Interesting features appear for compositions with $x \geqslant 2$ where the single domain size is large. The Kittel (1946) criterion for spherical particles gives a $22 \cdot 5$ μm critical diameter for $x = 2$. A more recent calculation by Ignatchenko and Zaharov (1964) for a quadratic prism-shaped particle gives the dimensions of the single domain particle for the above compositions as $c = 1$ mm and $c_y = 57$ μm (c_y being the side in the basal plane).

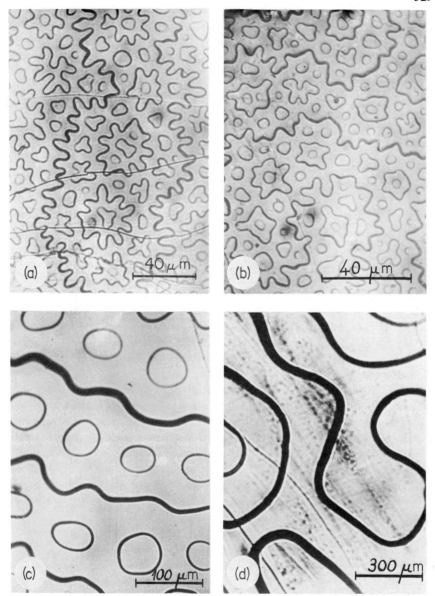

Figure 9.33 Typical domain configurations on the basal plane of crystals from the series $SrO \cdot (6-x)Fe_2O_3 \cdot xAl_2O_3$. (a) to (d) correspond to $x = 1$; $1 \cdot 5$; $1 \cdot 9$ and 2 respectively. The corresponding thicknesses of the samples are: (a) $1200\ \mu m$; (b) $1100\ \mu m$; (c) $1500\ \mu m$; (d) $1050\ \mu m$.

The magnetization processes for these compositions are similar to those described for barium and strontium ferrites. With the compositions containing a greater amount of Al it is worth noting the possibility of forming bubble domains in small bias fields. These fields also decrease as the Al content increases

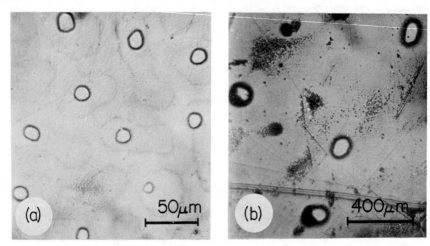

Figure 9.34 Bubble domains obtained in two strontium ferrites with $x = 1.9$ and 2.4 in bias fields of 600 Oe and 220 Oe respectively. The corresponding thicknesses of the specimens are 155 μm and 30 μm respectively.

due to the decrease of magnetic moment and consequently the magnetostatic energy. Figure 9.34(a) and (b) shows bubble domains for two compositions having $x = 1.9$ and 2.4 respectively. In some cases hollow bubbles (see Section 9.2.4.1) can be obtained as shown in Figure 9.35 for composition with $x = 1.9$; $H = 440$ Oe.

Interesting features of the magnetization processes appear in small single crystals with only a few domains. Thus for a small crystal, 1 mm^3, with $x = 2.4$ which has only 5 domains (Rosenberg *et al.*, 1968) movement of the walls starts in fields of about 20 Oe and saturation is reached in fields less than 100 Oe.

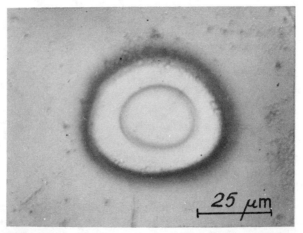

Figure 9.35 Ring domain (hollow bubble) on the composition with $x = 1.9$ in a bias field of 440 Oe. The thickness of the specimen is 300 μm.

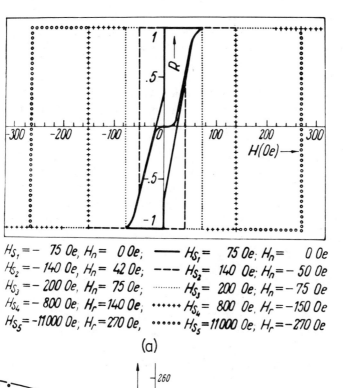

$H_{S_1} = -75\ Oe,\ H_n = 0\ Oe;$ —— $H_{S_1} = 75\ Oe;\ H_n = 0\ Oe$
$H_{S_2} = -140\ Oe,\ H_n = 42\ Oe;$ - - - $H_{S_2} = 140\ Oe;\ H_n = -50\ Oe$
$H_{S_3} = -200\ Oe,\ H_n = 75\ Oe;$ ········ $H_{S_3} = 200\ Oe;\ H_n = -75\ Oe$
$H_{S_4} = -800\ Oe,\ H_r = 140\ Oe;$ •••••• $H_{S_4} = 800\ Oe,\ H_r = -150\ Oe$
$H_{S_5} = -11000\ Oe,\ H_r = 270\ Oe,$ ••••• $H_{S_5} = 11000\ Oe;\ H_r = -270\ Oe$

(a)

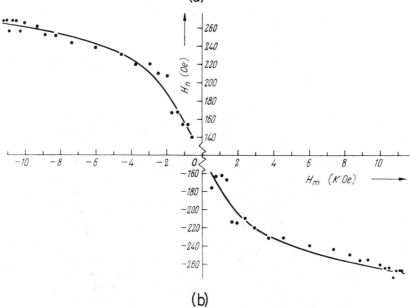

(b)

Figure 9.36 (a) Typical hysteresis loops of $SrO \cdot 4Fe_2O_3 \cdot 2Al_2O_3$ crystal for various magnetizing fields. H_{s_i} stands for the saturating fields or magnetizing fields, H_n the nucleation fields, H_r the reversal field (equivalent to the nucleation field) and $R = M/M_s$. (b) The dependence of H_n on the peak magnetizing field H_m: for $H > 10\,kOe$, H_n tends to a maximum constant value of about 270 Oe.

532

When the saturating field is reduced, reverse domains spontaneously nucleate. H_n (nucleation field) is a function of the magnetizing field H_m. The greater H_m the more H_n is shifted from the region of forward fields toward the opposite one. In this way the hysteresis loop has all the shapes discussed in Section 9.2.4.3 as a function of the magnetizing field. Such a hysteresis loop for a small single crystal with $x = 2$; $1.13 \times 0.66 \times 1.08$ mm^3 (where the dimension along the easy axis is given last) is illustrated in Figure 9.36(a). The dependence $H_n (H_m)$ is shown in Figure 9.36(b).

The explanation of such a behaviour is based on the assumption that structural imperfections of a chemical nature (local deviations from stoichiometry) are present inside the crystal and they produce local variations of the magneto-structural constants, which act as nucleation centres (Rosenberg et al., 1968).

So far we have discussed the structures obtained in fields parallel to the easy axis. But if strong (saturating) fields are applied normal to the easy axis and then removed, quite different patterns may subsequently be formed as illustrated in Figure 9.37 for barium ferrite. Such structures, known as honeycomb domain structures, have been investigated both theoretically (Kaczer and Gemperle, 1961; Kozlowski and Zietek, 1966; Müller, 1967) and experimentally (Kojima and Goto, 1962, 1964, 1965) and it has been found that in some circumstances they are energetically favourable. Honeycomb domain structures have been demonstrated in magnetoplumbite (Gemperle, 1964), barium and strontium ferrites (Kojima and Goto, 1965; Rosenberg et al., 1967) and ferrite aluminate with $x = 1$ (Carey and Isaac, 1964).

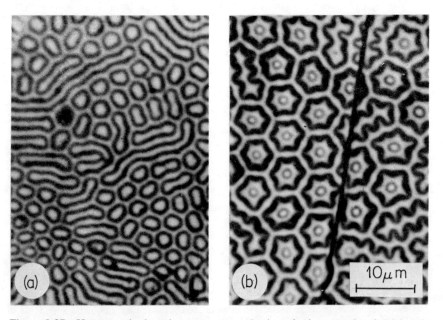

Figure 9.37 Honeycomb domain structure on barium ferrite crystals of thickness (a) 10 μm; (b) 75 μm.

9.4.1.2 Compounds with Z structure

The general chemical formula of Z compounds is $A_3^{2+}Me_2^{2+}Fe_{24}O_{41}$ where $A = $ Ba or Sr and Me means a divalent ion of Mg, Mn, Fe, Co, Ni, Cu, Zn or a combination of them. The domain structure was investigated for single crystals of the series $Ba_3(Zn_{2-x}Co_x)Fe_{24}O_{41}$ with $x = 0; 0.5$ and 2 by Rosenberg *et al.* (1964). For compounds with $x = 0$ there is a preferred axis while for $x > 0.5$ the basal plane is a preferred plane of magnetization.

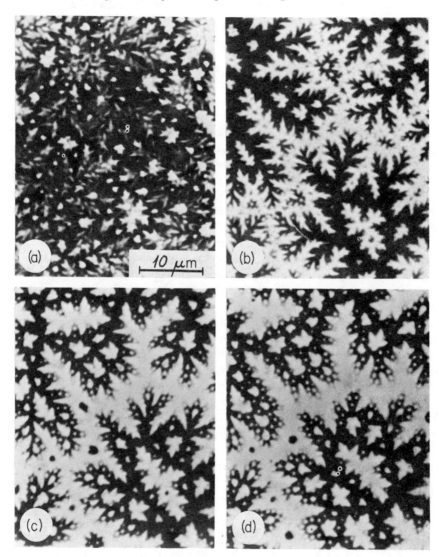

Figure 9.38 Domain patterns on the basal plane of a $Zn_{1.5}Co_{0.5}Z$ crystal as a function of the field applied along the easy axis: (a) $H = 0$ (b) $H = 700$ Oe (c) $H = 2200$ Oe (d) $H = 2800$ Oe.

Zn_2Z has an entirely similar structure to barium or strontium ferrites. For $x = 0.5$, i.e. $Zn_{1.5}Co_{0.5}Z$, though there is still preferred axis, due to the low values of the anisotropy constant (about one order of magnitude lower than for Zn_2Z) the domain structure differs significantly from that of Zn_2Z. The patterns on the basal plane consist of a continuous network of lines with irregular branches which cannot be divided into long continuous walls and discrete rings. This superficial structure is similar almost in all details to that of cobalt single crystals (Bates and Craik, 1962). The changes of the patterns in a magnetic field parallel to the easy axis is illustrated in Figure 9.38. The white regions were interpreted as small closure domains magnetized across the easy direction.

Thus the structure may be considered as intermediate between a free pole and closure domain structure as suggested by Bates and Craik for cobalt single crystals.

The most interesting domain structure is found in Co_2Z compounds which exhibit planar anisotropy. On surfaces parallel to the c axis the patterns consist of a set of nearly parallel walls directed across the c axis (Figure 9.39(a)) whereas on basal planes a great variety of domain shapes can be observed (Figure 9.39(b)). They are triangular, trapezoidal, half moon loops or just fine lines which stretch from one end of the surface to another. Occasionally spike domains are observed at the end of the specimen surface, very similar to those observed on low anisotropy thin films.

The structure of this material was interpreted as one of a 'sandwich' like form with uncharged Bloch walls parallel to the easy plane of magnetization as Kaczer (1962) proposed for infinite cylinders with planar anisotropy. Estimation of the equilibrium domain width l_0, using Equation (9.82) gives $l_0 = 3.3\,\mu m$ which agrees well with the experimental findings of $l_0 = 3.4\,\mu m$. Thus the model represented in Figure 9.12(b) for the domain structure of this compound seems very probable. The patterns observed on basal planes were explained by assuming local variations of the anisotropy in the hexagonal plane owing to the presence of strains, non-magnetic inclusions or other lattice defects which are effective in the case of low anisotropy materials (Rosenberg et al., 1964).

9.4.1.3 Compounds with Y structure

All known compounds having the chemical formula $BaMeFe_6O_{11}$ where Me is a divalent ion as in the previous paragraph, show planar anisotropy. It has been found that the domain structure is entirely similar to that of Co_2Z described in Section 9.4.1.2 (Verwell, 1967a, 1967b).

9.4.2 Orthoferrites

Domain structures in orthoferrite single crystals were first observed by Sherwood et al. (1959) in eleven compositions by means of the Faraday effect and the Bitter technique. Owing to the exceptionally low M_s and high H_K the domain structure consists of a set of 180° domains, the magnetization within

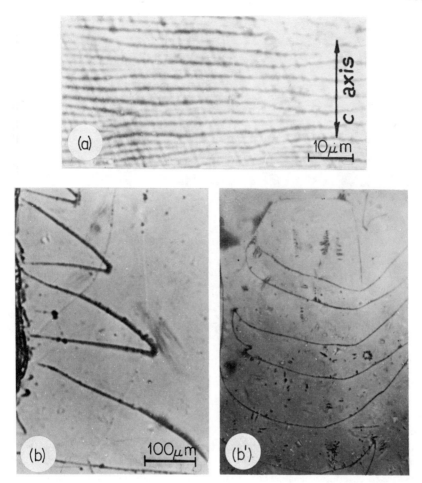

Figure 9.39 Domain patterns on Co_2Z single crystals with planar anisotropy: (a) axial plane (b) and (b') basal plane.

a domain being nearly uniform and aligned along the c axis. The domain wall patterns at opposite surfaces normal to the easy axis, in homogeneous samples, are mirror images indicating that the walls cut completely through the sample.

For thin platelet specimens the structure in the demagnetizing state consists of an array of strip domains, as illustrated in Figure 9.40. The dependence of l on c was recently theoretically analysed by Farztdinov *et al.* (1970a, 1970b) who found $l_0 \propto c^{\frac{1}{2}}$ as in the case of the Kittel structure, but so far there is no experimental evidence for such a dependence. It has been shown (Kurtzig and Shockley, 1968; Rossol, 1968, 1969) that in such samples planar parallel, equally spaced walls can be induced by the use of a properly applied magnetic field pattern, as obtained by passing a direct current through a set of parallel, equally spaced wires lying within a plane under the sample. Such structures

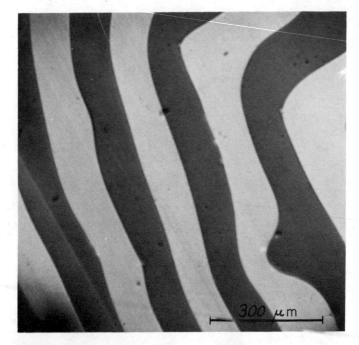

Figure 9.40   Strip domains in yttrium orthoferrite (63 μm thick) as shown by the Faraday effect (Craik).

are stable but the spacing between adjacent walls can be varied by at least a factor of two (Rossol, 1968, 1969). Thus it was not possible to determine which of these configurations is the lowest energy configuration for the sample.

Calculation of wall energy density for $ErFeO_3$ using Kittel's equation for such a configuration have shown that γ_w varies from 1·4 to 2·3 erg/cm². It is possible to calculate more reliable values for such structures using the method of Kurtzig and Shockley (1968) which consists of stretching the planar walls into sinusoidal corrugation by means of a sinusoidal magnetic field obtained by passing a direct current through a set of parallel, equally spaced wires lying under the sample and directed perpendicular to the walls. Opposite currents in adjacent wires give rise to an increase of the total wall area and energy. Calculation of the work exerted by the applied and demagnetizing magnetic field in forcing the controlled increase of the wall area in terms of the pressure due to the applied field P_a, the pressure due to the demagnetizing field P_d, and the pressure due to the domain wall surface tension P_w give the measure of γ_w. These pressures are reversible and are related at a static wall by $P_a + P_d + P_w = 0$.

The sample must be carefully prepared and free of stresses and other imperfections. For instance in the case of $ErFeO_3$, using 10 samples of various thicknesses, the values of γ_w ranged from 1·9 to 2·1 erg/cm². The precision is limited by the sample inhomogeneity and experimental uncertainty and not by inherent uncertainty of the method. Other methods and results are given in Chapter 13.

Domain structures of particular shapes appear in orthoferrite crystals in which the size along the easy axis is comparable with that across it. Owing to their low spontaneous magnetization the critical size for single domain particle is of the order of 10^{-2} cm (compared to 10^{-4} cm for barium ferrite). Therefore single crystals with sizes of around a millimetre, which can easily be handled, will contain only a small number of domains (Craik, 1970).

In Figure 9.41 very simple domain structures were obtained on $EuFeO_3$ simply by reducing the size of a single crystal to $1.5 \times 2.6 \times 2.8$ mm^2 and then

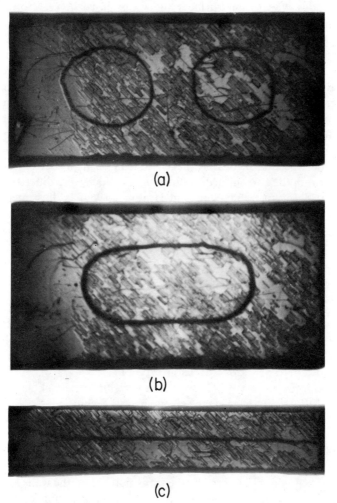

(a)

(b)

(c)

Figure 9.41 Simple domain structures on small $EuFeO_3$ single crystals, 2.8 mm thick obtained by thermal demagnetization. By reducing a dimension (normal to the easy axis) of the crystal toward the critical diameter for collapse, structures with cylindrical domains become unstable and simple structures with two equal domains separated by a straight wall parallel to the long dimension become favourable (Craik and Tănăsoiu).

thermally demagnetizing it. Structures consisting of a single elliptical domain or just two cylinders were achieved (Craik, 1970). If the dimension normal to the easy axis is reduced towards the critical diameter for collapse in the larger crystal, then the structure consists of two domains separated by a domain wall which runs parallel to the long axis of the surface (Figure 9.41(c)). It is assumed that such structures do not represent the minimum energy configuration.

The magnetization processes in such crystals are dependent on the physical state of their surfaces, particularly those perpendicular to the easy axis. In perfect crystals with virgin surfaces the walls move in very low fields and saturation is achieved in fields less than fifty oersteds, but once saturated they remain single domain and fields of the order of kilo-oersteds are required to reverse the magnetization by nucleation and growth of reverse domains. Multidomain structures are obtained in such crystals only by thermal demagnetization. But as soon as the crystal becomes strained, simply by mechanical polishing, the nucleation field is drastically lowered and domains may be spontaneously created by removing the saturating field. Moreover the nucleation field is a function of the saturating field as shown in Figure 9.42(a) for a crystal of $FuFeO_3$ mechanically polished on both surfaces perpendicular to the easy axis. The most peculiar feature is the asymmetry of the hysteresis loop created by the strains in the two surfaces of the crystals which were differently polished (Craik and McIntyre, 1966; Shur and Hrabrov, 1969) as illustrated in Figure 9.42(b).

The asymmetry of the loop depends upon the detailed surface treatment. So far there is no full explanation of this phenomenon.

9.4.3 Haematite

As discussed in Chapter 11 haematite, like the orthoferrites, is a canted-spin weak ferromagnet, with a Morin transition at $-10\,°C$. In natural crystals the transition is not so sharp and it has been found that domain walls are still present at all temperatures down to $-40\,°C$ even in those crystals for which the transition occurred at $-20\,°C$ (Gustard, 1967).

Magnetic domains in α-Fe_2O_3 were first identified by Blackman, Haigh and Lisgarten (1957, 1959) using the electron shadow technique (see also Williams et al., 1958; Kaye, 1961; Szymczak, 1970).

The domain patterns observed on (110) surfaces of synthetic haematite crystals at temperatures above the transition consist of long continuous walls occasionally forming closed loops but without any geometrical symmetry (Gallon, 1968). On cooling through the transition region the walls are shown to disappear either by a denucleation process or by a fading of the patterns (Gustard, 1967). An example of such a behaviour is illustrated in Figure 9.43.

Above the transition temperature the walls are easily moved in very low fields (<2 Oe) while at lower temperatures fields up to 100 Oe are required. On natural haematite crystal the patterns are less regular and more stable over a wide range of temperature below the transition region.

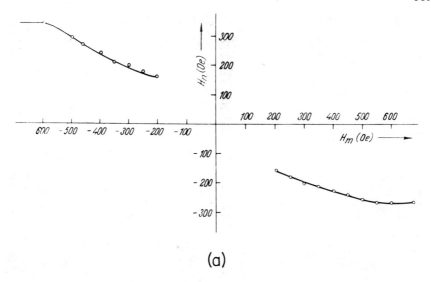

(a)

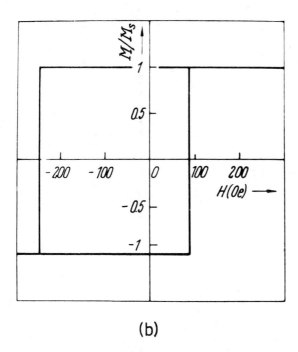

(b)

Figure 9.42 (a) The dependence of H_n on the peak magnetizing field for an $EuFeO_3$ single crystal with both basal planes identically polished. (b) Asymmetric hysteresis loop for the same crystal as in (a) but with the basal planes unequally polished, one with fine emery, the other with 1 μm diamond (Craik and Tănăsoiu).

540

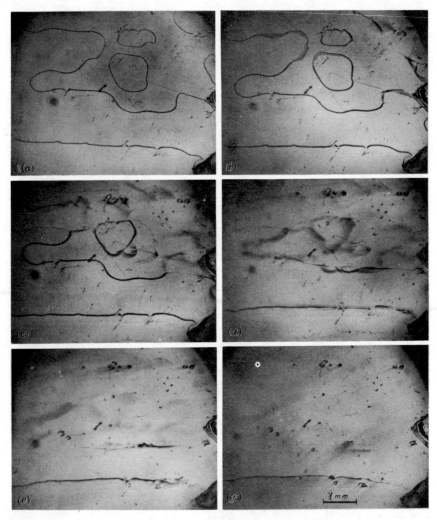

Figure 9.43   The domain structure on a (110) face of a synthetic haematite crystal at temperatures in the neighbourhood of the transition: (a) $-11.5\,°C$, (b) $-11.7\,°C$, (c) $-11.9\,°C$, (d) $-12.2\,°C$, (e) $-12.5\,°C$, (f) $-13.0\,°C$ (overall magnification $\times 17$) (Gallon, 1968).

By annealing natural crystals at suitably high temperatures for a long time the strains are partly removed and the behaviour of the domain patterns approaches that of synthetic crystals.

Eaton and Morrish (1969) reported a systematic investigation of the domain patterns on large highly perfect single crystals of pure and doped haematite. Their observations on specimens of various thicknesses (25–200 μm), first mechanically polished and then carefully annealed at 1000 °C to remove as much strains as possible showed that on surfaces parallel to the basal plane

[(111) plane] no patterns are formed while on any plane perpendicular to the basal plane a regular structure consisting of a set of walls lying parallel to the basal plane and regularly spaced 50–150 µm apart was observed (Figure 9.44).

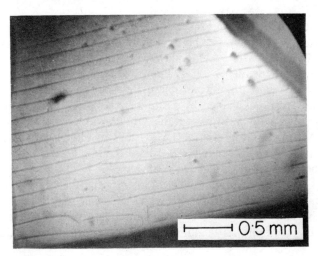

Figure 9.44 A 200 µm section of a pure haematite crystal. The surface is perpendicular to the basal plane. The edge at the bottom of the photograph lies in the basal plane (Eaton and Morrish, 1969).

When fields are applied in the basal plane the walls move up to 10 µm for fields less than 1 Oe, and in general all domain patterns are removed by about 25 Oe. The manner in which the patterns coalesced suggests the existence of only two directions of magnetization, i.e. the domains with magnetization oriented favourably with respect to the applied field are enlarged at the expense of those unfavourably oriented. Therefore it is assumed that the magnetizations of the adjacent domains are antiparallel and the domain walls are 180° (Eaton *et al.*, 1968, 1969).

An estimate of the wall energy made from domain patterns for thin sections (Figure 9.44) using the formula $\gamma_w = (1{\cdot}7m_D^2 l^2)/c$, where m_D is the weak Dzialoshinsky magnetization yields $\gamma_w = 10^{-3}$ erg/cm^2.

9.4.4 The influence of temperature on domain structure

Fundamental changes are to be expected in domain structure as a function of the temperature due to the changes of the parameters which define the magnitude and shape of the domains such as M_s, K and A.

For uniaxial crystals with Kittel-like domain structure, this is particularly obvious from the relation (9.60) written as

$$l_e = [2(AK)^{\frac{1}{2}}(1 + \mu^*)c/1{\cdot}7M_s^2]^{\frac{1}{2}} \qquad (9.102)$$

which gives the equilibrium structure at a given temperature T.

542

This problem has been investigated by different authors using mainly two methods: Faraday effect and a colloidal modified technique (Rosenberg *et al.*, 1970; Carey and Tănăsoiu, 1969; Florescu and Rosenberg, 1970; Gemperle *et al.*, 1963; Rossol, 1968; Szymczak, 1968).

The experimental observations carried out by Gemperle *et al.* (1963) on thin specimens of magnetoplumbite between $-150\,°C$ and $+380\,°C$ and by Kojima and Goto (1962) on barium ferrite between $-100\,°C$ and $+200\,°C$ have shown a nearly linear dependence of $l(T)$ in this temperature range. A thermal hysteresis was observed in magnetoplumbite especially in the low temperature region, qualitatively explained by assuming that new domains nucleate with difficulty at lower temperature when the domain width must decrease. A similar hysteresis was observed by Rossol (1968, 1969) on Thulium orthoferrite. From $l(T)$ it was possible to calculate $\gamma_w(T)$ using Equation (9.102) with $\gamma_w = 4(AK)^{\frac{1}{2}}$ and also the temperature dependence of A.

For thicker crystals with undulating domain walls, there has also been found to be a decrease of the wall amplitude with increasing T, as illustrated in Figure 9.45 for a barium ferrite single crystal (Szymczak, 1970). A similar behaviour was found in strontium ferrite aluminate with $x < 2$ (Florescu and Rosenberg, 1970).

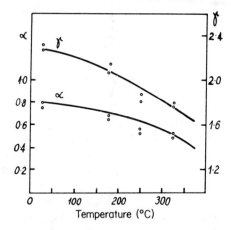

Figure 9.45 The dependence of amplitude α and period γ on temperature for barium ferrite (Scymczak, 1970).

Interesting features of the domain structure as a function of temperature appear in the case of crystals with a small number of domains as in the case of strontium ferrite aluminates with $x > 2$ (Rosenberg *et al.*, 1970). The demagnetized structures depend on whether this state is achieved by thermally or by a.c. demagnetization, the more stable being those obtained by a.c. demagnetization (Carey and Tănăsoiu, 1969).

The patterns are very stable up to the Curie point, when they suddenly disappear and on cooling down quite different patterns are formed (Rosenberg *et al.*,

1970). This was explained by assuming the existence of regions with different chemical compositions within the crystal, acting as traps for the walls.

9.4.5   Uniaxial ferrimagnetic materials: polycrystals

Although the majority of magnetic oxides for practical purposes are in the form of polycrystals, little attention has been paid to the study of their domain structure. The lack of information is partly explained by the difficulty encountered in revealing the actual domain structure in fine grain specimens by the usual techniques described in Section 9.3. To the knowledge of the authors the only works dealing with this problem are those due to Sixtus et al. (1956) and Craik (1959). In Sixtus's work the Bitter patterns on specimens having different grain sizes and different coercivities were examined by the usual optical microscopy. They found that grains with a thickness less than about 10 μm do not form domain structure, but their failure to observe domain structures in finer grains was primarily due to the limited resolution of optical microscopy technique. Craik, using the electron microscope technique, succeeded in observing domain structure in grains with thickness down to about 1 μm.

Generally it is to be expected that each grain in a polycrystal will have a structure typical of relatively large single crystals but somewhat modified by the interactions with the neighbouring grains. Figure 9.46(a) and (b) shows typical patterns on the axial planes of two crystals representing 180° domains lying parallel to the easy axis. In Figure 9.46(a) the walls run continuously from one grain to the other while in Figure 9.46(b), spike domains are formed at the boundary, due to the misorientation of the crystallites against the easy axis. In

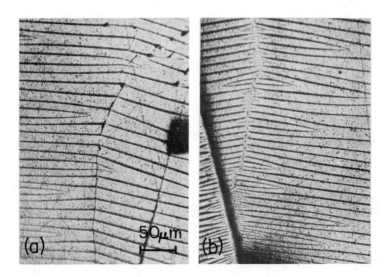

Figure 9.46 Domain patterns on axial planes in barium ferrite polycrystals with very large crystallites.

Figure 9·47 Domain structures on axial (a) and basal planes (b) in grain oriented barium ferrite by the electron microscope replica technique (Craik).

such specimens l is controlled not only by the thickness of the individual grain but also by that of the neighbouring grains with which it interacts.

In specimens with smaller grains the patterns frequently consist of only few straight boundaries. An example of such patterns on both lateral and basal plane in grain oriented barium ferrite is illustrated in Figure 9.47. The sizes of the particles with specific number of domains are shown in Table 9.4. It seems

Table 9.4 Domain width dependence on particle thickness in polycrystalline barium ferrite (from Craik, 1959)

Number of domains	Particle thickness (μm)	Domain width (μm)
1	<1·2	1·20
2	2·7	1·35
3	4·0	1·32
4	5·5	1·35
5	6·0	1·30
6	7·0	1·30
7	8·0	1·20

that the domain spacing is practically independent of the particle size and from these measurements the corresponding critical diameter for single domain particle is about 1 μm, while the critical diameter for non-interacting particle of barium ferrite has been found to be 0·5 μm (Shirk and Buessem, 1970). This difference can be ascribed to the interaction between particles in a polycrystalline specimen.

The particle interaction has also an effect on $l(c)$. Figure 9.48 shows the $l(c)$ dependence calculated by using Kittel Equation (9.60) and the experimental points (Craik, 1959). It can be seen that all points are above the theoretical curve calculated for an isolated particle.

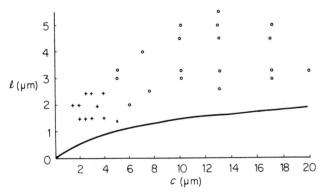

Figure 9.48 The dependence of the domain width on grain size in two specimens of polycrystalline barium ferrite (Craik).

Hysteresis loops in fine grain specimens are distinctly square with only slightly rounded corners (e.g. Craik, 1959). The main change in the magnetization takes place over a narrow field range (about 15 Oe). The individual grains are saturated at remanence and no domain structure can be observed. The small decrease of magnetization which occurs when the field is reduced from the high values to zero is accounted for by deviations from perfect orientation of the crystallites. The extent of the misorientation can be estimated from micrographs and generally the experimental findings are in good agreement with the observed value.

9.4.6 Cubic materials: single crystals

Though the variety of spinel ferrites is rather great the domain structures have been studied for only a small number of them. Difficulties are experienced in observing magnetic domains due to the rather poor contrast of powder patterns, owing to the low values of M_s and particularly K of this group of materials. But with great care and experimental skill domain structures of typical spinel ferrites have been successfully revealed.

9.4.6.1 Materials with positive anisotropy

In these materials M_s lying along the $\langle 100 \rangle$ easy axes, will give both 180° and 90° walls. If the specimen surface examined is parallel to the (100) and (110) crystallographic planes the domain patterns can be easily interpreted. From this group of materials only Co ferrite has been investigated (Carey and Isaac, 1963; Stoltz, 1966).

Examination of the surfaces parallel to (100) indicated that different regions of the same crystal surface show domain structure typical of uniaxial crystals. Each [100] type direction can become the easiest axis of magnetization in separate regions of the crystal where the magnetic anisotropy has a strong uniaxial component in the direction concerned. Basal patterns are identical with those on cobalt single crystals.

Stoltz showed that it was possible, by thermomagnetic annealing, to induce a single preferred direction of easy magnetization parallel to one [100] direction so that the remanent domain structure become characteristic of a uniaxial material with an anisotropy comparable to that of metallic cobalt.

9.4.6.2 Materials with negative anisotropy

With eight easy directions of magnetization along [111] axes the domain structures should be more complex than those described above. Now surfaces parallel to (110) will contain two easy axes of magnetization which intersect at angles of 70° 32' and 109° 28'. Therefore walls of 71° and 109° are expected to be demonstrated, as well as 180° walls. Spinel ferrites of this group on which domain structures were observed by various methods are Cu, Mn, Ni, Mn–Zn, Mg–Mn ferrites and magnetite.

To reveal the true domain structure in spinel ferrites great care must be taken in preparing strain free surfaces for observation. Such surfaces were successfully prepared by polishing with fine diamond polishing compound and then chemically etching.

9.4.6.2.1 Manganese ferrite. Observations of domain structure on (110) planes have shown the expected type of patterns with 180°, 109° and 71° walls (Bates *et al.*, 1958a; Craik and Griffiths, 1959; Kawado *et al.*, 1968) very similar to those observed on Ni crystals by Yamamoto and Iwata (1953). In some cases a pronounced curvature of the walls was observed and explained in terms of the presence of small varying stresses in the crystal which cause the anisotropy to vary from place to place. Stressed surface give more regular patterns which do not represent the true domain structure.

The main characteristics of the domain structure of Mn–Zn ferrite are similar to those of Mn ferrite (Craik and Griffiths, 1959).

9.4.6.2.2 Nickel ferrite. On natural (111) planes the domain patterns consist of a system of three groups of fundamental lines which intersect at angles of approximately 120°. Since such planes do not contain easy axes, very complicated closure domains are formed (Craik and Griffiths, 1959; Rosenberg *et al.*, 1966). Figure 9.49 shows typical detailed structures. On a (110) plane the

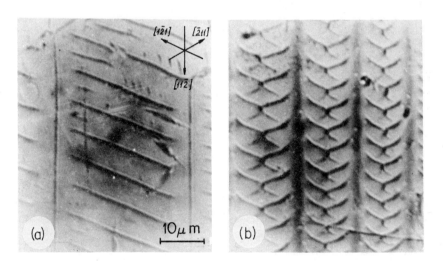

Figure 9.49 Typical details of the domain structures on (111) planes of nickel ferrite single crystal.

patterns consist of a system of nearly parallel lines directed along [001] (Figure 9.50) with 71° walls. The peculiar feature of this structure is the existence of faint walls alternating with more intense ones. A small magnetic field applied parallel to the walls caused the fine lines to disappear. At higher fields the walls rotate to a [111] direction and probably become of the 180° type.

Paulus (1960) investigating the domain structure of a nickel ferrite containing a 0·1 % Co, discovered a special type of 360° wall on a (100) plane.

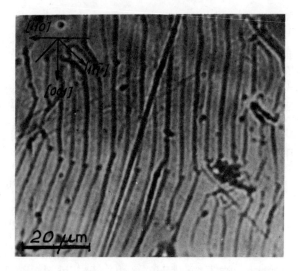

Figure 9.50 Domain patterns on (110) plane of nickel
ferrite single crystal.

The domain structure in nickel ferrite single crystals has also been investigated by a ferromagnetic resonance method (Pilshchikov and Syreev, 1969).
This consists in determining the type of the domain structure from the
dependence of the resonant frequencies for the longitudinal modes, since they
are not connected with the displacements of the domain walls. Measurements
have been made on spherical samples in the case of orientations along [100]
and [011] axes. The most probable structure found in this way is a perpendicular
plate-like domain structure (with boundaries perpendicular to the [011] axis),
for both orientations.

Similar measurements were carried out on YIG, MnMg-ferrite and Li-ferrite
single crystals (Dudkin and Pilshchikov, 1967). For YIG the structure is
entirely similar to Ni-ferrite while for MnMg and Li-ferrites the most probable
structure is a parallel domain structure both for field orientation along the
[011] axis and along [100].

Very simple domain structures can be obtained in frame-shaped samples cut
in such a way that each leg of the frame lies along one of the [111] easy direction
and the major face is a (110) plane as in Figure 9.51. The domain structure of
such a sample consists of four stationary walls, one at each corner and one
movable wall which goes all the way around the sample. Such simple structures
wese actually observed in a Ni–Fe-ferrite single crystal by Galt (1954). Owing
to imperfections of strains persisting even in very carefully prepared samples,
the walls along each leg were curved at some points but the main domain
structure consisted of only two domains in each limb.

9.4.6.2.3 Magnetite. Most observations have been made on natural magnetite
single crystals and only a few on synthetic crystals (e.g. Blackman *et al.*, 1963).

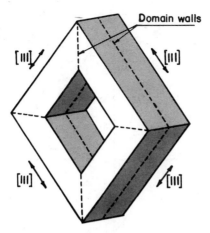

Figure 9.51 Single crystal sample of cubic ferrite with only one movable domain wall. Domain walls are shown dotted (Galt, 1954).

Plane and strain-free surfaces were obtained by electrolytic polishing (Soffel, 1965) or annealing in vacuum at high temperature (Vlasov and Bogdanov, 1964) after careful mechanical polishing. Etching at 600 °C in a mixture of boron and lead oxides also gives strain-free surfaces (Hanss, 1964).

On surfaces parallel to (111) complex patterns with closure domains are formed.

Domains of parallelogram shape with 180°, 109° and 71° walls are observed on planes parallel to (110). These are entirely similar to those described in detail by Yamamoto and Iwata (1953) for Ni single crystals.

The influence of elastic stresses was investigated by Bogdanov and Vlasov (1965, 1966a, 1966b), the stress σ being applied along the $[\bar{1}1\bar{1}]$ direction in the (110) plane. The structure in the absence of the stress consists of simple 180° domains magnetized along $[\bar{1}1\bar{1}]$ as shown in Figure 9.52(a). The domains start to move at $\sigma = 5.5 \, \text{kg/mm}^2$ and at the same time closure domains appear (Figure 9.52(b)). At $\sigma = 8.2 \, \text{kg/mm}^2$ the structure changed to give domains magnetized along $[\bar{1}1\bar{1}]$ and $[\bar{1}11]$ directions (Figure 9.52(c)). At $\sigma = 8.5 \, \text{kg/mm}^2$ all the domains changed and their magnetization made an angle of 71° with the stress direction (Figure 9.52(d)). Complicated structures with spike domains are formed if a tensile stress is applied along $[\bar{1}10]$ direction. All these results correspond to the positive magnetostriction of magnetite.

9.4.6.3 Ferrimagnetic garnets

Sufficiently thin sections of garnets are transparent in the visible portion of the spectrum thus allowing the domain structure to be investigated by the Faraday effect (Dillon, 1958).

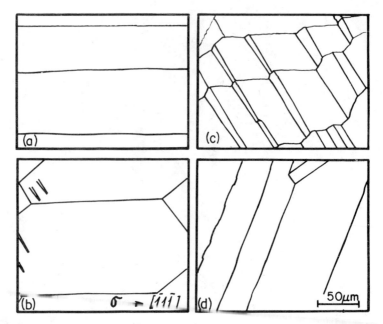

Figure 9.52 Drawing of the changes in the domain pattern of magnetite on (110) plane under the influence of a tensile stress applied along $[\bar{1}1\bar{1}]$ (Bogdanov and Vlasov, 1966).

(a) *Strained samples*

The domain structure has the form of irregular strips in which the magnetization is perpendicular to the plane of the sample and antiparallel in adjacent domains. Strains produced either by polishing or during crystal growth result in a uniaxial anisotropy which may override the cubic anisotropy. Such a strain structure is clearly seen in the central part of Figure 9.53 which is a GdIG slice 40 μm thick. There are large strain-free portions in this sample, containing in-plane domains with 180°, 109° and 71° walls. The surfaces of the sample were polished first with 6 μm and $\frac{1}{4}$ μm diamond paste and then carefully Syton polished down to the final thickness of 40 μm (Craik, private communication). If only one side of the sample is ground, the other being a growth face, the strain is generally less important (Smith and Williams, 1960, 1961; Lefever *et al.*, 1962).

Cylindrical, or 'bubble', domains can be produced in strained crystal slices (Nemchik, 1969) or epitaxial films with lattice mismatch: these are the basis of bubble domain devices and are described in Chapter 13.

(b) *Strain-free samples*

Using various etching and polishing procedures (Smith and Williams, 1960; Lefever *et al.*, 1965; Basterfield, 1968) essentially strain-free samples can be achieved, as indicated by typical domain structures for materials with negative cubic anisotropy. An example is illustrated in Figure 9.54 on an (110) plane of YIG slice 40 μm thick after two hours Syton polishing: only in-plane domains with 180°, 109° and 71° walls are present (Craik, private communication).

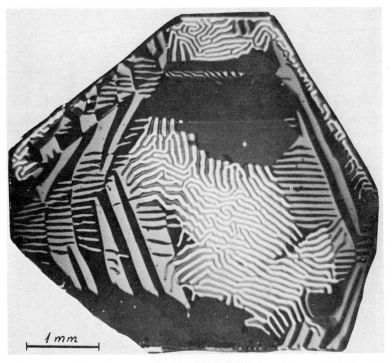

Figure 9.53 Domain structure of (211) GdIG slice, 40 μm thick, after fine mechanical polishing. The strained regions give rise to strip domains magnetized up and down (Craik).

Figure 9.54 Domain structure in a thin strain-free (110) section of YIG. The sample was Syton polished for two hours. Only in-plane domains characteristic for strain-free samples are present (Craik).

Bloch lines separate the black and white sections of 109° and 180° domain walls, as expected for materials in which free poles associated with the intersections of the Bloch wall with the surface of the crystal make a substantial contribution to the total wall energy.

The influence on domain structure of stresses introduced either during the crystal growth or applied along a certain direction, was studied in some detail by Lefever *et al.* (1962, 1965): the presence of a small amount of silicon as an impurity diminishes the effects of strain

Correlation between the surface and the bulk structure can be made by infrared microscopy (Enoch and White, 1971). A crystal of YIG, about 1·7 mm thick (Figure 9.55(a)) was polished on the two (110) faces with diamond paste of

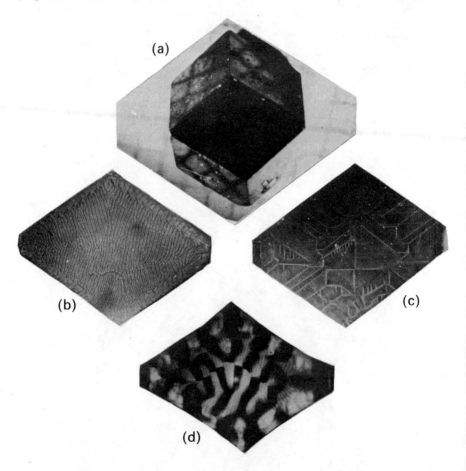

Figure 9.55 (a) YIG single crystal about 1·7 mm thick, the top and bottom surfaces being (110). (b) The domain structure on the top surface of the crystal by the colloid technique. (c) The domain structure of the bottom surface by the colloid technique. (d) The bulk domain structure by the infrared Faraday technique (Enoch and White, 1971).

$3\,\mu m$, $1\,\mu m$ and $\frac{1}{4}\,\mu m$ particle size. The colloid and infrared Faraday studies showed that: (a) patterns on the top surface indicated simple strip domains magnetized normal to the surface (Figure 9.55(b)) with spacing of about $12\,\mu m$; (b) the bulk domain structure consisted of 'ribbon' domains $120\,\mu m$ wide, running nearly parallel to the [001] axis, with the magnetization in the domains normal to the surface (Figure 9.55(c)); (c) strain-free structures with in-plane domains separated by $180°$, $109°$ and $71°$ walls are formed on the bottom surface (Figure 9.55(d)).

The existence of the in-plane structure on the bottom surface could be explained by assuming that the uniaxial anisotropy was induced during the crystal growth and not during the surface preparation by mechanical polishing.

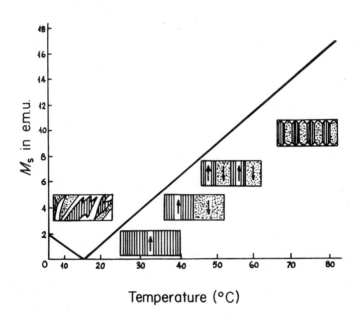

Figure 9.56 Illustration of typical domain structures in GdIG at different temperatures and thus at different values of M_s (Craik, 1970).

Interesting results were reported by Craik (1970) for strain-free specimens of GdIG in the range of the compensation temperature ($T_c = 12\,°C$). Small rectangular plates cut from thin (211) slices which contained only one [111] easy axis in this plane were investigated. Typical results are schematically illustrated in Figure 9.56. The most peculiar is the structure found in the narrow region of the compensation temperature where generally no domains are to be expected. With increasing temperature, structures with closure domains are formed. The width of the main domains decreases as the temperature increases.

9.4.7 Cubic materials: polycrystals

As for single crystals the main difficulty in observing domain structures in cubic polycrystalline specimens is connected with strain-free surface preparation, which is particularly troublesome in ferrites with low anisotropy. The general method of surface preparation (Craik, 1959) consists mainly of annealing the sample at a suitably high temperature after careful mechanical polishing (for a short time to avoid a detectable grain growth or a change in magnetic properties). This method has been used for polycrystalline nickel and cobalt ferrites (Craik, 1959), manganese zinc ferrites (Bates et al., 1958b) and magnesium-manganese ferrite (Knowles, 1960). A similar method was developed by Soffel (1965) for polycrystalline natural magnetite, in which case the specimens were annealed in vacuo for 60 hours.

The interpretation of the patterns obtained in cubic polycrystals is more difficult than in the uniaxial case due to the six or eight easy directions and the general random distribution of the crystallites.

Figure 9.57(a) shows patterns on polycrystalline cobalt ferrite with a very sharp definition, corresponding to the high K and low δ, which permits the portrayal of minute reverse domains at grain boundaries and other sub-micrometre features. Some of the smaller grains, as at A, appear to consist of only two domains. Patterns on some grains of a nickel ferrite specimen, with average diameter 12 μm (9.57(b)) indicate a finely-divided superficial closure structure reminiscent of those at (100) or (111) surfaces of nickel metal, while simpler patterns of walls some 2–3 μm apart probably represent the domain spacing within the crystallites. Craik also showed that the crystallites of a similar specimen with grain size 2·5 μm were either single-domain or contained only two domains.

A detailed investigation of domain behaviour in a single particle of about 50 μm in a bulk specimen of polycrystalline Mn–Mg-ferrite was made by Knowles (1960). The grain had a rectangular section with surface parallel to (110) plane and its domain structure consisted of a few rather parallel 180° walls. The magnetization of the grain was deduced from the total area of favourably and unfavourably magnetized domains at any given applied field. The hysteresis loop of this particular grain was compared with that for the entire specimen. Although the loop of the grain was rather similar to that of the whole sample it does not possess a well defined knee; this being explained by assuming that heavily stressed areas pinned the walls until a sufficiently large field forced them past the greatest stress gradient, when they moved forward irreversibly. A general treatment based on these observations gives loops for the whole specimen which correspond well with those measured.

A very peculiar hysteresis loop was found in a small grain of natural magnetite (about 50 μm) in a polycrystalline sample (Soffel, 1965), which was explained by the assumption that due to the negative anisotropy of magnetite, not only the 180° walls seen at the surface are formed, but 71° and 109° walls are likely to

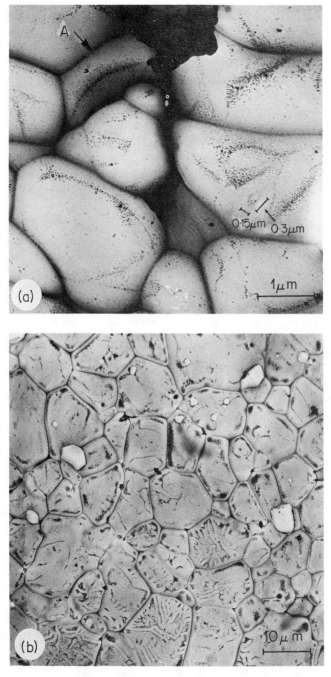

Figure 9.57 Typical domain patterns on (a) a fine grained cobalt ferrite and (b) nickel ferrite, obtained by electron microscope replica technique (D. J. Craik).

exist underneath the polished surface of the grain. Therefore, even in the demagnetized state, the grain apparently possesses a finite remanence.

9.4.8   Other oxides

9.4.8.1   *Chromium dioxide, CrO₂*

The domain structure of CrO_2 has been observed by the colloid technique on single crystals, 2–18 μm thick, formed epitaxially on highly polished TiO_2 single crystal substrates (Rodbell *et al.*, 1967). Patterns with zigzag domain walls were observed on the basal planes. The estimation of wall energy from the dependence of coercive force on sample thickness gives $\gamma_w = 0.4 \text{ erg/cm}^2$.

9.4.8.2   *Antiferromagnetic oxides*

Domain structures in antiferromagnetic oxides such as NiO, MnO, CoO, Cr_2O_3, have been reported by various authors in the last decade.

Since these crystals possess no magnetic moment, their domains differ basically from those found in ferromagnetic or ferrimagnetic oxides.

The most studied oxide from this group, theoretically and experimentally, is NiO and in this section we shall briefly consider the main results obtained on this type of oxide. A review paper on domain structures in antiferromagnetic oxides has been given by Farztdinov (1964d).

9.4.8.2.1   Domain walls. The particular type of domain boundary found in antiferromagnetic oxides is due primarily to the fact that the crystal lattice is deformed, either by contraction or by expansion along certain crystallographic axes, by the antiferromagnetic ordering below the Néel point.

Nickelous oxide is paramagnetic above $T_N = 523$ K with the cubic rock salt structure and becomes antiferromagnetic below T_N with a structure slightly distorted to rhombohedral. The amount of distortion increases as temperature decreases. This distortion consists of contraction of the original unit cell along one of the four [111] axes and causes a macroscopic crystal to be twinned below T_N and to be composed of many small regions each characterized by one of the four [111] contraction axes and each twinned to other such regions on (100) or (110) planes. These particular regions are antiferromagnetic domains and the twinning planes form a kind of antiferromagnetic domain wall. Consequently a twinning plane across which a change in contraction axis occurs is named a *T* (twinning) wall. A second type of antiferromagnetic wall called *S* (spin rotation) wall occurs within a region which possesses a given contraction axis but where there is a rather abrupt change in spin orientation within the (111) plane. Two types of *S* walls can be distinguished as shown in Figure 9.58: one in which the wall is parallel to the ferromagnetic sheets, denoted $S_{\parallel}$ and the other in which the wall is perpendicular to the ferromagnetic sheets and contains the rhombohedral axis, named $S_{\perp}$. The anisotropy in the (111) plane is very small and although a variety of *S* walls are possible, the most stable walls are formed

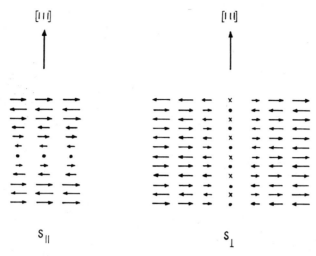

Figure 9.58 Two different types of S domain walls. In the $S_{\parallel}$ wall the magnetization direction changes by rotation of adjacent ferromagnetic sheets while in the $S_{\perp}$ wall the rotation takes place within the ferromagnetic sheet (Roth, 1960).

between regions where the angle between spins is 90°, 120° and 180°. Since exchange energy is involved in a S wall, the transition layer must have a certain thickness. Calculations made by Farztdinov (1964a, 1964b, 1964c) on wall thickness and wall energy for some antiferromagnetic oxides, taking into consideration higher order anisotropy constants, gave the results shown in Table 9.5.

Table 9.5 S wall energy density and width in antiferromagnetic monoxides (Farztdinov)

Oxide	δ (Å)	γ_w (erg/cm^2)
NiO	3000	0·15
MnO	600	0·5
Cr$_2$O$_3$	300	4·6

Following Slack (1960) there are 12 possible T walls in NiO, due to the contraction along one of the four diagonals of the original cubic lattice. A twinning plane is a reflection plane, in which the crystal on one side is a mirror image of that on the other. The only planes which satisfy this condition are those which bisect the angles between the cube diagonals, namely (100) and (110) planes, as illustrated for two simple cases in Figure 9.59 for NiO. The four differently oriented rhombohedral regions into which a cube can distort are designated as crystal regions of the following types: I—[111]; II—[111]; III—[111] and IV—[111]. Thus the 12 possible T walls are those given in Table 9.6.

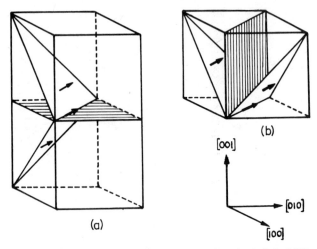

Figure 9.59 Two types of simple T walls (shaded) in NiO
(Roth, 1960).

Table 9.6 Possible crystallographic planes for T walls
in NiO (Slack, 1960)

Region pair	Twinning plane	
I–II	(001)	(110)
I–III	(010)	(101)
I–IV	(100)	(011)
II–III	(100)	(01$\bar{1}$)
II–IV	(010)	(10$\bar{1}$)
III–IV	(001)	(1$\bar{1}$0)

Though a T wall is expected to have no thickness, calculations made by Yamada (1960) by taking into consideration the superexchange energy between neighbours along [100], the exchange energy between neighbours along [110], the elastic energy, the magnetic dipolar energy and a minor anisotropy energy, have shown that T walls have a finite thickness. The results for NiO and MnO both for wall thickness and for the wall energy associated with them are given in Table 9.7.

T walls are caused to move either by the application of a small mechanical pressure or, in well annealed crystals with few T walls, by magnetic fields from

Table 9·7 T wall energy density and width in
antiferromagnetic monoxides (Yamada, 1966)

Oxide	δ (Å)	γ_w (erg/cm^2)
NiO	80	4
MnO	9	20

5 kOe to 20 kOe. Slack found that minimum stress for T wall motion, expressed in terms of pressure corresponds to about 240 dynes/cm^2. The energy dissipated in moving the T walls, in magnetic fields, expressed also in terms of pressure, is given by

$$P_w = (1/2)H^2\rho(\partial\bar{\chi}(g)/\partial g) \tag{9.103}$$

where ρ is the density of NiO, g is the fractional volume of the crystal that is of type I,

$$\bar{\chi}(g) = (1/2)[\chi_{\perp 1} + (1/2)(\chi_{\perp 2} + \chi_{\parallel})] + (1/6)[\chi_{\perp 1} - (1/2)(\chi_{\perp 2} + \chi_{\parallel})]$$
$$\times [9 - 32g(1 - g)]^{\frac{1}{2}} \tag{9.104}$$

and $\chi_\perp$ and $\chi_\parallel$ are the susceptibilities measured perpendicular and parallel to the preferred spin direction respectively. They start to move at $P = 100$ dynes/cm^2 and the whole crystal become a single type at about 4000 dynes/cm^2, values which are close to those found mechanically.

9.4.8.2.2 Domain observation. The antiferromagnetic domains can be made visible by methods which may completely differ from those applied in the ferrimagnetic materials.

(a) *Neutron diffraction.* The scattering of the neutrons by the (111) planes depends on the crystallographic orientation of the ferromagnetic layers but is independent of the direction of the magnetization in a given layer. The determination of the magnetic reflections from the (111) planes gives directly the distribution of the I–IV type regions in a NiO. Information on S domains distribution can be obtained from the variation of intensities of reflections from planes not parallel to (111) but from the (113) reflection which has been found to be exceedingly sensitive. Thus for instance a 90° rotation of spins in (111) plane gives a 97% reduction of the intensity in the (113) reflection.

(b) *Optical observation.* For specimens less than 100 μm thick the domain structure can be observed directly by the conventional transmitted light technique. An example of the most simple T domain structure observed in thin (111) sections of NiO is shown in Figure 9.60. Using this method, Kondoh (1962, 1963) observed a fine structure of S domains within a T domain in NiO.

For thicker crystals the specular reflection technique can be applied. It consists of studying the reflected light from the very slight facial angles arising from the rhombohedral distortion: from 2·5′ to 12′ for NiO. Slack (1960) was able to deduce the T domain structure for an entire crystal of NiO, polished on all six surfaces, as in Figure 9.61.

(c) *X-ray technique.* T domain structures can also be observed by X-ray back reflection or by the Berg–Barrett technique. The last fails to reveal any T structure if the angle of the relative tilt is very small: in such cases the first method has proved successful. Saito (1962) studied the interaction of T walls with crystal defects. A departure of T walls from twin planes in NiO was also

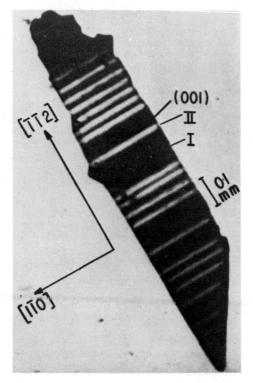

Figure 9.60 Simple T domain structure in a NiO crystal section as seen by transmitted light (Roth, 1960).

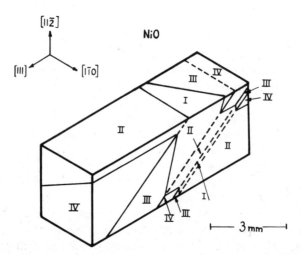

Figure 9.61 T domain structure in a NiO single crystal deduced by the specular reflection technique (Slack, 1960).

detected and attributed to the crystalline and magnetic defects present in the crystal.

Yamada *et al.* (1966a, 1966b) investigated the S domain structure in NiO by the Berg–Barrett method. Figure 9.62 shows a typical S domain structure as obtained by X-ray diffraction topograph. The behaviour of S domains in fields up to 10 kOe along various crystallographic directions, indicated that the spin directions in NiO are [112] within the (111) plane, neglecting the possible slight

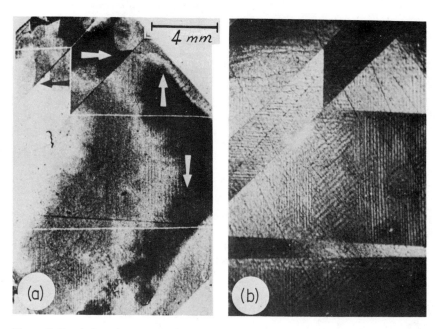

Figure 9.62 A domain structure in NiO observed (a) by X-ray diffraction topograph and (b) by birefringence (Yamada *et al.*, 1966).

deviation from this plane. The accompanied spontaneous magnetostriction was found to be $(9 \pm 3) \cdot 10^{-5}$, in fairly good agreement with the theoretical figure of $2 \cdot 4 \cdot 10^{-4}$. The small difference is explained by assuming that even in well annealed NiO crystals a good number of imperfections still exist and the amount of spontaneous magnetostriction changes locally according to the strain distribution produced by the crystal imperfections. Thus the value of the magnetostriction obtained by measuring overall strain may become smaller as compared with the value obtained by measuring local strain, especially if the S structure is formed to release the strain field associated with the crystal imperfections.

The X-ray method has also been successfully applied to CoO at low temperatures by Saito *et al.* (1966). They discovered a new rhombohedral deformation in addition to the well known tetragonal one, when the crystal was cooled down to $-150\,°C$, accompanied by a new type of domain structure, called the R

structure. The R domains are formed as a result of the change in the direction of the trigonal axis of the antiferromagnetic spin arrangement, and therefore they are distinguished by the direction of the rhombohedral axis. Since four equivalent rhombohedral spin arrangements are possible in each T domain, twelve kinds of R domains appear in the three kinds of T domains. Examples are given in Figure 9.63.

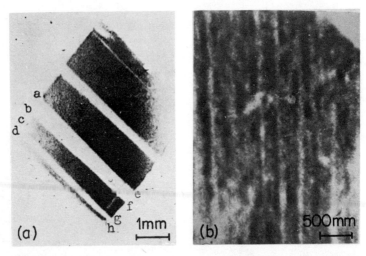

Figure 9.63 (a) Low temperature Berg Barrett photograph of T walls in a CoO crystal. (b) Diffraction photograph of fairly large R domains in a CoO crystal (Kondoh, 1962, 1963).

9.5 DOMAIN WALL DYNAMICS

9.5.1 Theory

9.5.1.1 Introduction

Since magnetic oxides are mostly applied in high frequency a.c. or pulsed fields, domain wall dynamics is important for both scientific and practical purposes. This section deals only with those aspects of the domain wall dynamics which could be put in a direct connection with a known domain structure. As in the foregoing the most reliable results on domain wall dynamics are provided by experiments on single crystals. Domain wall dynamics is investigated in conditions which correspond either to small or to large amplitude motion.

9.5.1.2 Effective mass (inertia) of a moving domain wall

Döring (1948) showed that, to a good approximation, the energy of a wall moving with velocity v differs from that of the wall at rest by a term $\Delta\gamma_w$, proportional to v^2. From $\Delta\gamma_w = (1/2)m_{ef}v^2$ one can define an effective mass m_{ef}, describing the inertia of the moving wall.

The expression of the effective mass in a rather general form was obtained by Sandikov (1964) for a Bloch wall in uniaxial crystals and a 90° wall in cubic crystals. When the normal to the wall is also perpendicular to the directions of easy magnetization in the neighbouring domains, the effective mass is

$$m_{\text{ef}} = (\gamma_w/K_1\delta)/(8\pi\gamma^2\delta) \qquad (9.105)$$

where $\gamma = (ge)/(2mc)$ is the gyromagnetic ratio (with spectroscopic splitting factor g).

When $K_2 \ll K_1$,

$$\gamma_w = 4K_1\delta \quad \text{and} \quad m_{\text{ef}} = (2\pi\gamma^2\delta)^{-1} \qquad (9.106)$$

For a 90° wall in a cubic material with positive anisotropy and the normal to the wall parallel to [100], one has

$$m_{\text{ef}} = (8\pi\gamma^2\delta)^{-1}[1 + (5/6)\lambda(1 + K_2/5K_1)]^{-1} \qquad (9.107)$$

where $\lambda = K_1/2\pi M_s^2$.

For $\lambda \ll 1$ one obtains Döring's result

$$m_{\text{ef}} = (8\pi\gamma^2\delta)^{-1} \qquad (9.108)$$

In the case of periodic fields the analysis of Rado (1951) indicates that inertial behaviour is obtained at frequencies ω for which

$$\gamma H \ll \omega \ll \gamma(2K_1/M_s) \qquad (9.109)$$

i.e. ω must lie between the Larmor frequencies in the applied field H and the anisotropy field $(2K_1/M_s)$. This requires the absence of any interaction between the spin precession in the external field and the overall rotation of the spins at the frequency of natural ferromagnetic resonance. A second condition for the inertia of the wall to be manifest is $\lambda \ll 1$.

In Sandikov's paper it was shown that $\lambda \ll 1$ is not in fact a necessary condition, and this is important for uniaxial crystals where usually $\lambda \gg 1$. In this case, by neglecting K_2, the analytical expression (9.106) for m_{ef} may be obtained.

9.5.1.3 Domain wall damping

During domain wall movement, energy is dissipated owing to eddy currents or relaxation mechanisms. The resulting damping can be generally expressed in terms of a viscous damping parameter β to which corresponds a rate of energy dissipated per second and per unit area of the wall equal to βv. The relatively high resistivity of the magnetic oxides investigated makes it possible to neglect the influence of the eddy currents and to consider the damping parameter as providing important information about the different types of relaxation mechanisms in these materials.

Details of the relaxation mechanisms are described in Chapter 10. We give only the expressions of the damping parameter β in terms of the Landau–Lifshitz's (1935) and Gilbert and Kelly's (1955) damping parameters λ and α, as follows

$$\beta = \lambda(2\gamma^2\delta)^{-1} \quad \text{or} \quad \beta = \alpha M_s(2\gamma\delta)^{-1} \qquad (9.110)$$

9.5.1.4 Domain wall stiffness

In low fields domain wall motion is impeded by the imperfections, which give rise to the finite value of the initial reversible susceptibility χ_{rev}. In many cases, the interaction between domain walls and imperfections may be satisfactorily expressed by a term characterizing the domain wall stiffness, proportional to the domain wall displacement z. For unit surface of a planar 180° wall, a simple computation gives a stiffness coefficient $c = (4M_s^2)/(\chi_{rev}l)$ where l is the domain width and χ_{rev} is defined per unit surface of the crystal.

9.5.1.5 Small amplitude motion

The equation of motion of a wall field applied parallel to the easy axis is that of a damped harmonic oscillator. For a 180° wall:

$$m_{ef}\ddot{z} + \beta\dot{z} + cz = 2M_sH \tag{9.111}$$

where z and H are time dependent.

Denoting the wall-resonant frequency by $\omega_0^2 = c/m_{ef}$ and the relaxation frequency by $\omega_1 = c/\beta$, a standard calculation gives for the displacement in a sinusoidal field $H = H_0 \sin \omega t$,

$$z = (2M_sH_0/c)[(1 - \omega^2/\omega_0^2)^2 + (\omega/\omega_1)^2]^{-\frac{1}{2}} \sin(\omega t - \varphi) \tag{9.112}$$

where the loss angle φ is given by

$$\varphi = \text{arctg}\{(\omega/\omega_1)/[1 - (\omega/\omega_0)^2]\} \tag{9.113}$$

For the single wall complex susceptibility $\chi = \chi' - i\chi''$, denoting $4M_s^2/cl$ by χ_0, the real and imaginary parts are

$$\chi' = \chi_0[1 - (\omega/\omega_0)^2]\{[1 - (\omega/\omega_0)^2]^2 + (\omega/\omega_1)^2\}^{-1} \tag{9.114}$$

and

$$\chi'' = \chi_0(\omega/\omega_1)\{[1 - (\omega/\omega_0)^2]^2 + (\omega/\omega_1)^2\}^{-1} \tag{9.115}$$

The frequency dependence of χ' and χ'' is characteristic for the domain wall resonance.

At frequencies low enough compared to the wall-resonant frequency ω_0, the motion of the domain wall behaves as a simple relaxation with the real and imaginary parts of the susceptibility given by

$$\chi' = \chi_0[1 + (\omega/\omega_1)^2]^{-1} \tag{9.116}$$

and

$$\chi'' = \chi_0(\omega/\omega_1)[1 + (\omega/\omega_1)^2]^{-1} \tag{9.117}$$

with $\varphi = \text{arctg}\,(\omega/\omega_1)$.

The velocity of the wall is, for domain wall relaxation

$$z = z_0\omega[1 + (\omega/\omega_1)^2]^{-1} \cos(\omega t - \phi) \tag{9.118}$$

where $z_0 = 2M_sH_0/c$ is the wall displacement in a static field H_0.

The displacement for wall relaxation is given by

$$z = z_0[1 + (\omega/\omega_1)^2]^{-\frac{1}{2}} \sin(\omega t - \varphi) \qquad (9.119)$$

and the damping parameter may be written as

$$\beta = \frac{2M_s H_0}{\omega_1 z_0} \qquad (9.120)$$

It is often useful to investigate the relaxation in pulsed fields, i.e. when $H = 0$ for $t < 0$ and $H = H_0$ for $0 < t < T$, for an ideally rectangular pulse.

In this case the equation of motion becomes

$$\beta \dot{z} + cz = 2M_s H \qquad (9.121)$$

and for $0 < t < T$, one obtains

$$z = z_0[1 - \exp(-\omega_1 t)] \qquad (9.122)$$

with a velocity

$$\dot{z} = [(2M_s H_0)/\beta] \exp(-\omega_1 t) \qquad (9.123)$$

A condition which has to be taken into account for small amplitude reversible domain wall motion is that the maximum of the displacement does not exceed the half of the width 2Δ of the potential well in which the wall is located at equilibrium. This condition requires a pulse duration $T \leqslant (\beta/c) \ln[z_0/(z_0 - \Delta)]$ and H_0 must be higher than $(c\Delta)/(2M_s)$. A case of interest for some experimental purposes is that of linearly and parabolic increasing fields, $H = at$ and $H = at^2$. For the former, one obtains

$$z = [(2M_s \beta a)/c^2][\exp(-\omega_1 t) - 1] + (2M_s a/c)t \qquad (9.124)$$

and

$$\dot{z} = (2M_s a/c)[1 - \exp(-\omega_1 t)] \qquad (9.125)$$

For $H = at^2$, the corresponding solutions of Equation (9.121) are

$$z = (2M_s a/c)(t^2 - 2\tau t + 2\tau^2) - (4M_s a\tau^2/c) \exp(-t/\tau) \qquad (9.126)$$

with $\tau = \omega_1^{-1} = \beta/c$, and

$$\dot{z} = (4M_s a/c)(t - \tau) + (4M_s a\tau/c) \exp(-t/\tau) \qquad (9.127)$$

which gives asymptotically

$$\dot{z} = (4M_s a/c)(t - \tau) \qquad (9.128)$$

9.5.1.6 Large amplitude motion

In materials with a well-defined critical field for wall displacement the large amplitude motion in pulsed fields is an irreversible one. The stiffness term becomes independent of the wall position, being well described by a constant field pressure $2M_sH_c$, opposite to the pressure exerted by the applied field on the unit area of the moving wall. Thus, the equation of motion reduces to

$$\beta\dot{z} = 2M_s(H - H_c) \qquad (9.129)$$

i.e. motion with constant velocity for a given value of $H > H_c$.

More careful examinations of the wall motion (Enz, 1964; Palmer and Willoughby, 1967; Feldtkeller, 1968; etc.) have shown that the wall contracts during its movement, the amount of contraction depending upon the value of the field, resulting in a non-linear field dependence of the velocity at higher values of the applied field.

Equation (9.129) is valid only for the large amplitude motion of a planar domain wall with a constant area. But, as has been shown in Section 9.4.1.1, even in the static regime the magnetization processes do not take place wholly by parallel wall displacement so that a more peculiar behaviour than that described by Equation (9.129) is to be expected at higher fields.

9.5.2 Experiment

9.5.2.1 Cubic ferrites and garnets

The rather scarce experimental data on single crystals with a well known domain structure were obtained by measuring the voltage induced in the secondary winding of a picture-frame cut single crystal by the displacement of a single domain wall under the influence of a pulsed field, produced by the primary winding. Usually, Equation (9.129) is applied for large amplitude motion and the velocity obtained from the induced voltage varies linearly with the applied field. For a moving planar wall and n turns of the secondary winding, the velocity is expressed in terms of the induced voltage V (in volts), by

$$\dot{z} = 10^8(v/n)/(8\pi M_s d) \qquad (9.130)$$

where d is the width of the planar surface of the wall.

Using Equation (9.129) and the field dependence of the wall velocity at various temperatures, the values and temperature dependences of β and H_c can be obtained.

9.5.2.2 Hexagonal ferrites and orthoferrites

The only single crystals of hexagonal ferrites investigated in pulsed fields were $BaFe_{12}O_{19}$, $BaFe_{11.3}Al_{0.7}O_{19}$ and $BaMg_2Fe_{16}O_{27}$ (Asti et al., 1965, 1967, 1968). Linearly and parabolically increasing fields were used in these investigations.

As long as the walls are moving without sensible changes in shape or total area, Equation (9.121) is expected to hold well.

The term characterizing the wall stiffness is very important in these materials, generally highly-anisotropic, owing to the demagnetizing field H_D which cannot be avoided as in the case of the picture-frame cut cubic ferrites or garnets.

H_D is proportional to the first approximation to the magnetization, i.e. to the wall displacement, exceeding by far the other contributions to the domain wall stiffness.

The results of Asti and coworkers on the above materials are only in broad agreement with the known theories. For a highly-anisotropic single crystal, magnetized along the c axis in linear and parabolic time-dependent pulsed fields, one has to use, owing to the high demagnetizing effects, equations like (9.125) and (9.127) for a relaxation. In this case, the stiffness constant is $c = 4M_s^2 N/l$, where N is the demagnetizing factor of the crystal, approximately assumed as field independent. In this case, for $t \gg \tau$, the asymptotic values of the domain wall velocity are $(la)/(2NM_s)$ for $H = at$ and $(la/2NM_s)t$ for $H = at^2$ giving no information about the damping parameter β.

Asti and coworkers used the instantaneous values of z, measured in their experiments to compute the instantaneous value of the demagnetizing field H_D and applied Equation (9.129) in which H was replaced by $H - H_D$. They obtained a parabolic increase of the wall velocity with the internal field $H - H_D$ for rates of increase $dH/dt < 1.6$ kOe/s and a fairly constant value for $1.6 < dH/dt < 5$ kOe/s.

Such behaviour corresponds, for high internal fields, to low values of the mobility ($u < 10$ cm/s Oe) i.e. to unusually high value of the damping parameter as compared to the cubic ferrites. Umebayashi and Ishikawa (1965) have investigated the motion of a single domain wall in a $YFeO_3$ single crystal ($0.6 \times 1.2 \times 5.6$ mm^3) with two domains only. At low fields the wall velocity increases nearly linearly in time, reaching a flat maximum above which Equation (9.129) applies quite well. The linear rise was explained as the result of pinning centres at the surface and inside the crystal. A thermally activated process is imagined to explain the releasing of the wall from these centres.

The relatively good transparency of thin platelets of orthoferrites permits the use of magneto-optical (Faraday) methods. Rossol (1968, 1969) developed a stroboscopic technique for this. It involves holding a domain wall in an equilibrium position by a combination of external magnetic fields and magnetostatic forces. An additional a.c. field is applied to modulate the equilibrium position the image of which is focused onto a calibrated micrometer slide. The frequency dependence of both the amplitude and phase of the wall displacement are measured. The experimental data for Y, Eu, Ho, Er, Tm and Lu orthoferrites (Rossol, 1968, 1969, 1970) in the temperature range 77–350 K fit (9.122) quite well showing that, within the frequency range used, the displacements occur according to the simple relaxation model. Mobilities in the range 300–50,000 cm s^{-1} Oe^{-1} were measured.

Seitchik et al. (1971) showed, by optical methods, for stripe domains in relatively thick platelets of Tm and Dy orthoferrites, that (9.122) applied and that z_0 depended linearly on H at least up to 5 Oe. The asymptotic value of the mobility is, by (9.123), $u = 2M_s/\beta$ and could be determined from the initial slope of Equation (9.122) or equivalently from a measurement of ω and z_0/H, since at $t = 0$, $z = z_0$. Mobilities as high as 6,000 cm s^{-1} Oe^{-1} were measured.

568

The magneto-optical methods have very great prospects for the future investigation of the domain wall dynamics in a great number of magnetic oxides including, besides orthoferrites, the hexagonal ferrites and garnets.

Other methods used recently in the investigation of domain wall dynamics of bubble domains are referred to in Chapter 13.

REFERENCES

Aharoni, A., 1959, *J. Appl. Phys.*, **30**, 70 S.
Andrä, W., 1956, *Ann. d. Phys.*, **17**, 233.
Andrä, W., 1959, *Ann. d. Phys.*, **3**, 334.
Andrä, W. and E. Schwabe, 1955, *Ann. d. Phys.*, **17**, 55.
Asti, G., M. Colombo, M. Giudici and A. Levialdi, 1965, *J. Appl. Phys.*, **36**, 3581.
Asti, G., M. Colombo, M. Giudici and A. Levialdi, 1967, *J. Appl. Phys.*, **38**, 2195.
Asti, G., F. Conti and C. M. Maggi, 1968, *J. Appl. Phys.*, **39**, 2039.
Basterfield, J., 1968, *J. Appl. Phys.*, **39**, 5521.
Basterfield, J. and M. J. Prescott, 1968, *J. Appl. Phys.*, **38**, 3190.
Bates, L. F. and D. J. Craik, 1962, *J. Phys. Soc. Japan*, **17 Suppl. B-I**, 535.
Bates, L. F. and S. Spivey, 1964, *Brit. J. Appl. Phys.*, **15**, 705.
Bates, L. F., D. J. Craik and P. M. Griffiths, 1958a, *Proc. Roy. Soc.*, **71**, 789.
Bates, L. F., J. Clow, D. J. Craik and P. M. Griffiths, 1958b, *Proc. Roy. Soc.*, **72**, 224.
Bates, L. F., D. J. Craik, P. M. Griffiths and E. D. Isaac, 1959, *Proc. Roy. Soc.*, **A253**, 1.
Bean, C. P. and R. W. De Blois, 1962, private communication.
Bergman, W. H., 1956, *Z. Angew. Phys.*, **11**, 559.
Blackman, M. and B. Gustard, 1962, *Nature (London)*, **193**, 360.
Blackman, M., G. Haigh and N. D. Lisgarten, 1957, *Nature (London)*, **179**, 1288.
Blackman, M., G. Haigh and N. D. Lisgarten, 1959, *Proc. Roy. Soc.*, **A251**, 117
Blackman, M., G. Haigh and N. D. Lisgarten, 1963, *Proc. Phys. Soc.*, **81**, 244.
Bloch, F., 1932, *Z. Phys.*, **74**, 295.
Bobek, A. H., 1967, *Bell System. Techn. J.*, **46**, 1901.
Bobeck, A. H., R. F. Fischer, A. J. Perneski, J. P. Remeika and L. G. van Uitert, 1969, *I.E.E.E. Trans. Mag.*, **MAG-5**, 544.
Boersch, H. and M. Lambeck, 1961, *Z. für Phys.*, **165**, 176.
Bogdanov, A. A. and Ya. A. Vlasov, 1965, *Izv. Akad. Nauk (USSR) Ser. Fiz. Zemli*, **1**, 49.
Bogdanov, A. A. and Ya. A. Vlasov, 1966a, *Izv. Akad. Nauk (USSR) Ser. Fiz. Zemli*, **1**, 42.
Bogdanov, A. A. and Ya. A. Vlasov, 1966b, *Izv. Akad. Nauk (USSR) Ser. Fiz. Zemli*, **9**, 53.
Brown, W. F. Jr., 1945, *Rev. Mod. Phys.*, **17**, 15.
Brown, W. F. Jr., 1957, *Phys. Rev.*, **105**, 1479.
Brown, W. F., 1963, *Micromagnetics* (Interscience: New York).
Carey, R. and E. D. Isaac, 1963, *Proc. Phys. Soc.*, **81**, 741.
Carey, R. and E. D. Isaac, 1964, *Phys. Letters*, **8**, 239.
Carey, R. and E. D. Isaac, 1966, *Magnetic Domains and the Techniques for their Observation* (Academic Press: New York).
Carey, R. and C. Tănăsoiu, 1969, *Phys. Stat. Sol.*, **35**, K115.
Charap, S. H. and J. M. Nemchik, 1969, *I.E.E.E. Trans. Mag.*, **MAG-5**, 566.
Cockayne, B. and D. S. Robertson, 1964, *Brit. J. Appl. Phys.*, **15**, 643.
Conger, R. L. and G. H. Moore, 1963, *J. Appl. Phys.*, **34**, 1213.
Cooper, P. V. and D. J. Craik, 1973, *J. Phys. D.*, **6**, 1393.
Craik, D. J., 1956, *Proc. Phys. Soc.*, **B69**, 647.
Craik, D. J., 1959, Ph.D. Thesis, Nottingham.
Craik, D. J., 1966, *Brit. J. Appl. Phys.*, **17**, 873.

Craik, D. J., 1967a, *Phys. Letters*, **25A**, 126.

Craik, D. J., 1967b, *J. Appl. Phys.*, **38**, 931.

Craik, D. J., 1970, *Contemp. Phys.*, **11**, 65.

Craik, D. J. and P. V. Cooper, 1970, *Phys. Letters*, **33A**, 411.

Craik, D. J. and P. M. Griffiths, 1958, *Brit. J. Appl. Phys.*, **9**, 279.

Craik, D. J. and P. M. Griffiths, 1959, *Proc. Roy. Soc.*, **73**, 1.

Craik, D. J. and D. A. McIntyre, 1966, *Phys. Letters*, **21**, 288.

Craik, D. J. and D. A. McIntyre, 1967, *Proc. Roy. Soc.*, **A302**, 99.

Craik, D. J. and D. A. McIntyre, 1969a, *I.E.E.E. Trans. Mag.*, **MAG-5**, 378.

Craik, D. J. and D. A. McIntyre, 1969b, *Proc. Roy. Soc.*, **A313**, 97.

Craik, D. J. and R. S. Tebble, 1961, *Repts. Progr. Phys.*, **24**, 116.

Craik, D. J. and R. S. Tebble, 1965, *Ferromagnetism and Ferromagnetic Domains* (North-Holland: Amsterdam).

Craik, D. J., P. V. Cooper and W. F. Druyvesteyn, 1971, *Phys. Lett.*, **34A**, 244.

De Bitetto, D. J., 1964, *J. Appl. Phys.*, **35**, 3482.

De Johnge, F. A., W. F. Druyvesteyn and A. G. H. Verhulst, 1971, *J. Appl. Phys.*, **42**, no. 4, 1270.

Dillon, J. F. Jr., 1958, *J. Appl. Phys.*, **29**, 1286.

Dillon, J. F. Jr., 1963, 'Domains and domain walls', in *Magnetism*, Vol. 3 (Academic Press: New York).

Dillon, J. F. Jr. and H. E. Earl Jr., 1959, *J. Appl. Phys.*, **30**, 202.

Dreschel, W., 1961, *Z. für Phys.*, **164**, 324.

Drokin, A. I. and B. V. Beznosikov, 1964, *Kristalographia (U.S.S.R.)*, **9**, 427.

Drokin, A. I. and V. D. Dylgerov, 1961, *Fiz. Tverdovo Tela*, **3**, 553.

Drokin, A. I. and V. D. Dylgerov, 1963, *Fiz. Metal. Metalovedenie*, **15**, 128.

Druyvesteyn, W. F. and J. W. F. Dorleyn, 1971, *Philips Res. Repts.*, **26**, 11.

Dudkin, V. I. and A. I. Pilshchikov, 1967a, *Zh. Eksp. Teor. Fiz.*, **52**, 677. (English translation, *Sov. Phys. JETP*, **25**, 444 (1967).)

Dudkin, V. I. and A. I. Pilshchikov, 1967b, **53**, 56. (English translation, *Sov. Phys. JETP*, **26**, 38 (1968).)

Dylgerov, V. D. and I. A. Drokin, 1960, *Kristalographia (U.S.S.R.)*, **5**, 945.

Dylgerov, V. D. and I. A. Drokin, 1965, *Kristalographia (U.S.S.R.)*, **7**, 465.

Dzialoshinskii, I. E., 1957, *J. Expt. Theor. Phys.*, **32**, 1547.

Eaton, J. A. and A. H. Morrish, 1969, *J. Appl. Phys.*, **40**, 3180.

Eaton, J. A., A. H. Morrish and C. W. Searle, 1968, *Phys. Letters*, **26A**, 520.

Elmore, W. C., 1937, *Phys. Rev.*, **51**, 982.

Elmore, W. C., 1938, *Phys. Rev.*, **54**, 309.

Enoch, P. D. and R. M. Lambert, 1970, *J. Phys. E. Sci. Instr.*, **3**, 728.

Enoch, R. D. and E. A. D. White, 1971, *J. Mat. Sci.*, **6**, 263.

Enz, U., 1964, *Helv. Phys. Acta*, **37**, 245.

Farztdinov, M. M., 1964a, *Fiz. Metal. Metallov.*, **19**, 809 (*Phys. Met. Metallography*, **19**, 10 (1964)).

Farztdinov, M. M., 1964b, *Fiz. Met. Metallov.*, **19**, 641 (*Phys. Met. Metallography*, **19**, 1 (1964)).

Farztdinov, M. M., 1964c, *Fiz. Met. Metallov.*, **19**, 321 (*Phys. Met. Metallography*, **19**, 1 (1964)).

Farztdinov, M. M., 1964d, *Uspehi Fiz. Nauk*, **84**, 611.

Farztdinov, M. M., 1965, *Sov. Phys. Uspehii*, **7**, 855.

Farztdinov, M. M. and S. D. Malginova, 1970, *Fiz. Tverdovo Tela*, **12**, 2955.

Farztdinov, M. M., S. D. Malginova and A. A. Halfina, 1970, *Izv. Akad. Nauk (U.S.S.R.) Ser. Fiz.*, **34**, 1104.

Feldtkeller, E., 1968, *Phys. Stat. Sol.*, **27**, 161.

Florescu, V. and M. Rosenberg, 1970, *Japan J. Appl. Phys.*, **9**, 217.

570

Forlani, F. and N. Minnaja, 1969, *J. Appl. Phys.*, **40**, 1092.

Fowler, C. A. and E. M. Fryer, 1954a, *J. Opt. Soc. Am.*, **44**, 256.

Fowler, C. A. and E. M. Fryer, 1954b, *Phys. Rev.*, **94**, 52.

Fowler, C. A., E. M. Fryer, B. L. Brandt and R. A. Isaacson, 1963, *J. Appl. Phys.*, **34**, 2064.

Frei, E. H., S. Shtrikman and D. Treves, 1957, *Phys. Rev.*, **106**, 446.

Gallon, T. E., 1968, *Proc. Roy. Soc.*, **A303**, 525.

Galt, J. K., 1952, *Phys. Rev.*, **85**, 664.

Galt, J. K., 1954, *Bell System Techn. J.*, **33**, 1023.

Garrood, J. R., 1962, *Proc. Phys. Soc.*, **79**, 1252.

Gemperle, R., 1964, *Phys. Stat. Sol.*, **6**, 89.

Gemperle, R., E. V. Shtoltz and M. Zeleny, 1963, *Phys. Stat. Sol.*, **3**, 2015.

Gianola, U. F., D. H. Smith, A. A. Thiele and L. G. van Uitert, 1969, *I.E.E.E. Trans. Mag.*, **MAG-5**, 558.

Gilbert, T. L. and J. M. Kelly, 1955, *Proc. Conf. Mag. and Magnetic Materials*, *A.I.E.E.*, New York, 253.

Goodenough, J. B., 1956, *Phys. Rev.*, **102**, 356.

Goto, K., 1966, *Japan J. Appl. Phys.*, **5**, 117.

Grundy, P. J., 1965, *Brit. J. Appl. Phys.*, **16**, 409.

Grundy, P. J. and R. S. Tebble, 1968, *Advances in Phys.*, **17**, 153.

Gustard, B., 1967, *Proc. Roy. Soc.*, **A297**, 269.

Hagedorn, F. D. and E. M. Gyorgy, 1961, *J. Appl. Phys.*, **32**, Supple. 3, 282.

Hale, M. E., H. W. Fuller and H. Rubinstein, 1959, *J. Appl. Phys.*, **30**, 789.

Hanss, R., 1964, *Science (U.S.A.)*, **146**, 398.

Harper, H. and R. W. Teale, 1967, *J. Phys. Chem. Solids*, **28**, 1781.

Ignatchenko, V. A. and I. V. Zaharov, 1964, *Izv. Acad. Nauk (U.S.S.R.) Ser. Fiz.*, **28**, 568.

Isaac, E. D., 1959, *Proc. Phys. Soc.*, **74**, 786.

Kaczer, J., 1962, *Czech. J. Phys.*, **B.12**, 354.

Kaczer, J., 1964, *J. Exptl. Theor. Phys. (U.S.S.R.)*, **46**, 1787. (English translation, *Sov. Phys. J.E.T.P.*, **19**, 1204 (1964).)

Kaczer, J. and R. Gemperle, 1960, *Czech. J. Phys.*, **B10**, 505 and 614.

Kaczer, J. and R. Gemperele, 1961, *Czech. J. Phys.*, **B11**, 510.

Kandaurova, G. S., 1964, *Izv. Vishih Zaved. Phys.*, **5**, 12.

Kandaurova, G. S., 1967, *Izv. Vishih Zaved. Phys.*, **1**, 81.

Kandaurova, G. S., 1968, *Sov. Sol. State Phys.*, **10**, 2311.

Kandaurova, G. S. and Ya. S. Shur, 1967, *Repts. Inst. Phys. Metal. Sverdlovsk*, **26**, 29.

Kawado, S., T. Maruyama and Z. Ishii, 1968, *J. Phys. Soc. Japan*, **24**, 208.

Kaye, G., 1961, *Proc. Phys. Soc.*, **78**, 869.

Kittel, C., 1946, *Phys. Rev.*, **70**, 965.

Kittel, C., 1949a, *Rev. Mod. Phys.*, **21**, 541.

Kittel, C., 1949b, *Rev. Mod. Phys.*, **21**, 1527.

Kittel, C. and J. K. Galt, 1956, 'Ferromagnetic domain theory', in *Solid State Physics*, Vol. 3, p. 437 (Academic Press: New York).

Knowles, J. E., 1960, *Proc. Phys. Soc.*, **75**, 885.

Kocinski, J., 1958, *Acta Phys. Pol.*, **17**, 283.

Kojima, H. and K. Goto, 1962, *J. Phys. Soc. Japan*, **17** Suppl. B-I, 201.

Kojima, H. and K. Goto, 1962, *J. Phys. Soc. Japan*, **17**, 584.

Kojima, H. and K. Goto, 1964, *Proc. Int. Conf. Mag. Nottingham*, 727.

Kojima, H. and K. Goto, 1965a, *J. Appl. Phys.*, **36**, 538.

Kojima, H. and K. Goto, 1965b, *J. Appl. Phys.*, **36**, 539.

Kondoh, H., 1962, *J. Phys. Soc. Japan*, **17**, 1316.

Kondoh, H., 1963, *J. Phys. Soc. Japan*, **18**, 595.

Kondoh, H. and T. Takeda, 1964, *J. Phys. Soc. Japan*, **19**, 2041.

Kondorski, E. I., 1952, *Izv. Akad. Nauk (U.S.S.R.) Ser. Fiz.*, **16**, 398.

Kooy, C., 1958, *Philips Techn. Rev.*, **19**, 286.
Kooy, C. and U. Enz, 1960, *Philips Res. Repts.*, **15**, 7.
Kozlowski, G. and W. Zietek, 1965, *J. Appl. Phys.*, **36**, 2162.
Kozlowski, G. and W. Zietek, 1966, *Acta Phys. Pol.*, **29**, 261.
Kranz, J. and H. C. Thomas, 1963, *Naturwissenschaften*, **50**, 326.
Kurtzig, A. J. and W. Shockley, 1968a, *I.E.E.E. Trans. Mag.*, **MAG-4**, 426.
Kurtzig, A. J. and W. Shockley, 1968b, *J. Appl. Phys.*, **39**, 5619.
Landau, L. and E. Lifshitz, 1935, *Phys. Z. Sowjetunion*, **8**, 153.
Lang, A. R., 1959, *Acta Cryst.*, **12**, 249.
Lefever, R. A., A. B. Chase and K. A. Wickersheim, 1962, *J. Appl. Phys.*, **33**, 2249.
Lefever, R. A., K. A. Wickersheim and A. B. Chase, 1965, *J. Phys. Chem. Solids*, **26**, 1529.
Lifshitz, E., 1944, *J. Phys. U.S.S.R.*, **8**, 337.
Lilley, B. A., 1950, *Phil. Mag.*, **41**, 792.
Lo, D. S., G. F. Sauter and W. J. Simon, 1969, *J. Appl. Phys.*, **40**, 5402.
Malek, Z. and V. Kambersky, 1958, *Czech. J. Phys.*, **8**, 416.
Martin, T. J. and J. C. Anderson, 1966, *I.E.E.E. Trans. Mag.*, **MAG-2**, 446.
Mayer, L., 1957, *J. Appl. Phys.*, **28**, 975.
Merz, K. M., 1960, *J. Appl. Phys.*, **31**, 147.
Mitzek, A. I., 1966, *Fiz. Metal. Metallov.*, **22**, 481.
Moriya, T., 1960, *Phys. Rev.*, **120**, 91.
Müller, M. W., 1967, *Phys. Rev.*, **162**, 423.
Néel, L., 1944a, *Cahiers de Phys.*, **25**, 19.
Néel, L., 1944b, *J. Phys. Rad.*, **5**, 241.
Néel, L., 1944c, *J. Phys. Rad.*, **5**, 265.
Nemchik, J., 1969, *J. Appl. Phys.*, **40**, 1086.
Palmer, W. and R. A. Willoughby, 1967, *I.B.M. J. Res. Develop.*, **11**, 284.
Patel, J. R., K. A. Jackson and J. F. Dillon Jr., 1968, *J. Appl. Phys.*, **39**, 3767.
Paulus, M., 1960a, *Comp. Rend.*, **250**, 2332.
Paulus, M. M., 1960b, *Compt. Rend.*, **250**, 1213.
Pearson, R. F., 1957, *Proc. Phys. Soc.*, **70**, (448B), 441.
Perneski, A. J., 1969, *I.E.E.E. Trans. Mag.*, **MAG-5**, 554.
Pilshchikov, A. I. and N. E. Syreev, 1969, *Zh. Eksp. Teor. Fiz.*, **57**, 1940. (English translation, *Sov. Phys. J.E.T.P.*, **30**, 1050 (1970).)
Polcarova, M. and J. Kaczer, 1967, *Phys. Stat. Sol.*, **21**, 635.
Polcarova, M. and A. R. Lang, 1962, *Appl. Phys. Letters*, **1**, 13.
Prutton, M., 1959, *Phil. Mag.*, **4**, 1063.
Rado, G. T., 1951, *Phys. Rev.*, **83**, 821.
Remaut, G., P. Delavignette, A. Lagasse and S. Amelincks, 1964a, *J. Appl. Phys.*, **35**, 1351.
Remaut, G., A. Lagasse and S. Amelincks, 1964b, *Phys. Stat. Sol.*, **7**, 497.
Rhodes, P. and G. Rowlands, 1954, *Proc. Leeds Phil. Soc.*, **6**, 191.
Rodbell, D. S., R. C. De Vries, W. D. Barber and R. W. De Blois, 1967, *J. Appl. Phys.*, **38**, 4542.
Roessler, B., 1967, *Phys. Stat. Sol.*, **20**, 713.
Rosenberg, M., 1968, *Report of the Int. Conf. on Mag. Oxides, Bucharest*, Sept. (1968).
Rosenberg, M. and C. Tănăsoiu, 1963a, *Rev. Phys. Buch.*, **8**, 384.
Rosenberg, M. and C. Tănăsoiu, 1963b, *Phys. Stat. Sol.*, **3**, 1790.
Rosenberg, M., C. Tănăsoiu and C. Rusu, 1964a, *Phys. Stat. Sol.*, **6**, 639.
Rosenberg, M., C. Tănăsoiu and C. Rusu, 1964b, *Phys. Stat. Sol.*, **6**, 141.
Rosenberg, M., C. Tănăsoiu and C. Rusu, 1965, *Phys. Stat. Sol.*, **10**, 613.
Rosenberg, M., C. Tănăsoiu and V. Florescu, 1966a, *J. Appl. Phys.*, **37**, 3826.
Rosenberg, M., C. Tănăsoiu and L. Nowicki, 1966b, *Phys. Stat. Sol.*, **14**, 499.
Rosenberg, M., C. Tănăsoiu and V. Florescu, 1967, *Phys. Stat. Sol.*, **21**, 197.
Rosenberg, M., C. Tănăsoiu and V. Florescu, 1968, *J. Appl. Phys.*, **39**, 879.
Rosenberg, M., C. Tănăsoiu and V. Florescu, 1970, *I.E.E.E. Trans. Mag.*, **MAG-6**, 207.

Rossol, F. C., 1968, *J. Appl. Phys.*, **39**, 5263.
Rossol, F. C., 1969a, *J. Appl. Phys.*, **40**, 1082.
Rossol, F. C., 1969b, *I.E.E.E. Trans. Mag.*, **MAG-5**, 562.
Rossol, F. C., 1970, *Phys. Rev. Letters*, **24**, 1021.
Roth, W. L., 1960, *J. Appl. Phys.*, **31**, 2000.
Saito, S., 1962, *J. Phys. Soc. Japan*, **17**, 1287.
Saito, S., K. Nakahigashi and Y. Shimomura, 1966, *J. Phys. Soc. Japan*, **21**, 850.
Sandikov, V., 1964, *Izv. AN S.S.S.R.*, **28**, 584.
Schwartze, W., 1957, *Ann. d. Phys.*, **19**, 322.
Schwink, C. and H. Spreen, 1965, *Phys. Stat. Sol.*, **10**, 57.
Seitchik, J. A., W. D. Doyle and G. K. Goldberg, 1971, *J. Appl. Phys.*, **42**, no. 4, 1272.
Sherwood, R. C., J. P. Remeika and H. G. Williams, 1959, *J. Appl. Phys.*, **30**, 217.
Shirk, B. T. and W. R. Buessem, 1970, *J. Amer. Ceram. Soc.*, **53**, 192.
Shirobokov, M., 1939, *Doklad. Akad. Nauk U.S.S.R.*, **24**, 426.
Shirobokov, M., 1945, *J. Exptl. Theoret. Phys. (U.S.S.R.)*, **15**, 57.
Shockley, W., 1948, *Phys. Rev.*, **73**, 1246.
Shur, Ya. S. and V. I. Hrabrov, 1969, *J. Expt. Theoret. Phys.*, **57**, 1899.
Shur, Ya. S., A. A. Glaser, Yu. N. Dragochanski, V. A. Zaikova and G. S. Kandaurova, 1964, *Izv. Akad. Nauk (U.S.S.R.)*, **28**, 553.
Silcox, J., 1963, *Phil. Mag.*, **8**, 7.
Sixtus, K. J., K. J. Kronenberg and R. K. Tenzer, 1956, *J. Appl. Phys.*, **27**, 1051.
Slack, G. A., 1960, *J. Appl. Phys.*, **31**, 1571.
Smith, A. W. and G. W. Williams, 1960, *Canad. J. Phys.*, **38**, 1187.
Smith, A. W. and G. W. Williams, 1961, *Canad. J. Phys.*, **39**, 768.
Soffel, H., 1963, *Z. Geophysik*, **29**, 21.
Soffel, H., 1964, *Z. Geophysik*, **30**, 45.
Soffel, H., 1965, *Z. Geophysik*, **31**, 345.
Spivak, G. V., N. G. Kanavina, I. S. Sbitnikova and T. N. Dombrovskaya, 1955, *Dokl. Akad. Nauk (U.S.S.R.)*, **105**, 706 and 965.
Stoltz, E. V., 1966, *Fiz. Tverdovo Tela.*, **8**, 3147.
Sugiura, I., 1951, *Busseiron Kenkyu*, **38**, 106.
Szymczak, R., 1965a, *Bull. Acad. Pol. Sci.*, **13**, 111.
Szymczak, R., 1965b, *Arkiv. Elektr.*, **14**, 185.
Szymczak, R., 1966, *Archiv. Elektr.*, **15**, 477.
Szymczak, R., 1968, *Electron Technology*, **1**, 5.
Szymczak, R., 1970, *Inst. Electr. Technology Repts. Poland*, **21**, 1.
Thiele, A. A., 1969, *Bell System. Techn. J.*, **48**, 3287.
Träuble, H., O. Boser, H. Kronmüller and A. Seger, 1965, *Phys. Stat. Sol.*, **10**, 283.
Umebayashi, H. and Y. Ishikawa, 1965, *J. Phys. Soc. Japan*, **20**, 2193.
van Loef, J. J. and A. Broese van Groenou, 1964, *Proc. Int. Conf. Mag. Nottingham*, 646.
van Uitert, L. G., R. C. Sherwood, W. A. Bonner, W. H. Grodkiewicz, L. Pictorski and G. Zydzik, 1970, *Mat. Res. Bull.*, **5**, 153.
Verwell, J., 1967a, *J. Appl. Phys.*, **38**, 1111.
Verwell, J., 1967b, *Z. angew. Phys.*, **23**, 200.
Vlasov, Ya. A. and A. A. Bogdanov, 1964, *Izv. Akad. Nauk (U.S.S.R.) Ser. Geophys.*, **3**, 386.
Wanas, M. A., 1967, *J. Appl. Phys.*, **38**, 1019.
Williams, H. J., P. M. Bozorth and W. Shockley, 1949, *Phys. Rev.*, **75**, 155.
Williams, H. J., R. C. Sherwood and J. P. Remeika, 1958, *J. Appl. Phys.*, **29**, 1772.
Yamada, T., 1963, *J. Phys. Soc. Japan*, **18**, 520.
Yamada, T., 1966a, *J. Phys. Soc. Japan*, **21**, 650.
Yamada, T., 1966b, *J. Phys. Soc. Japan*, **21**, 664.
Yamada, T., S. Saito and Y. Shimomura, 1966, *J. Phys. Soc. Japan*, **21**, 672.
Yamamoto, M. and T. Iwata, 1953, *Sci. Rep. Res. Inst. Tohoku Univ.*, **A5**, 433.

ACKNOWLEDGEMENTS

In preparing this chapter the authors have benefited from helpful discussions with Dr. D. J. Craik to whom they are greatly indebted.

Permission from authors and publishers to reproduce the figures used in this chapter is gratefully acknowledged.

10 *Microwave resonance and relaxation*

C. E. PATTON

10.1 INTRODUCTION

In a typical ferromagnetic resonance (FMR) experiment, a sample is placed in a uniform magnetic field large enough to magnetize it parallel to the field direction. If the magnetization is disturbed slightly from this equilibrium position, it does not return directly but precesses about the field direction. Energy loss associated with the magnetization motion causes the precession to be damped, or undergo relaxation, with eventual alignment along the field direction. To study this phenomenon, a small sinusoidal field is applied perpendicular to the static field. This sinusoidal field excites precessional motion but unless the frequency ω is nearly equal to the precessional frequency ω_u, the energy coupled into the precessing magnetization will be small. If $\omega \approx \omega_u$, the coupling is large and the amplitude of the precession is limited only by the damping of the system. This simple description of FMR contains the three basic ingredients common to resonance phenomena: precession, resonant response and relaxation.

The first observations of FMR were reported by Griffiths (1946) for electrolytically deposited films of iron, cobalt and nickel. The resonances were observed to occur for values of the static external field H_0 much lower than expected from the usual resonance relation for electrons. The apparent discrepancy was first explained by assigning extremely large g values to the magnetic electrons. This approach, however, yielded different g values, depending on sample shape and orientation. The understanding of FMR was advanced considerably when Kittel (1948) pointed out the role of demagnetizing fields in determining the resonance condition; the internal fields at an electronic site in the sample can be quite different from the applied fields. The usual resonance condition with $g \approx 2$ is satisfied for most materials if demagnetizing fields are taken into account. (The fact that the g value determined by microwave data is generally greater than 2, while static determinations give $g < 2$, has been explained by Kittel (1949) and others (Polder, 1948, 1949; van Vleck, 1949).)

The resonances observed in the early experiments were very broad and the results were understood only vaguely. The substantial progress in the

575

understanding of microwave phenomena began with the discovery of YIG (Bertaut and Forrat, 1956; Geller and Gilleo, 1957a, 1957b) with a cubic structure, no disorder and excellent insulating properties. Sparks (1964) has cited Kittel's classic remark that YIG is to FMR research what the fruit fly is to genetics research. The gradual realization of the important role played by spinwaves, the short wavelength precessional modes in the magnetic system, and by impurity ions in the physical damping mechanisms added in-depth understanding of relaxation to the general picture of microwave phenomena. Finally, the prediction and observation of spin-wave instability for specific spin-wave modes (Suhl, 1956a, 1956b; Schlömann et al., 1960a) has led to a multitude of experimental and theoretical work pertaining to spin-wave relaxation mechanisms and damping.

The objective of this chapter is to provide a broad survey of microwave phenomena touched on above. The emphasis is on physical concepts. Detailed theoretical developments are avoided. Much of the material has been covered at length in one or more of several comprehensive books on microwave ferrites (Sparks, 1964; Lax and Button, 1962; Soohoo, 1960; von Aulock, 1965; Gurevich, 1960; Smit and Wijn, 1959). The intent here is to draw on these works and the technical literature and use selected areas to develop a unified picture of microwave phenomena. Certain areas are touched on only briefly and others are neglected entirely, in the interest of both brevity and continuity. Except for the basic concepts, detailed presentations are reserved for areas in which recent advances have been made and which have been discussed only briefly in scattered articles in the recent technical literature.

Most of the discussion will embrace a macroscopic model in describing microwave phenomena. The magnetic material is viewed as homogeneous with a given M_s. As long as the wavelengths of the magnetic disturbances are long compared to the atomic spacings, the model is quite adequate and an atomistic picture need not be considered in detail. The details of the magnetic ordering will not be considered. Even though microwave oxide materials have ferrimagnetic ordering, for most microwave phenomena the materials can be viewed simply as ferromagnets with a net magnetization. Except near a compensation point, the phenomena peculiar to ferrimagnets occur at high frequencies, out of the microwave range, and will not be considered here (see Lax and Button, 1962, for a review of such phenomena). Experimental techniques for microwave measurements will not be discussed. Basic methods are reviewed in several books (Lax and Button, 1962; Soohoo, 1960) as well as in recent technical articles (see Green and Kohane, 1964; Patton and Kohane, 1971).

10.2 NORMAL MODES OF MAGNETIC SYSTEMS

Before a detailed discussion of microwave resonance and relaxation, it is important to have some understanding of the various types of modes which exist and can be excited in magnetic materials. In general, these are classified according to the approximations used in treating the dipole and exchange fields.

In most situations of practical interest, the dipolar fields are handled in the quasistatic approximation in which the fields induced by the magnetization are assumed to propagate instantaneously to all parts of the sample. In very large samples, however, the propagation time for the fields cannot always be neglected and the quasistatic approximation is not valid. Another approximation neglects the dipolar fields generated by the magnetization divergence at the surfaces of the sample. If the mode wavelength is small compared to the sample dimensions, such fields oscillate very rapidly over the sample surface and are essentially zero inside the material. For wavelengths of the order of the sample dimensions, these fields must be considered explicitly. For most cases of interest, the mode wavelength is much larger than unit cell dimensions so that the exchange interaction can be handled in a continuum approximation.

For the modes which are used most frequently in discussions of microwave phenomena, the spin-wave modes, the quasistatic approximation is invoked and surface dipole fields are neglected. Spin waves are considered, first using a simple intuitive model and then from an analytic quantum theory approach. Macroscopic spin-wave theory, which is relied on rather heavily in subsequent sections, will then be summarized, with emphasis on its similarity to the quantum theory analysis. Finally, the other types of normal modes will be briefly discussed.

In any quantitative discussion involving spin-waves, one of the most important considerations is the nature of the normal modes. As for phonons and electrons in solids, only definite states described by a characteristic dispersion relation $\omega_{\mathbf{k}} = \omega(\mathbf{k})$ are allowed. The quantity $\mathbf{k}$ is the spin-wave wave vector and $\omega_{\mathbf{k}}$ the spin-wave frequency. Three energy terms contribute, in general, to the spin-wave energy $\hbar\omega_{\mathbf{k}}$ and hence to the normal mode frequency $\omega_{\mathbf{k}}$ at a specified $\mathbf{k}$. The field or Zeeman energy, the exchange energy and the magnetostatic or dipolar energy which is simply the self energy of the system of spins in their own dipolar field.

Consider first the role of Zeeman and exchange energies:

$$(\hbar\omega)_{\text{Zeeman}} = g\beta \sum_i \mathbf{S}_i \cdot \mathbf{H} \qquad (10.1\text{a})$$

$$(\hbar\omega)_{\text{exch}} = -2J \sum_{i,j} \mathbf{S}_i \cdot \mathbf{S}_j \qquad (10.1\text{b})$$

$\mathbf{H}$ is the internal magnetic field, and J is the exchange integral. The j summation in the exchange term is over the nearest neighbours of spin S_i and the i summations are over the N spins of the system. The ground state is that in which every spin in the sample is aligned parallel to the applied static field. Both the Zeeman and exchange energy are minimum for this configuration. One initial guess as to the first excited state would be to flip one spin antiparallel to the magnetic field. However, this state corresponds to a very high exchange energy. A judicious twist of the system, in which each spin is tipped slightly away from the static field direction, gives a much smaller exchange energy. The twisted configuration also represents an eigenmode while a single flipped spin does not.

An appropriate twist is shown schematically in Figure 10.1. Each spin is tipped at the same angle β with respect to the applied field. For small β the increase in Zeeman energy is small. This disturbance on the spin system is called a spin wave and is a normal mode of the system. The shortest distance between parallel spins (except in the plane containing all parallel spins) defines the propagation direction and the wavelength $\lambda_{\mathbf{k}}$. The wave vector $\mathbf{k}$ has magnitude $2\pi/\lambda_{\mathbf{k}}$ and is directed along the propagation direction.

β = Angle of precession cone

δ = Angle between adjacent spins

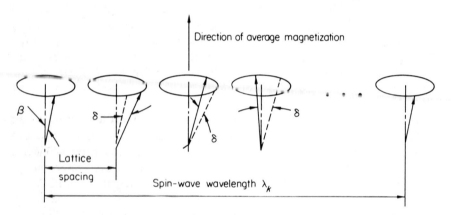

Figure 10.1 Spin configuration (spatial distribution) for spin-wave mode.

Using the above qualitative spin-wave picture, the general shape of the dispersion relation $\omega_{\mathbf{k}} = \omega(\mathbf{k})$ can be obtained. The exchange energy of a spin-wave is given by

$$(\hbar\omega_{\mathbf{k}})_{\text{exch}} \propto S_i \cdot S_{i+1} \propto -\cos\delta \approx -(1 - \tfrac{1}{2}\delta^2) \tag{10.2}$$

where δ is the angular deviation between adjacent spins. Since δ is inversely proportional to $\lambda_{\mathbf{k}}$, $(\delta \approx 2\pi a/\lambda_{\mathbf{k}}$ if a is the spin separation distance) and k is also inversely proportional to λ_k, δ is proportional to k. Thus, the exchange energy of the spin wave is

$$(\hbar\omega_{\mathbf{k}})_{\text{exch}} = Bk^2 \tag{10.3}$$

where constant terms have been neglected. The coefficient B is given by $2SJa^2$, where S is the electron spin. Rigorous calculations show that for a single magnon corresponding to a spin-wave $(\omega_{\mathbf{k}}, \mathbf{k})$ of energy $\hbar\omega_{\mathbf{k}}$, each spin is tipped by an angle β such that the total change of the magnetic moment in the field direction is the same as if one spin were completely flipped. The total magnon energy is then

$$\hbar\omega_\mathbf{k} = g\beta H + Bk^2 \tag{10.4}$$

The dispersion relation, however, is still incomplete. The importance of the magnetostatic energy can be demonstrated by considering modes with $\mathbf{k}$ in different directions with respect to the applied field and average magnetization. Figure 10.2 shows spin-wave modes with $\mathbf{k}$ parallel to and perpendicular to

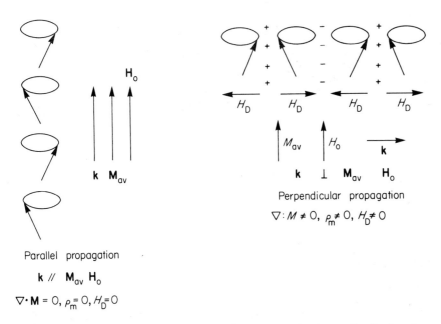

Parallel propagation

$\mathbf{k} \parallel \mathbf{M}_{av} H_o$

$\nabla \cdot \mathbf{M} = 0,\ \rho_m = 0, H_D = 0$

Perpendicular propagation

$\nabla : M \neq 0,\ \rho_m \neq 0, H_D \neq 0$

Figure 10.2 Schematic representation of magnetostatic charge distribution for spin-wave modes.

the static field. For $\mathbf{k} \perp \mathbf{H}$, the magnetization $\mathbf{M}$ has a non-zero divergence and dipolar fields are generated. These fields may be viewed as arising from magnetic charges $\rho_m = 4\pi\nabla\cdot\mathbf{M}$, as in the domain analyses in Chapter 9. One can see that $\nabla\cdot\mathbf{M}$ is zero for $\mathbf{k}$ parallel to $\mathbf{H}$ and the dispersion relation is essentially given by (10.4). The dipolar fields generated by the non-zero magnetization divergence for $\mathbf{k}$ and $\mathbf{H}$ not parallel have two effects. (1) The fields tend to raise the energy of the spin-system. (2) The fields alter the spin motions from a circular precession to an elliptical precession cone. When the dipolar interaction is taken rigorously into account, the dispersion relation is given by

$$\omega_\mathbf{k}/\gamma = [(H + Dk^2)(H_i + Dk^2 + \omega_m \sin^2\theta_k)]^{\frac{1}{2}} \tag{10.5}$$

where $\omega_m = 4\pi\gamma M_s$ and $D = B/\hbar$. Here $\gamma = g\beta/\hbar$ is the gyromagnetic ratio $(1.76 \times 10^7 \mathrm{Oe}^{-1}\mathrm{s}^{-1}$ for $g = 2)$ and θ_k is the angle between $\mathbf{k}$ and $\mathbf{H}$. Note that

(10.5) reduces to the simple relation of (10.4) for $\theta_k = 0$. The total dispersion relation is sketched in Figure 10.3, including Zeeman, exchange and volume magnetostatic energies. The magnitude of **k** is denoted by k. Curves for $0 < \theta_k < \pi/2$ are contained in the region between the $\theta_k = 0$ and $\theta_k = \pi/2$ dispersion curves. Considered collectively, these curves represent a quasi-continuum of magnon states which is called the spin-wave manifold. The increase in ω_k with k is due to exchange, the increase with θ_k is due to dipolar interactions, and the field dependent term γH is the Zeeman contribution.

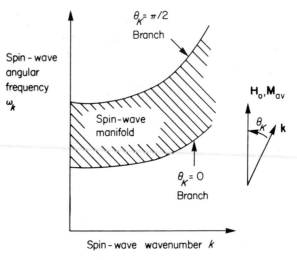

Figure 10.3 Spin-wave manifold.

A rigorous derivation of (10.5) starts with a macroscopic equation of motion for the magnetization, including dipolar terms (Soohoo, 1960, p. 229) or uses quantum theory. Both are extremely useful techniques. Many of the concepts which come out of the quantum approach are useful in relating magnons to physical properties and in discussing relaxation mechanisms. Following Spark's (1964) review of the original Holstein–Primakoff (1940) quantum treatment, the basic elements of this approach are outlined below.

The Holstein–Primakoff treatment is based on the Hamiltonian of (10.1) but with the dipole–dipole term

$$\frac{1}{2} \sum_{j(i \neq j)} \left(\frac{4\mu_0^2}{r_{ij}^5} \right) [r_{ij}^2 (\mathbf{S}_i \cdot \mathbf{S}_j) - 3(\mathbf{r}_{ij} \cdot \mathbf{S}_i)(\mathbf{r}_{ij} \cdot \mathbf{S}_j)] \qquad (10.6)$$

included, where $\mathbf{r}_{ij}$ is a vector connecting the spatial coordinates of $\mathbf{S}_i$ and $\mathbf{S}_j$, and $\mu_0 = \frac{1}{2}g\beta$. Note that each term in (10.6) is essentially $\mathbf{H}_{ij} \cdot \mathbf{S}_j$, where H_{ij} is the dipole field of $\mathbf{S}_i$ at $\mathbf{S}_j$. Propagation times for the dipole field are neglected. This is the quasistatic approximation. The Hamiltonian is diagonalized using three transformations to obtain an harmonic oscillator form

$$\mathcal{H} = \sum_k \hbar\omega_k a_k^\dagger a_k + \cdots \tag{10.7}$$

where the $a_k^\dagger$ and a_k are magnon creation and annihilation operators which are closely related to harmonic oscillator creation and annihilation operators. These operators have commutation relations

$$[a_i, a_j^\dagger] = \delta_{ij}; \qquad [a_i, a_j] = 0; \qquad [a_i^\dagger, a_j^\dagger] = 0 \tag{10.8}$$

Operating on a magnon state $|n_i\rangle$, shorthand for $|n_i, n_2, \ldots, n_i, \ldots, n_N\rangle$ and corresponding to a reduction in S_{iz} from S by n_i units of μ_0, a_i and $a_i^\dagger$ follow the relations

$$a_i^\dagger|n_i\rangle = (n_i + 1)^{\frac{1}{2}}|n_i + 1\rangle, \qquad a_i|n_i\rangle = (n_i)^{\frac{1}{2}}|n_i - 1\rangle \tag{10.9}$$

The first transformation is given by

$$S_i^+ = S_{ix} + iS_{iy} = (2S)^{\frac{1}{2}}\left(1 - \frac{a_i^\dagger a_i}{2S}\right)a_i \tag{10.10a}$$

$$S_i^- = S_{ix} - iS_{iy} = (2S)^{\frac{1}{2}}a_i^\dagger\left(1 - \frac{a_i^\dagger a_i}{2S}\right) \tag{10.10b}$$

$$S_{iz} = S - a_i^\dagger a_i \tag{10.10c}$$

and results in a Hamiltonian of the form

$$\mathcal{H} = \sum_{i,j\,i\neq j} f(r_{ij})a_i^\dagger a_j + \cdots \tag{10.11}$$

To diagonalize the $a_i^\dagger a_j$ terms, a Fourier transformation

$$a_i = N^{-\frac{1}{2}}\sum_k e^{-i\mathbf{k}\cdot\mathbf{r}_i}b_k \tag{10.12a}$$

$$a_i^\dagger = N^{-\frac{1}{2}}\sum_k e^{i\mathbf{k}\cdot\mathbf{r}_i}b_k^\dagger \tag{10.12b}$$

is used. The $a_i^\dagger a_j$ terms each relate to two different electronic sites. The b_k relate to *all* sites and the transformation results in terms of the form $b_k^\dagger b_k$ with a single subscript. The b_k and $b_k^\dagger$ operators satisfy the same commutation relations given in (10.8). This transformation results in a Hamiltonian of the form

$$\mathcal{H}/\hbar = \sum_k A_k b_k^\dagger b_k + \frac{1}{2}\sum_k (B_k b_k b_{-k} + B_k^* b_{-k}^\dagger b_k^\dagger) + \cdots \tag{10.13}$$

with

$$A_k = \gamma Dk^2 + \gamma H + \frac{1}{2}\omega_m \sin^2\theta_k \tag{10.14}$$

and

$$B_k = \frac{1}{2}\omega_m \sin^2\theta_k\, e^{-i2\phi_k} \tag{10.15}$$

ϕ_k is the angle between the x axis and the projection of $\mathbf{k}$ on the x-y plane. Although the first term is in diagonal form, pairs of creation or annihilation operators for k and $-k$ magnons appear in the second term. A final transformation

$$b_k = U_k c_k - V_k c_k^\dagger \tag{10.16}$$

is used to diagonalize these terms. The U_k and V_k are chosen so that the c_k operators satisfy the commutation relations in (10.8). The transformation is given by

$$U_k = \cosh \mu_k, \qquad V_k = e^{i v_k} \sinh \mu_k \tag{10.17}$$

with

$$\tanh 2\mu_k = |B_k|/A_k \quad \text{and} \quad v_k = 2\phi_k \tag{10.18}$$

The diagonalized Hamiltonian is now

$$\mathscr{H} = \sum_k \hbar \omega_k c_k^\dagger c_k \tag{10.19}$$

with ω_k given by

$$\omega_k/\gamma = [(H + Dk^2)(H + Dk^2 + 4\pi M_s \sin^2 \theta_k)]^{\frac{1}{2}} \tag{10.20}$$

While (10.20) gives the correct dispersion relation, the approximate relation (10.14) is usually adequate. This is because the coefficient B_k of the non-diagonal terms which are present after the second transformation is usually much smaller than the coefficient A_k for the diagonal terms. The third transformation is related to the spin-wave ellipticity and is important whenever $H + Dk^2 \gg 4\pi M_s \sin^2 \theta_k$ is not satisfied or when the spin-wave ellipticity plays an explicit role in the phenomena of interest.

The macroscopic approach closely follows the quantum theory. Instead of creation and annihilation operators, one deals with spin-wave amplitudes α_k which represent the direction cosines of that part of the magnetization associated with a particular spin wave at $\mathbf{k}$ and ω_k. A brief outline is given here and, in Section 10.4, the approach is extended to explain high power phenomena. The starting point for the macroscopic theory is the magnetic torque equation $d\mathbf{M}/dt = -\gamma(\mathbf{M} \times \mathbf{H})$. In terms of a unit vector $\boldsymbol{\alpha}(\mathbf{r}) = \mathbf{M}(\mathbf{r})/M_s$, the torque equation becomes

$$d\boldsymbol{\alpha}(\mathbf{r})/dt = -\gamma\boldsymbol{\alpha}(\mathbf{r}) \times \mathbf{H}(\mathbf{r}) \tag{10.21}$$

The $\boldsymbol{\alpha}(\mathbf{r})$ vectors in the classical theory correspond to the a_i operators in the quantum theory. The spin-wave amplitudes $\boldsymbol{\alpha}_k$ are defined by a Fourier transform

$$\boldsymbol{\alpha}(\mathbf{r}) = \sum_k \boldsymbol{\alpha}_k e^{i\mathbf{k}\cdot\mathbf{r}} \tag{10.22}$$

similar to (10.12a and 10.12b). The internal field $\mathbf{H}(\mathbf{r})$ contains the static internal field H, assumed to be z directed, an effective field due to exchange (Herring and Kittel, 1951)

$$\mathbf{H}_{ex} = D\nabla^2 \boldsymbol{\alpha}(\mathbf{r}) \tag{10.23}$$

and an effective field due to dipolar interactions $H_{dip}(\mathbf{r})$

$$H_{dip}(\mathbf{r}) = -4\pi M_s \sum_{k \neq 0} \frac{\mathbf{k}(\mathbf{k} \cdot \boldsymbol{\alpha}_k)}{k^2} e^{i\mathbf{k}\cdot\mathbf{r}} \qquad (10.24)$$

The $k = 0$ term in H_{dip} is usually included in the static field H as the uniform demagnetizing field. If the above expressions are substituted in the torque equation, the linear terms in the α_k give an equation of motion

$$\dot{\alpha}_k = i\{A_k\alpha_k + B_k e^{2i\phi_k}\alpha^*_{-k}\} \qquad (10.25)$$

where a complex scalar amplitude $\alpha_k = \alpha_k^x + i\alpha_k^y$ has been introduced. A_k is given by (10.14). This B_k is slightly modified from that given in (10.15).

$$B_k = \tfrac{1}{2}\omega_m \sin^2 \theta_k \qquad (10.26)$$

The last Holstein–Primakoff transformation in the form $b_k = U_k\alpha_k + V_k\alpha^*_{-k}$ is all that is required to diagonalize (10.25) to give $\dot{b}_k = i\omega_k b_k$ with ω_k given by (10.20). The classical macroscopic theory is quite similar to the quantum theory. The relevant terms in the Hamiltonian are replaced by equivalent field terms in the torque equation, and the same transformations are used to diagonalize the spin-wave amplitude equation-of-motion.

It is important to show how the intuitive spin-wave picture and the quantitative theories are related. First consider the simple picture of a spin-wave consisting of a collection of precessing spins, each tipped slightly from the precession axis. This picture is expressed mathematically by the second transformation (10.12). The b_k, which are linear combinations of the a_i operators, operate on all lattice sites to tip the spins with a strength which varies as $e^{-\mathbf{k}\cdot\mathbf{r}_i}$. Thus, the b_k are spin-wave operators and describe collective modes for the magnetic system. The spin-wave ellipticity, justified by plausibility arguments (Figure 10.2), also comes out of the formalism simply by expressing S_{ix} and S_{iy} in terms of the c_k, calculating $\langle S_{ix}^2 \rangle / \langle S_{iy}^2 \rangle$, and considering terms for the wave vector of interest. The result is

$$\left(\frac{\langle S_{i,\text{minor}}^2 \rangle}{\langle S_{i,\text{major}}^2 \rangle} \right)^{\frac{1}{2}} = \left(\frac{A_k - |B_k|}{A_k + |B_k|} \right)^{\frac{1}{2}} \qquad (10.27)$$

When $A_k \gg |B_k|$ is satisfied, the precession follows a circular orbit. In general, however, the precession is elliptical. One final result which can be obtained from the formalism is the relation between the magnetic moment and the magnon occupation numbers. It is clear that the excitation of spin-wave modes causes a reduction in the average magnetization. Specifically

$$M_z = M_s V - 2\mu_0 \sum_k n_k \qquad (10.28)$$

and

$$M = M_s V - 2\mu_0 \sum_{k \neq 0} n_k \qquad (10.29)$$

Recall that $n_k = 1$ is equivalent to flipping a single spin antiparallel to the field direction. Thus, non-zero occupation numbers for magnon states decrease M_z. If $k \neq 0$ the net average magnetization M is also reduced.

The above spin-wave models, particularly the microscopic model with single electronic sites in a simple cubic lattice and nearest neighbour interactions only, bear little resemblance to real magnetic oxide crystals. In applying the theory to real materials, a phenomenological approach in which the exchange parameter D is based on experiment, has proved to be highly successful. The fact that many oxides are ferrimagnetic, not ferromagnetic, also appears to be relatively unimportant. The spin-wave analysis also appears to be applicable to metals and, in particular, to thin films, even though conduction electrons are not considered in the Holstein–Primakoff treatment. In short, the basic results appear to be generally applicable to a wide variety of materials and are essentially model independent. For most applications, the details of the crystal structure, electronic configurations, exchange interactions etc. need not be explicitly taken into account.

As indicated in the introduction to this section, the spin-wave modes discussed above represent only one category of modes which are important in describing microwave phenomena. Following Schlömann and Joseph (1968), the different types of modes can be classified according to the way in which the electromagnetic (dipolar) and exchange interactions are taken into account. His classification scheme is shown in Figure 10.4. The mode type (if named), dispersion

Classification of magnetic normal modes

Approximation used for: Dipole – dipole interaction → Neglect Exchange interaction ↓	Neglect	Quasistatic approximation	Rigorous treatment
Neglect	$\omega = \gamma H$ (Zeeman energy only)	$\omega = \gamma [H(H + 4\pi M_s \sin^2 \theta)]^{1/2}$ Magnetostatic modes	Magnetodynamic modes
Continuum approximation	$\omega = \gamma(H + Dk^2)$ Exchange modes	$\omega = \gamma [(H + Dk^2)(H + Dk^2 + 4\pi M_s \sin^2 \theta)]^{1/2}$ Spin – wave modes	Not analysed

Figure 10.4   Classification of normal modes for magnetic systems.

relation and approximations used are indicated. The simplest type (not named) is the one where both exchange and electromagnetic interactions are neglected completely. The only important energy term is the Zeeman term and $\omega_k = \gamma H$ is the dispersion relation. The most complicated case in which exchange is taken into account in the continuum approximation and the electromagnetic interaction treated using the full set of Maxwell's equations has never been solved satisfactorily and this type of mode is also not named.

For magnetostatic modes, the electromagnetic interaction is handled in the quasistatic approximation and exchange is neglected. These approximations are valid as long as (1) the propagation time for the dipolar fields across the sample is negligible compared to the period of the microwave excitation and (2) the wavelengths of the magnetic modes are long enough so that the exchange energy is negligible. The range of k over which the magnetostatic mode approximations are valid is typically $1 \text{ cm}^{-1} < k < 10^3 \text{ cm}^{-1}$. In magnetostatic mode calculations, the dipole fields generated by the non-zero magnetization divergence at the sample surfaces must be taken into account explicitly. For very high k modes, where λ_k is much less than the sample size, such surface fields average to zero and can be neglected (as in the spin-wave case—surface contributions to the dipolar term in the Hamiltonian were neglected in (10.5)). For $k < 10^3 \text{ cm}^{-1}$, however, λ_k and typical ferrite dimensions may be comparable and surface fields do not have a zero spatial average. Thus, most magnetostatic mode calculations reduce to a boundary value problem involving Maxwell's equations with displacement currents neglected (quasistatic approximation) and an equation of motion for the magnetization neglecting exchange. For very large samples, where electromagnetic propagation times must be taken into account, the full set of Maxwell's equations are employed and the resultant modes are known as magnetodynamic modes.

If the exchange interaction is taken into account in the continuum approximation (represented by a term Dk^2 in the dispersion relation) but the electromagnetic interaction is neglected, the dispersion relation is the same as for spin-wave modes propagating parallel to the average magnetization direction with $\nabla \cdot M = 0$. If the dipolar fields are taken into account in the quasistatic approximation, the spin-wave modes discussed above are obtained. For such modes in which exchange is important, $\lambda_k \ll$ sample size is usually satisfied so that only dipolar fields generated in the volume of the sample need be considered. Surface dipolar fields average to zero and can be neglected.

Finally, it is important to mention the much discussed uniform mode and how it fits into the above scheme. By definition, a uniform mode has $\lambda_k = \infty$. Because all samples are finite, no mode wavelength can really be infinite. Unless the sample is very small, however, a uniform mode is a very good approximation to the realizable lowest order magnetostatic mode. In such cases, the $\lambda_k = \infty$ assumption is usually invoked because of the analytic simplicity which results. In the next section, uniform precession resonance will be considered in detail.

10.3 RESONANCE AND RELAXATION

The precessional motion of the magnetization which is the basis for resonance follows directly from the classical magnetic torque equation

$$(d\mathbf{M}/dt)_{\text{precession}} = -\gamma(\mathbf{M} \times \mathbf{H}) \tag{10.30}$$

Equation (10.30) corresponds to that for electronic precession. The gyromagnetic ratio γ is taken as a positive quantity. The magnetization response to a microwave field $\mathbf{h}$ can be calculated in terms of a permeability tensor $\boldsymbol{\mu}$ or susceptibility tensor $\boldsymbol{\chi}$,

$$\mathbf{m} = \boldsymbol{\chi} \cdot \mathbf{h} \tag{10.31a}$$

$$\mathbf{b} = \mathbf{h} + 4\pi\mathbf{m} = \boldsymbol{\mu} \cdot \mathbf{h} \tag{10.31b}$$

As shown by the early analyses by Kittel (1948) and Polder (1949), the particular choice of $\mathbf{h}$ is extremely important. If one considers the response in terms of the internal field $\mathbf{h}$, which includes microwave demangetizing fields due to the sample shape, one obtains the well known Polder tensor

$$\mathbf{m} = \begin{pmatrix} \chi & -i\kappa \\ i\kappa & \chi \end{pmatrix} \mathbf{h} \tag{10.32}$$

Equation (10.32) describes the response $\mathbf{m} = (m_x, m_y)$ to an internal microwave field $\mathbf{h} = (h_x, h_y)$ for a sample saturated in the z direction. Note that $b_z = h_z$, so that $\chi_{zz} = 0$ and $\mu_{zz} = 1$. The tensor is invariant to a rotation about the z axis, and thus expresses the required symmetry for an isotropic material. Polder showed that the tensor components are given by

$$4\pi\chi = \omega_m\omega_0/(\omega_0^2 - \omega^2) \tag{10.33a}$$

and

$$4\pi\kappa = -\omega_m\omega/(\omega_0^2 - \omega^2) \tag{10.33b}$$

where $\omega_m = \gamma 4\pi M_s$ and $\omega_0 = \gamma H$. (H is the static internal field.) Equations (10.33a and 10.33b) are valid for small precession angles, so that $m_z \approx 0$ and $M_z \approx M_s$ are satisfied to first order. All the microwave quantities (i.e. $m_{x,y}, h_{x,y}$) contain an implicit $e^{i\omega t}$ time dependence.

If we evaluate the response in terms of external field quantities, $\mathbf{h}_e$ (applied microwave field) and $\mathbf{H}_0$ (z directed static field), the sample shape enters into the calculation. The result is

$$\mathbf{m} = \begin{pmatrix} \chi_e^{(x)} & -i\kappa_e \\ i\kappa_e & \chi_e^{(y)} \end{pmatrix} \mathbf{h}_e \tag{10.34}$$

with

$$4\pi\chi_e^{(x,y)} = \frac{\omega_m\omega_{y,x}}{\omega_x\omega_y - \omega^2} \tag{10.35a}$$

and

$$4\pi\kappa_e = \frac{-\omega_m\omega}{\omega_x\omega_y - \omega^2} \qquad (10.35b)$$

where

$$\omega_{x,y} = \gamma[H_0 + 4\pi M_s(N_{x,y} - N_z)] \qquad (10.35c)$$

The subscript on χ_e and κ_e denotes that the response is in terms of the applied field. χ_e is usually called the external susceptibility tensor. $N_{x,y,z}$ are the sample demagnetizing factors. The diagonal elements of the external susceptibility tensor are not equal, except for the special case of rotational z axis symmetry $(N_x = N_y)$. The resonance frequency is given by

$$\omega_{res}^{(Kittel)} = \sqrt{\omega_x\omega_y}$$

$$= \sqrt{(\omega_0 + N_x\omega_m)(\omega_0 + N_y\omega_m)} \qquad (10.36)$$

This is the famous Kittel resonance condition (Kittel, 1948). It demonstrates the explicit role played by the microwave demagnetizing fields, through N_x and N_y, in determining the resonance condition.

The external susceptibility tensor χ_e is related to the internal susceptibility tensor χ through the demagnetizing tensor $\mathbf{N}$, according to

$$\chi_e = (\mathbf{I} - 4\pi\chi\mathbf{N})^{-1}\chi \qquad (10.37)$$

$(\mathbf{I} - 4\pi\chi\mathbf{N})^{-1}$ is simply the inverse of $(\mathbf{I} - 4\pi\chi\mathbf{N})$, $\mathbf{I}$ is a unit matrix and $\mathbf{N}$ is diagonal. In the hypothetical limiting case of an infinite magnetic medium with $\mathbf{N} = 0$, χ_e and χ are identical. For any sample of finite extent, however, it is necessary to clearly specify which kind of susceptibility is under consideration. Most of the discussion in this chapter will be in terms of χ_e, the external susceptibility. Thus, the response is directly related to the applied external fields, both microwave and static. In most cases of interest the effect of the sample on the field distribution can be neglected, so that the applied fields are the same as those with the sample absent. Another advantage of the external formulation is that the role of geometry (sample shape) is shown explicitly.

One serious difficulty with the above expressions is that the magnetization response diverges at $\omega = \omega_{res}$. The basic problem is that the torque equation contains no loss term. Once the uniform precession mode is excited by the microwave field to some amplitude, (10.30) predicts that the motion will continue indefinitely at the same amplitude if the exciting field is turned off. In real materials, coupling between the precessing spin system and the lattice provides a way for the precession motion to be damped. In the absence of a driving field the magnetization will gradually relax to the equilibrium position, along the static field for an isotropic specimen.

There are a number of ways in which damping or relaxation can be introduced into the formulation. One widely used technique, originally proposed by

Landau and Lifshitz (1935), is to add a phenomenological damping term on the right-hand side of (10.30). They suggested an equation of motion of the form

$$(d\mathbf{M}/dt)_{\text{total}} = -\gamma(\mathbf{M} \times \mathbf{H}) - (\alpha\gamma/M_s)[\mathbf{M} \times (\mathbf{M} \times \mathbf{H})] \qquad (10.38)$$

Their L–L damping term relaxes $\mathbf{M}$ toward the static field direction at a rate proportional to the precessional component of $d\mathbf{M}/dt$ only,

$$(d\mathbf{M}/dt)_{\text{L–L}} = (\alpha/M_s)\mathbf{M} \times (d\mathbf{M}/dt)_{\text{precession}} \qquad (10.39)$$

Gilbert (1955) later proposed that the relaxation should occur at a rate proportional to the *total* $(d\mathbf{M}/dt)$. If the damping is small ($\alpha^2 \ll 1$) so that the total relaxation in a hypothetical decay experiment occurs over many precession cycles, $(d\mathbf{M}/dt)_{\text{total}} \approx (d\mathbf{M}/dt)_{\text{precession}}$ and the two approaches are essentially equivalent (see Iida (1963) for a detailed discussion of the differences). A somewhat different approach is to simply replace the natural resonance frequency, ω_{res}, by a complex frequency, $\Omega_{\text{res}} = \omega_{\text{res}} + i\eta_0$. The imaginary term results in an exponential decay in the transient response and a finite response for steady state resonance excitation.

These purely phenomenological forms of the damping are not always strictly compatible with many physical processes which have been found to contribute to the relaxation in the first place. One important example of this, to be considered shortly, is two-magnon scattering. If the uniform precession amplitude is relaxed by an energy transfer from uniform mode magnons ($k = 0$) to spin-waves ($k \neq 0$), the average magnetization is decreased. In the Landau–Lifshitz or Gilbert equations, however, $d\mathbf{M}/dt$ is always normal to M, so that $|\mathbf{M}|$ is conserved. A number of formulations have been introduced to resolve such difficulties and make the equation-of-motion approach strictly compatible with physical relaxation processes. Bloembergen (1956) adapted Bloch's NMR equations (1946) to FMR:

$$dM_z/dt = -\gamma(\mathbf{M} \times \mathbf{H})_z - (M_z - M)/T_1 \qquad (10.40a)$$

$$dM_{x,y}/dt = -\gamma(\mathbf{M} \times \mathbf{H})_{x,y} - M_{x,y}/T_2 \qquad (10.40b)$$

T_1 is a spin–lattice relaxation time for the relaxation of M_z and T_2 is a spin–spin relaxation time for the transverse components of $\mathbf{M}$. Note that spin–spin interactions, which conserve the sum total of magnon occupation numbers, cannot contribute to the relaxation of M_z (10.28). They can relax the total $|\mathbf{M}|$ (10.29). Fletcher *et al.* (1960) and Callen (1958) have developed more elaborate formulations which separate the uniform precession and the spin-wave contributions to the total magnetization, and derive relaxation parameters with explicit contributions from physical processes. Such approaches are discussed in detail by Lax and Button (1962). In the following discussion, the three basic approaches mentioned above are used to establish some of the basic features of relaxation and to point out the important differences between the formulations.

One natural definition of relaxation comes from the analysis of transient response after the microwave field in a resonance experiment is switched off.

In the simple precession picture, the magnetization will simply spiral in toward the static field direction with eventual alignment. For an ellipsoidal sample, the magnetization response predicted by the Landau–Lifshitz (L–L), and Bloch–Bloembergen (B–B) approaches, is given by $m(t) \approx \exp(i\Omega_0 t)$ with $\Omega_0 = \omega_{res}^{(Kittel)} + i\eta_0$. The precession frequency, $\omega_{res}^{(Kittel)}$, is the same as obtained without damping. The decay rate, expressed in terms of the damping parameters in the L–L or B–B equations, is

$$\eta_0^{(L-L)} = \alpha(\omega_x + \omega_y)/2 \tag{10.41a}$$

$$\eta_0^{(B-B)} = 1/T_2 \tag{10.41b}$$

Second order terms in the loss were neglected ($\alpha^2 \ll 1$, $1/T_2^2 \ll \omega_{res}^2$) in obtaining this and the results to follow. Equation (10.41) illustrates one very important point: α is not simply proportional to η_0 or T_2^{-1}, which are usually considered as intrinsic relaxation parameters. α also depends on the value of $(\omega_x + \omega_y)$. At a given resonance frequency this term can be strongly dependent on the shape of the sample (Patton, 1968, 1971c).

To deal with resonance phenomena, it is necessary to evaluate the various susceptibility expressions including loss. Expressions for $\chi_e^{(x,y)}$ and κ_e which follow from the L–L and B–B equations, and from complex frequency substitutions for $\omega_{x,y}$ and $\omega_{res} = \sqrt{\omega_x \omega_y}$ in Equations (10.35), are given in Table 10.1. It is necessary to replace $\omega_{x,y}$ by $\omega_{x,y} + iq$, with

$$q = 2\eta_0\sqrt{\omega_x \omega_y}/(\omega_x + \omega_y) \tag{10.42}$$

in order for this substitution to be compatible with the $\omega_{res}^{(Kittel)}$ ($= \sqrt{\omega_x \omega_y}$) replacement, $\sqrt{\omega_x \omega_y} + i\eta_0$.

In reference to Table 10.1, first note the points of similarity between the different sets of expressions. All three describe resonance response, centred at $\omega = \omega_{res}$. For a sharp resonance, neither the field (frequency fixed) or frequency (field fixed) change appreciably over the width of the resonance, so that the contribution of the numerator terms to the shape of the line can be disregarded.

Table 10.1

	$4\pi\chi_e$	$4\pi\kappa_e$
L–L	$\dfrac{\omega_m(\omega_{y,x} + i\alpha\omega)}{\omega_x\omega_y - \omega^2 + i\alpha\omega(\omega_x + \omega_y)}$	$\dfrac{-\omega_m\omega}{\omega_x\omega_y - \omega^2 + i\alpha\omega(\omega_x + \omega_y)}$
B–B	$\dfrac{\omega_m\omega_{y,x}}{\omega_x\omega_y - \omega^2 + 2i\omega/T_2}$	$\dfrac{\omega_m(-\omega + i/T_2)}{\omega_x\omega_y - \omega^2 + 2i\omega/T_2}$
C.F.	$\dfrac{\omega_m[\omega_y + 2i\eta_0\sqrt{\omega_x\omega_y}/(\omega_x + \omega_y)]}{\omega_x\omega_y - \omega^2 + 2i\eta_0\sqrt{\omega_x\omega_y}}$	$\dfrac{-\omega_m\omega}{\omega_x\omega_y - \omega^2 + 2i\eta_0\sqrt{\omega_x\omega_y}}$

The response is determined mainly by the resonance denominators. Since $\omega \approx \sqrt{\omega_x \omega_y}$ is a good approximation in the damping terms, all three denominators give basically the same response, with

$$\eta_0 = 1/T_2 = \alpha(\omega_x + \omega_y)/2 \qquad (10.43)$$

This relation between the different types of damping parameters is the same as that given for transient response by (10.41). Hence as far as the shape of the resonance is concerned, all three formulations are equivalent and are consistent with the intuitive transient decay definition of relaxation. This point is extremely important, for example, in discussing the meaning of resonance linewidths. Resonance aside, the L–L and the C.F. expressions are exactly the same (both numerator *and* denominator) if α and η_0 are related by

$$\alpha\omega = 2\eta_0\sqrt{\omega_x \omega_y}/(\omega_x + \omega_y) \qquad (10.44)$$

For samples with rotational symmetry, this reduces to $\alpha\omega = \eta_0$. This simple relation is often invoked in discussions of FMR, but the general relation is actually that in (10.44).

The big differences appear to be between the B–B expressions and the other two types. The imaginary term in the numerator has been transferred from χ_e to κ_e. The differences are not as great as they appear, however, if the actual response to a microwave field is examined. Recall that χ_e and κ_e describe the response to the x and y components of $\mathbf{h}_e$. Consider instead the response to circularly polarized fields, rotating either in the proper sense to excite resonance (Larmor excitation) or in the opposite sense (anti-Larmor excitation). If the sample has rotational symmetry, the magnetization response for these two senses of circular polarized excitation is also circularly polarized in the appropriate sense, with

$$|m|_L = (\chi_e - \kappa_e)|\mathbf{h}_0| \quad \text{(Larmor)} \qquad (10.45a)$$

$$|m|_A = (\chi_e + \kappa_e)|\mathbf{h}_0| \quad \text{(anti-Larmor)} \qquad (10.45b)$$

Expressions for $(\chi_e \mp \kappa_e)$, which characterize the Larmor and anti-Larmor response, are listed in Table 10.2. These expressions follow from those in Table 10.1, after some manipulation and simplification. If only the Larmor response is considered, the three formulations give exactly the same result

Table 10.2

	$4\pi(\chi_e - \kappa_e)$ Larmor response	$4\pi(\chi_e + \kappa_e)$ anti-Larmor response
L–L	$\dfrac{\omega_m[\omega_{res} - \omega - i\alpha\omega]}{(\omega_{res} - \omega)^2 + \alpha^2\omega^2}$	$\dfrac{\omega_m(\omega_{res} + \omega - i\alpha\omega)}{(\omega_{res} + \omega)^2 + \alpha^2\omega^2}$
B–B	$\dfrac{\omega_m(\omega_{res} - \omega - i/T_2)}{(\omega_{res} - \omega)^2 + 1/T_2^2}$	$\dfrac{\omega_m(\omega_{res} + \omega + i/T_2)}{(\omega_{res} + \omega)^2 + 1/T_2^2}$
C.F.	$\dfrac{\omega_m(\omega_{res} - \omega - i\eta_0)}{(\omega_{res} - \omega)^2 + \eta_0^2}$	$\dfrac{\omega_m(\omega_{res} + \omega - i\eta_0)}{(\omega_{res} + \omega)^2 + \eta_0^2}$

with $\eta_0 = \alpha\omega = T_2^{-1}$. The $\eta_0 = \alpha\omega$ result follows in general for rotational symmetry, as already discussed. The root of the problem is the anti-Larmor response. The B–B formulation results in a positive imaginary part for $(\chi_e + \kappa_e)$. A positive sign is unreasonable, from a physical point of view, because the energy loss is proportional to $-4\pi Im(\chi_e + \kappa_e)$.* This quantity is negative for the anti-Larmor B–B term. This problem, however, is usually of no serious practical concern. For most experiments linear excitation is used, and the Larmor excitation is the dominant term in the response. When it is necessary to study a response which is predominantly anti-Larmor in character, the other approaches should be invoked.

The physical origin of the relaxation is quite complicated. There are many possible physical relaxation channels by which energy contained in the uniform precession mode (or other spin-wave modes) may be dissipated. One convenient way to describe the various channels is in terms of scattering interactions. The special case of two-magnon scattering has already been mentioned. Scattering involving three or more magnons, magnons and phonons, etc., must also be considered. The contribution of such interactions to the line-broadening was mentioned even in the early papers of Akhieser (1946) and Polder (1949). For complicated scattering sequences where many processes may be operative, the concept of a relaxation time $(T = 1/\eta)$ for each interaction is convenient in describing the total relaxation. The individual scattering interactions which may contribute are summarized schematically in Figure 10.5. The uniform mode

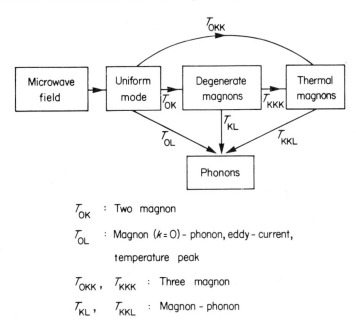

T_{OK} : Two magnon

T_{OL} : Magnon $(k=0)$ - phonon, eddy - current,

       temperature peak

T_{OKK}, T_{KKK} : Three magnon

T_{KL}, T_{KKL} : Magnon - phonon

Figure 10.5 Relaxation scheme, in terms of scattering processes, for the uniform precession.

* *Im* indicating the imaginary part.

resonance is excited by the microwave field and uniform mode magnons eventually relax into phonons (lattice heating) by various channels. A similar scheme was the basis of the Fletcher et al. (1960) analysis mentioned earlier. (Thermal magnons were not considered.) Various schemes have also been proposed for measuring the individual relaxation times indicated in Figure 10.5. A frequency modulation technique was developed by Fletcher et al. (1960) to separate T_{k1}, T_{01} and T_{0k} terms in a relaxation scheme involving only these processes. In addition, the different frequency, temperature etc. dependences of the various relaxation times provide a means of separating the different contributions to the relaxation. In many cases, the situation is simplified because one contribution dominates.

The details of various processes which contribute to the total relaxation will be covered in subsequent sections. To give an overall picture, the mechanisms are tabulated in Figure 10.5. Two-magnon relaxation is particularly important in polycrystals, where porosity or the anisotropy in randomly oriented crystallites can make a large contribution to the resonance loss. In single crystals, a somewhat smaller two-magnon contribution due to surface roughness can be important. There are numerous types of magnon–phonon processes. Three such processes give rise to loss which exhibits a peak as a function of temperature; valence exchange, slowly-relaxing-impurity and rapidly-relaxing-impurity. Valence exchange involves electron hopping between Fe^{2+} and Fe^{3+} ions. In impurity relaxation, some impurity atom in the lattice is the source of the loss. Magnon–phonon scattering induced by magnetostriction has been recently proposed as a possible mechanism for the loss in polycrystals. Eddy current loss can be important in ferrites with high conductivity. Three-magnon processes are generally invoked to explain the loss in high purity polished single crystals for which temperature-peak and two-magnon contributions are negligible.

The experimental parameter commonly used to characterize the microwave losses in ferrites is not the relaxation rate or time, but the resonance linewidth. The original linewidth parameter was simply the fieldwidth at half height of the resonance absorption curve. It is a convenient and easily accessible parameter. As increasingly sophisticated microwave phenomena were discovered, however, a number of different linewidth parameters were introduced to characterize these phenomena with loss parameters which were compatible with the original linewidth concept. The most important are the frequency swept linewidth, spin-wave linewidth and effective linewidth.

The half-power linewidth is derived from the shape of the resonance absorption curve, which is simply a plot of the negative imaginary part of $\chi_e^{(x,y)}$ as a function of static field. This curve has a peak near the field value at which $\omega = \omega_{res}$ is satisfied and falls monotonically to zero on either side. Consider the Larmor response given in Table 10.2. The absorption curve is described by

$$-Im\,4\pi(\chi_e - \kappa_e) = \frac{\omega_m \eta_0}{(\omega_{res} - \omega)^2 + \eta_0^2} \tag{10.46}$$

The peak in the absorption at $\omega_{\text{res}} = \omega$ is equal to ω_{m}/η_0. The location of the half power points is obtained by solving

$$\sqrt{\omega_x \omega_y} - \omega = \eta_0 \qquad (10.47)$$

If the experiment is done by sweeping frequency and keeping the field constant, the frequency linewidth is given by

$$\Delta\omega \doteq 2\eta_0 \qquad (10.48)$$

Thus, the frequency linewidth is a direct measure of the relaxation rate η_0 associated with resonance.

The more convenient experimental approach is to sweep the field and keep the frequency constant, thereby obtaining the half-power linewidth in field units. But, because of the non-linear relation between ω_{res} and H_0, this field linewidth cannot be used as a direct measure of η_0. (ω_{res} and H_0 are connected through the Kittel resonance condition involving a square-root relationship.) If the resonance peak is reasonably narrow so that the slope of ω_{res} versus H_0 is relatively constant in the vicinity of resonance, the two linewidths are related by

$$\Delta\omega \approx \frac{\partial\sqrt{\omega_x \omega_y}}{\partial H} \Delta H \qquad (10.49)$$

$$\Delta\omega/\gamma \approx \frac{(\omega_x + \omega_y)}{2\sqrt{\omega_x \omega_y}} \Delta H \qquad (10.50)$$

The distinction between ΔH and $\Delta\omega/\gamma$ is most significant for thin planar samples magnetized in-plane. For YIG with $4\pi M_s = 1750$ G, at 1 GHz, the frequency linewidth $\Delta\omega/\gamma$ (in field units) is greater than ΔH by a factor of three. The difference between $\Delta\omega/\gamma$ and ΔH for planar geometry and the interpretation of linewidth data in terms of physical relaxation has been discussed extensively by Patton (1968, 1971c).

Under conditions where ω_{res} is linear in H_0, the above distinctions vanish and $\Delta\omega/\gamma$ and ΔH are essentially equivalent. This situation applies to samples with rotational symmetry about the direction of H_0 or resonance at sufficiently high frequency so that H_0 (at resonance) $\gg 4\pi M$ is satisfied. Such conditions apply to most of the work to be discussed in subsequent sections, so that no further distinction need be made between frequency and field linewidths.

A single linewidth parameter is not always adequate in describing the relaxation. Some of the processes, namely those which involve magnon–magnon interactions, are field dependent. The resonance linewidth is a measure of the loss for only one particular bias condition, resonance, and it provides absolutely no information about the field dependences of the relaxation mechanisms. As will be considered in Section 10.6, such field dependences can provide important clues to the origin of the loss. To describe the field dependence of the relaxation rate, an effective linewidth parameter

$$\Delta H_{eff} = 2\eta_0(H_0)/\gamma_{eff}(H_0) \tag{10.51}$$

is used (Kohane and Schlömann, 1968; Vrehen, 1968; Patton, 1969a). The effective gyromagnetic ratio γ_{eff} is on the order of γ but modified somewhat by the presence of loss. It will be discussed shortly. Essentially, ΔH_{eff} provides an easy way to express the field dependence of the relaxation rate in linewidth units which are compatible with conventional linewidth results.

An effective line shift parameter S_{eff} can be used to describe the effect of relaxation on the real part of the uniform precession frequency (Vrehen, 1968). Recall that η_0 may be interpreted as the imaginary part of the complex uniform mode frequency ω_{res}. Relaxation introduces a change in the real part ω_{res} as well. This change from the Kittel frequency, ω_{res} in (10.36) can be expressed in terms of S_{eff} according to

$$\omega_{res}^{(exp)}/\gamma =$$

$$\sqrt{[H_0 + S_{eff} + 4\pi M_s(N_x - N_z)][H_0 + S_{eff} + 4\pi M_s(N_y - N_z)]} \tag{10.52}$$

If $N_x = N_y = N_z$ (spheres), S_{eff} is given by

$$S_{eff} = \omega_{res}^{(exp)}/\gamma - H_0 \tag{10.53}$$

Describing the mode frequency shift in terms of a lineshift is somewhat artificial because the basic physical parameter which determines the mode natural precession frequency in a given field is the gyromagnetic ratio. A description in terms of an effective gyromagnetic ratio γ_{eff} seems more appropriate, particularly since γ comes into the definition of ΔH_{eff}. If γ_{eff} is defined by

$$\omega_{res}^{(exp)} = \gamma_{eff}\sqrt{[H_0 + 4\pi M_s(N_x - N_z)][H_0 + 4\pi M_s(N_y - N_z)]} \tag{10.54}$$

the simple relation

$$\frac{\omega_{res}^{(exp)}}{\omega_{res}^{(Kittel)}} = \frac{\gamma_{eff}}{\gamma} \tag{10.55}$$

is satisfied. The relative change in the actual ω_{res} from the Kittel frequency is the same as the relative change in the gyromagnetic ratio from the intrinsic value γ in the absence of relaxation.

The parameters $\Delta H_{eff}(H_0)$ and $S_{eff}(H_0)$ have now been defined in terms of the more fundamental relaxation rate η_0 and effective gyromagnetic ratio. η_0 and γ_{eff} can be easily determined from the susceptibility expressions given in Table 10.1 or 10.2, using measured susceptibility values obtained by standard microwave techniques. The susceptibility which is measured in the usual linearly polarized microwave experiment, separated into its real and negative imaginary parts, $\chi_e = \chi_e' - i\chi_e''$, can be written as:

$$4\pi\chi_e' = \frac{\omega_m}{2}\left[\frac{\omega_{res} - \omega}{(\omega_{res} - \omega)^2 + \eta_0^2} + \frac{\omega_{res} + \omega}{(\omega_{res} + \omega)^2 + \eta_0^2}\right] \quad (10.56a)$$

$$4\pi\chi_e'' = \frac{\omega_m}{2}\left[\frac{\eta_0}{(\omega_{res} - \omega)^2 + \eta_0^2} + \frac{\eta_0}{(\omega_{res} + \omega)^2 + \eta_0^2}\right] \quad (10.56b)$$

In (10.56a and b), the expressions have been separated into their Larmor and anti-Larmor terms, by combining the expressions for $(\chi_e - \kappa_e)$ and $(\chi_e + \kappa_e)$ in Table 10.2, and including a one-half factor. Given measured values of χ_e' and χ_e'' versus field, the only remaining problem is to calculate $\eta_0(H_0)$ and $\omega_{res}(H_0)$. Two techniques have been used. One (Vrehen, 1968) involves the use of circularly polarized microwave fields to measure the Larmor or anti-Larmor contributions to X_e' and χ_e'' separately. Then, η_0 and ω_{res} are easily calculated using the Larmor or anti-Larmor parts of (10.56a and b). One problem with this approach is that the experimental arrangement is quite complicated and adjustment may be very tedious. In the other approach, linearly polarized fields are used (Patton, 1969a) to measure χ_e' and χ_e'', and solutions for η_0 and ω_{res} are obtained from (10.56a and b) using an iterative technique. Note that ω_{res} is contained implicitly in ω_m because of the γ factor (γ_{eff}).

The first technique avoids the assumption of equal η_0 and γ_{eff} values for the Larmor and anti-Larmor terms in χ_e' and χ_e''. For some physical relaxation processes, η_0 for the two terms might be quite different. Two-magnon relaxation, for example, should never contribute significantly to the anti-Larmor η_0. On the other hand, the anti-Larmor terms can be neglected near resonance, where two-magnon relaxation is most important. Consequently, the two approaches should yield about the same result. Once $\omega_{res}^{(exp)}$ and η_0 are determined, ΔH_{eff} is obtained from (10.51) and (10.55). The field dependence of ΔH_{eff} for relaxation contributions from various sources will be considered in subsequent sections.

The linewidths discussed so far are related to resonance. One very important linewidth parameter, the spin-wave linewidth ΔH_k, has to do with non-resonant excitation at high power. The basic experiment consists of pumping microwave energy into the spin system and observing the change in susceptibility with power level. When the input power exceeds the rate at which specific modes can relax to thermal equilibrium, the susceptibility changes sharply. The threshold microwave field amplitude h_{crit} at the break point can be related to the relaxation rate η_k for the spin-wave mode with the lowest threshold under the prescribed experimental conditions. Determining h_{crit} is equivalent to measuring η_k for the mode. The spin-wave linewidth ΔH_k is defined as

$$\Delta H_k = 2\eta_k/\gamma \quad (10.57)$$

similar to the definition of ΔH_{eff} in terms of the off-resonant η_0. Of course, ΔH_k is a linewidth only in the frequency swept linewidth sense. It is unambiguously related to the relaxation rate η_k. By changing the various experimental parameters, such as sample shape, pump field polarization, field and/or frequency, the wave vector k of the mode with the lowest threshold can be varied over a wide range. In this way ΔH_k as a function of k may be obtained, thus providing additional information by which to sort out the various possible contributions to the relaxation. High power effects will be considered in the next section.

10.4 HIGH POWER PHENOMENA

At low microwave power levels the magnetization response can be adequately described by a field independent susceptibility; i.e. the response is linear. At high power levels, coupling between the microwave field and spin-wave modes can lead to a non-linear response. The discovery of non-linear phenomena in ferrites and the understanding of the behaviour in terms of spin-wave interactions have provided new insight into microwave relaxation processes.

The early experiments were based on FMR data at high signal levels. At high power levels, the large uniform precession amplitude should result in a decrease in M_z from the saturation value M_s as well as a microwave field dependent susceptibility. Predictions based on the Bloch–Bloembergen equations (10.40a and b), indicate that the reduction in M_z from M_s and in the susceptibility at resonance χ''_{res} from its low power value χ''_0 should have the same form

$$\frac{\chi''_{res}}{\chi''_0} = \frac{M_z}{M_s} = (1 + \tfrac{1}{4}\gamma^2 |\mathbf{h}_0|^2 T_1 T_2)^{-1} \qquad (10.58)$$

Experiments by Bloembergen and Damon (1952), Damon (1953) and Bloembergen and Wang (1954), however, showed that for FMR in nickel ferrite the decrease in χ''_{res} occurs at much lower fields than that for M_z. This effect is often called premature saturation of the main resonance. Data (after Damon (1953)) are shown in Figure 10.6. In addition, an absorption peak at fields less

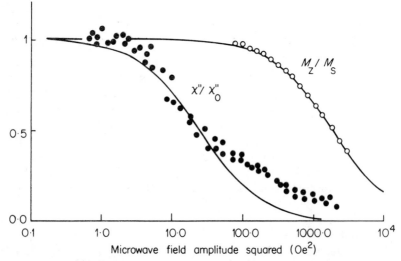

Figure 10.6 Susceptibility and average magnetization versus microwave field amplitude for nickel ferrite (after Damon (1953)).

than that required for resonance was observed at power levels comparable to the level for saturation of χ''_{res}. This effect is often labelled 'subsidiary absorption'. Both effects are shown in Figure 10.7 (after Bloembergen and Wang (1954)). The dotted line indicates the shape of the absorption curve at low power and the solid line shows the change at high power. Notice the subsidiary absorption peak centred around 2000 Oe. The excitation frequency in this experiment was 9 GHz.

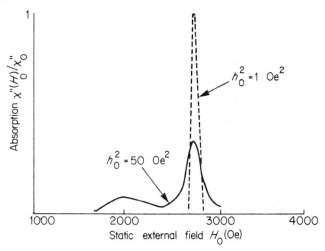

Figure 10.7 Resonance absorption curves at low power (dotted line) and high power (solid curve) (after Bloembergen and Wang (1954)).

These initial observations led to a detailed study of the microwave properties of ferrites at high power levels and the development of important techniques for studying spin-wave relaxation. The origin of the peculiar high power effects was first explained by Suhl (1956a, 1956b, 1957, 1958). He proposed that the parametric excitation of spin-waves by the uniform mode, because of the dipole–dipole interaction, was the cause of the observed behaviour. At low power levels the uniform precession amplitude is small so that the spin-wave amplitudes stay essentially at their thermal values. As the power level is increased, the uniform precession amplitude grows and more and more energy is pumped into the spin-wave modes. At some critical power level, the power input to the modes exceeds the rate at which the energy can be lost due to relaxation and some spin-wave amplitude(s) begin to increase beyond their thermal values. The power into any spin-wave mode is proportional to some coupling coefficient P_k multiplied by the mode occupation number n_k. The power out of the mode due to relaxation, given by the usual transition probability formula, is proportional to $\eta_k(n_k - \bar{n}_k)$, where $\bar{n}_k$ is the thermal equilibrium occupation number. The time dependence of n_k for this energy flow situation is

$e^{\kappa t}$ with $\kappa = P_k - \eta_k$. If P_k exceeds η_k, κ is positive and the mode occupation number grows exponentially. The power level or field at which $P_k = \eta_k$ is satisfied is the spin-wave instability threshold. Of course, the mode amplitude cannot increase indefinitely. At some point the amplitude (or occupation number) is limited by other interactions. In the case of subsidiary absorption, for example, this is in the form of a back reaction of the spin-wave modes on the uniform precession.

The early data and theory were concerned only with transverse microwave excitation. Schlömann *et al.* (1960a) and Morgenthaler (1960) independently proposed that a microwave field parallel to the static field could also result in the parametric excitation of spin waves and associated instability effects. In many respects, the parallel pump instability is easier to understand physically than resonance saturation or subsidiary absorption. Direct coupling between the parallel pump microwave field and certain spin-wave modes is possible because of the spin-wave ellipticity. A single precessing spin with an elliptical precession cone is shown in Figure 10.8. If the magnitude of **M** is constant, the

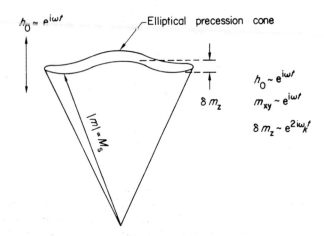

Figure 10.8 Schematic representation of elliptical precession cone for spin-waves which leads to the parallel pump instability.

elliptical orbit results in a wobble in the z component denoted by δm_z, with an $\exp(2i\omega_k t)$ time dependence. Consequently, a spin-wave mode at ω_k can couple to the parallel pump microwave field at $\omega = 2\omega_k$. The amplitude of δm_z (and the coupling) is zero at $\theta_k = 0$, where the ellipticity is zero and maximum at $\theta_k = \pi/2$ (maximum ellipticity). Since the ellipticity is essentially a result of the dipole–dipole energy term in the dispersion relation, $|\delta m_z| \approx \sin^2 \theta_k$ is a reasonable assumption for the θ_k dependence of the coupling strength. From the more detailed theory, the rate of energy input to spin waves at $\omega_k = \omega/2$ is proportional to

$$G = (\omega_m/\omega)(\gamma h_0/2) \sin^2 \theta_k \qquad (10.59)$$

The threshold field h_c is the value of h_0 at which G is equal to the relaxation rate η_k. In terms of the spin-wave linewidth $\Delta H_k = 2\eta_k/\gamma$, the threshold is given by

$$h_c = (\omega/\omega_m)\Delta H_k/\sin^2 \theta_k$$

Note that the threshold diverges at $\theta_k = 0$ because the ellipticity is then zero.

The above discussion outlines the basic ideas behind instability phenomena. The exact form of the coupling strength is needed to calculate thresholds for the different processes and for determining ΔH_k from threshold data. As in basic spin-wave theory, two different approaches are commonly used, a transition probability method based on quantum theory, and a macroscopic approach. In the latter, instability effects are obtained simply by including terms due to the applied microwave field $\mathbf{h}_0$ or the potentially large uniform precession amplitude α_0 excited by the transverse components of $\mathbf{h}_0$ in the spin-wave analysis. This technique was used in the original Suhl theory and in the parallel pump theory of Schlömann. In addition, Patton (1969b) has recently extended the Suhl–Schlömann theories to include more complicated microwave field configurations.

Macroscopic spin-wave theory was outlined in Section 10.2. In order to treat instability phenomena, it is necessary to include the microwave field in $\mathbf{H}(\mathbf{r})$, and to treat the uniform precession amplitude α_0 somewhat differently from the α_k with $k \neq 0$. In the spin-wave theory, all amplitudes were assumed small so that higher order products of the α_k could be neglected. Only linear terms in the α_k were considered. If the uniform mode is excited directly by the microwave field, however, α_0 may be large and terms of the form $\alpha_0\alpha_k$ cannot be neglected. The linearization of the equation of motion must be carried out only with respect to the α_k with $k \neq 0$. The equation of motion for the b_k is changed to the form

$$b_k = i\{\omega_k b_k + F_k b_k + G_k b_{-k}\} \tag{10.60}$$

The F_k term leads only to a small modulation of the spin-wave frequency which may be neglected. Spin-wave instability effects come from the G_k mixing term which couples the b_k and b_{-k} amplitudes. The different instability effects are separated by considering different parts of G_k. To sort out the different processes, G_k may be expanded in a series

$$G_k = \sum_n G_k^{(n)} e^{in\omega t} \tag{10.61}$$

with n ranging from -2 to $+2$. Parallel pumping and subsidiary absorption come from the $n = 1$ term in G_k. The $\exp(i\omega t)$ time dependence for $G_k^{(1)}$ comes from terms which are linear in α_0 or h (but not both). The related instabilities are often called first order processes. Resonance saturation, in the cases where a first order process is not allowed at resonance, comes from the $n = 2$ term. The $\exp(2i\omega t)$ time dependence comes from terms which are quadratic in α_0 (second order process). (For complete expressions, see Schlömann, 1959a or Patton, 1969b.) The remaining brief analysis will be concerned with first order processes.

If the F_k term in the equation of motion is neglected and only the $G_k^{(1)}$ contribution to G_k considered, the simplified equation becomes

$$\dot{b}_k = i\{\omega_k b_k + G_k^{(1)} e^{i\omega t} b_{-k}^*\} \qquad (10.62)$$

At this point, it is necessary to introduce spin-wave relaxation into the formulation by replacing ω_k with $\Omega_k = \omega_k + i\eta_k$, one method discussed in the previous section. It is also convenient to rewrite the equation in terms of a time dependent amplitude $b_{k0}(t)$,

$$b_k = b_{k0}(t) e^{i\omega t/2} \qquad (10.63)$$

and examine the conditions under which a stationary solution for the b_{k0} is permitted. By combining the differential equations for b_{k0} and b_{-k0}^* which result from (10.62) and (10.63), one obtains

$$\left\{\frac{d^2}{dt^2} + 2\eta_k \frac{d}{dt} + [(\omega_k - \omega/2)^2 + \eta_k^2 - |G_k^{(1)}|^2]\right\} b_{k0} = 0 \qquad (10.64)$$

with a solution, $b_{k0} \propto \exp(\kappa t)$.

$$\kappa = \eta_k \pm [|G_k^{(1)}|^2 - (\omega_k - \omega/2)^2]^{\frac{1}{2}} \qquad (10.65)$$

The stationary solution corresponds to $\kappa = 0$ and the smallest value of $|G_k^{(1)}|$ for which this solution is permitted occurs at $\omega_k = \omega/2$. (Recall that the parallel pump instability occurs at $\omega_k = \omega/2$ from the physical arguments.) The quantity $|G_k^{(1)}|$ is also proportional to the microwave field amplitude, so that the condition $|G_k^{(1)}| = \eta_k$ ($\kappa = 0$ for $\omega_k = \omega/2$ and $|G_k^{(1)}| = \eta_k$) determines the threshold h_c for spin-wave $(\omega/2, \mathbf{k})$. For $h_0 > h_c$, the analysis predicts an exponential growth of b_{k0} with time and to obtain the steady state solution above threshold, it is necessary to consider additional terms in the equation of motion.

Detailed results for general first-order processes are given by Patton (1969b). If a reduced coupling parameter W is defined by

$$G_k^{(1)} = \frac{\omega_m}{\omega} \frac{\gamma h_0}{2} e^{2i\phi_k} W \qquad (10.66)$$

the threshold h_c is simply given by

$$h_c = \frac{\omega}{\omega_m} \frac{\Delta H_k}{|W|} \qquad (10.67)$$

The experimentally observed threshold h_{crit} for a first-order instability corresponds to the minimum value of h_c with respect to $\mathbf{k}$ for a given sample geometry, microwave field configuration and applied static field. Thus, h_{crit} can be obtained by minimizing $|\Delta H_k/W|$ with respect to ϕ_k and θ_k, subject to the requirement that k be real.

$$h_{crit} = (\omega/\omega_m)[\Delta H_k/W]_{min} \qquad (10.68)$$

For a given ω_k, θ_k and k are related through the dispersion relation. It is important to realize that the range of allowed k or θ_k depends on the biasing condition. Consider the dispersion diagrams in Figure 10.9. For very large bias fields such that $\gamma H > \omega/2$ is satisfied (Figure 10.9(a)), there are no modes at $\omega_k = \omega/2$ and the threshold is infinite. Over an intermediate field range (Figure 10.9(b)), given by

$$\gamma H < \omega/2 < \gamma\sqrt{H(H + 4\pi M)}$$

the allowed modes extend from $k = 0$ to some maximum value. The upper limit on θ_k is less than $\pi/2$. At lower fields (Figure 10.9(c)), the smallest k allowed is non-zero and θ_k extends over the full range.

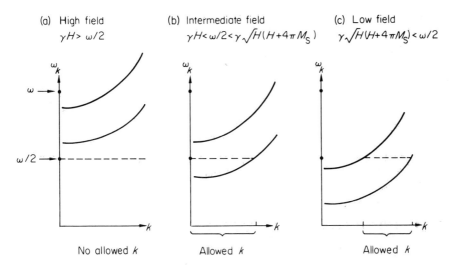

Figure 10.9 Position of spin-wave manifold with respect to potentially unstable modes at $\omega_k = \omega/2$ for different bias fields.

In the intermediate field range, the minimum $(\Delta H_k/W)$ which should determine the observed threshold h_{crit} often occurs at the maximum θ_k allowed or $k = 0$. This introduces some difficulty in interpreting experimental threshold in terms of the theory because surface dipole terms were not taken into account either in the spin-wave theory or in the instability analysis. For long wavelengths ($k \approx 0$), such terms are important and lead to magnetostatic modes. Schlömann and Joseph (1961) have considered the problem of magnetostatic mode instability in detail and showed that the spin-wave approach is still approximately correct.

The coupling parameter W for first order processes and an arbitrary pump field is given by Patton (1969b).

$$W = \sin\theta_k \cos\theta_k\{q_L[f(\theta_k) + \omega/2]\,e^{-i\phi_k} + q_A^*[f(\theta_k) - \omega/2]\,e^{i\phi_k}\}$$
$$+ a_z \sin^2\theta_k \tag{10.69}$$

The $f(\theta_k)$ function is simply

$$f(\theta_k) = \gamma(H + Dk^2) = \tfrac{1}{2}[(\omega_m^2 \sin^4 \theta_k + \omega^2)^{\frac{1}{2}} - \omega_m \sin^2 \theta_k] \qquad (10.70)$$

and the requirement of real k is equivalent to $f(\theta_k) - \gamma H > 0$. The $q_{L,A}$ are normalized Larmor and anti-Larmor contributions to the uniform precession amplitude α_0,

$$\alpha_0 = (\gamma h_0/2)(q_L\, e^{i\omega t} + q_A\, e^{-i\omega t}) \qquad (10.70a)$$

$$q_L = \frac{(\omega + \omega_y)a_x\, e^{i\delta_x} + i(\omega + \omega_x)a_y\, e^{i\delta_y}}{\omega_x \omega_y - \omega^2} \qquad (10.70b)$$

$$q_A = \frac{(\omega - \omega_y)a_x\, e^{-i\delta_x} + i(\omega - \omega_x)a_y\, e^{-i\delta_y}}{\omega_x \omega_y - \omega^2} \qquad (10.70c)$$

The $a_{x,y,z}$ and $\delta_{x,y}$ describe the microwave field polarization according to

$$h_0^x = h_0 a_x \cos(\omega t + \delta_x) \qquad (10.71a)$$

$$h_0^y = h_0 a_y \cos(\omega t + \delta_y) \qquad (10.71b)$$

$$h_0^z = h_0 a_z \cos \omega t \qquad (10.71c)$$

The ω_x and ω_y are given in Section 10.3. Note that the transverse components of h_0 enter into W only through the q_L and q_A terms in (10.64) so that for transverse excitation the coupling which leads to instability is only through the uniform precession. The parallel pump component of h_0 enters directly as the $a_z \sin^2 \theta_k$ term in W.

The simplest case, parallel pumping, yields $W = \sin^2 \theta_k$, independent of ϕ_k. If the possible θ_k dependence of ΔH_k is neglected, the threshold is given by (10.66) with $\theta_k = \pi/2$,

$$h_{\text{crit}} = \omega \Delta H_k/\omega_m \qquad (10.72)$$

for $\omega/2 > \gamma\sqrt{H(H + 4\pi M)}$ and k is given by $f(\pi/2) = \gamma(H + Dk^2)$. In the intermediate field range given by (10.68), the threshold is given by

$$h_{\text{crit}} = \frac{\omega\, \Delta H_k}{\omega_m \sin^2 \theta_k} \qquad (10.73)$$

with θ_k specified by $f(\theta_k) = \gamma H$ $(k = 0)$.

For transverse pumping, Larmor and anti-Larmor components of the uniform precession amplitude must first be evaluated for the particular microwave field polarization and sample shape under consideration. The original Suhl theory considered a circular polarized microwave field rotating in a Larmor sense and samples with rotational symmetry about the static field (and magnetization) direction ($\omega_x = \omega_y$). In this case, the microwave field is given by (10.71a, b and c) with $a_x = a_y = \tfrac{1}{2}$, $\delta_x = a_z = 0$, and $\delta_y = -\pi/2$, so that

$q_A = 0$ and

$$q_L = (\omega_{res} - \omega)^{-1} \tag{10.74}$$

$$W = \frac{\sin\theta_k \cos\theta_k[f(\theta_k) + \omega/2]\,e^{-i\phi_k}}{(\omega_{res} - \omega)} \tag{10.75}$$

The maximum value of $|W|$ occurs near $\theta_k = \pi/4$, independent of ϕ_k. For a Suhl process, therefore, the unstable modes are directed at approximately 45° to the magnetization direction. For samples which do not have rotational symmetry, the anti-Larmor component of the uniform precession amplitude is non-zero, even for Larmor microwave excitation. This anti-Larmor term introduces a ϕ_k dependence for $|W|$ and results in a slight reduction in the threshold. Excitation with a linearly polarized microwave field, which has Larmor and anti-Larmor components of equal amplitude, also yields an anti-Larmor contribution to α_0, introduces a ϕ_k dependence in $|W|$, and lowers the threshold. For such situations, the unstable modes which determine h_{crit} have specific values of ϕ_k.

The details of first-order processes for other pump configurations follow arguments similar to those presented above (see Patton, 1969b). Second order processes have been discussed in detail by Suhl (1957) and Schlömann (1959a). A summary of the important processes is given in Figure 10.10. The nature of the mode to which the instability is attributed is given in each case. By measuring the threshold h_{crit} as a function of static field for one or more of these configurations, one obtains a so-called butterfly curve (h_{crit} versus H_0) from which the wave vector dependence of ΔH_k can be determined. Butterfly curves and spin-wave linewidth will be discussed in more detail at the end of this section.

Configuration		Spin-wave angles for $k > 0$
H, M	Parallel pumping h_0 (1st order)	$\theta_k \approx \pi/2$, no ϕ_k dependence
H, M	Transverse Larmor pumping h_0 (1st order)	$\theta_k \approx \pi/4$, no ϕ_k dependence
H, M	Transverse linear pumping h_0 (1st order)	$\theta_k \approx \pi/4$, $\phi_k = \pi$ (or $\pm\pi/2$), depending on geometry and field (usually $\phi_k = \pi$)
H, M	Oblique linear pumping h_0 (1st order)	θ_k from $\pi/2$ to $\pi/4$ depending on pump angle, $\phi_k = \pi$
$H = H_{res}$ M	Resonance saturation h_0 (2nd order)	$\phi_k = 0$

Figure 10.10 Summary of the various instability processes.

The basic interactions which lead to instability effects have been described, both from a physical point of view and using macroscopic theory. Very little has been said about the changes in the observable microwave properties which accompany the onset of instability effects or the processes which determine these properties above threshold. The behaviour of the microwave susceptibility above threshold was considered in the original Suhl theory for subsidiary absorption and conditions under which it is separate from or coincides with the main resonance. (Recall that the field range for subsidiary absorption is limited to $H < \omega/2\gamma$. If $H_{\text{res}} < \omega/2\gamma$ is also satisfied, a first-order instability is possible at resonance.) The first order threshold was derived by considering a spin-wave equation of motion of the form

$$\dot{b}_k = i\{\Omega_k b_k + f_k^* \alpha_0 b_{-k}^*\} \tag{10.76}$$

where the uniform precession amplitude α_0 contained in the $G_k^{(1)}$ of (10.62) is displayed explicitly. The steady state response is obtained by simultaneously solving this equation and the uniform precession equation-of-motion, including feedback terms from the b_k, for α_0. Following Schlömann's treatment (1959a), the uniform precession equation is given by

$$\dot{\alpha}_0 = i\left\{\Omega_0 \alpha_0 + \tfrac{1}{2}\sum_k f_k b_k b_{-k} - \gamma h_0\right\} \tag{10.77}$$

If the sum term is excluded, (10.77) gives the Larmor resonance response described in Section 10.3. Above threshold, feedback to α_0 from the spin-wave modes determines the microwave susceptibility. (For details of the analysis, refer to Schlömann's treatment.)

In essence, the result of the feedback above threshold leads to a uniform precession amplitude $|\alpha_0|$ which is independent of the driving field. For a sample biased at resonance, this means that the imaginary part of the susceptibility χ'' is an inverse function of the microwave field amplitude. In terms of the low power susceptibility χ_0'', the susceptibility decline is given by

$$\chi'' = \chi_0'' h_{\text{crit}}/|h_0| \tag{10.78}$$

above threshold. Note that this result applies only if a first order process is allowed at resonance. For subsidiary absorption, where the first order process is allowed only for fields somewhat less than required for resonance, the behaviour of the susceptibility above threshold comes from the change in phase of α_0 with microwave field. If α_0 is described by

$$\alpha_0 = a_{00}\, e^{i\delta} \tag{10.79}$$

the imaginary part of the susceptibility is given by $\chi'' = (Ma_{00}/|h_0|)\sin\delta$. The final result is

$$\chi'' = \frac{Ma_{00}}{|h_0|^2}\sqrt{|h_0|^2 - h_{\text{crit}}^2 + \left(\frac{a_{00}\,\Delta H}{2}\right)^2} \tag{10.80}$$

for $|h_0| > h_{crit}$. When $|h_0|$ exceeds h_{crit}, χ'' increases sharply, reaches a maximum near $|h_0| = 2h_{crit}$ (if the ΔH term is neglected) and finally decreases to zero at large $|h_0|$.

These predicted theoretical dependences have been confirmed by microwave data on a number of single crystal materials. Data for a first order instability at resonance, due to Suhl (1956a) for Ni ferrite and Mn ferrite at 4 GHz are shown in Figure 10.11. The Ni ferrite data show poor agreement, presumably

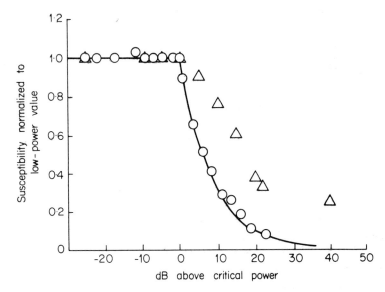

Figure 10.11 Susceptibility above threshold for resonance saturation in Ni ferrite (triangles) and Mn ferrite (open circles) (after Suhl (1965a)).

due to material inhomogeneities. Single crystal Mn ferrite data for subsidiary absorption at 9·3 GHz (Suhl, 1956a; data due to H. Scovil) in Figure 10.12 are in good agreement with the theory near threshold. At very high power levels the measured susceptibility does not decay as rapidly as predicted. Suhl suggested that higher order instability processes possibly enhance χ'' at high power levels.

Schlömann (1959a) and Schlömann et al. (1960b) have developed a general theory which, among other things, is used to study the so-called fold-over of the resonance curve at high power (Anderson and Suhl, 1955; Weiss, 1958) and the influence of crystalline anisotropy for second order processes. Other references on the steady state behaviour above threshold include Schlömann (1962a, 1962b, 1963), Schlömann and Green (1959, 1963), Green and Schlömann (1962), Kohane and Schlömann (1963), LeGall (1964, 1967, 1968), LeGall et al. (1969), Monosov (1969), Monosov and Surin (1967), Monosov et al. (1968), Desormiere and Milot (1967), Desormiere et al. (1969), Morgenthaler (1964),

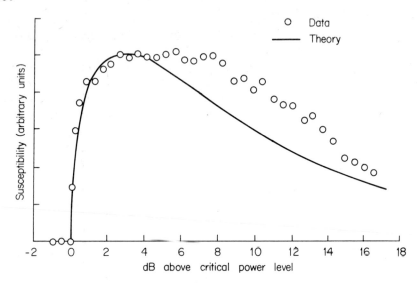

Figure 10.12   Susceptibility above threshold for subsidiary absorption in Mn ferrite single crystal (after Suhl (1956a)).

Joseph *et al.* (1966), Churkin and Chelishcher (1967), Kitaev and Fedoseeva (1968), Starobinets and Gurerich (1968) and Wang *et al.* (1968).

One of the most important results of the early observations and subsequent theories has been an extremely useful and versatile technique for studying microwave relaxation. Specifically, the application of instability theory to threshold data yields the spin-wave relaxation rate or spin-wave linewidth as a function of wave-vector $\mathbf{k}$. The concluding part of this section is concerned with the extraction of ΔH_k from h_{crit} data and the interpretation of ΔH_k results.

Butterfly curve data for transverse and parallel pumping in single crystal YIG at 9·1 GHz and 300 K are shown in Figure 10.13. The minimum threshold for transverse pumping occurs near 2000 Oe, the location of the subsidiary absorption peak in Figure 10.7. The minimum h_{crit} corresponds to the instability of modes with $k \approx 0$. At lower fields, the instability is for non-zero $\mathbf{k}$ and $\theta_k = \pi/2$ (parallel pumping) or $\theta_k \approx \pi/4$ (transverse pumping). The field at the minimum for transverse pumping is higher than that for parallel pumping because at $\mathbf{k} \approx 0$ the $\theta_k = \pi/4$ position on the dispersion curve falls at lower frequency than the $\theta_k = \pi/2$ position, thus requiring higher bias fields to satisfy the $\omega_k = \omega/2$ requirement. In both cases the thresholds become very large at about 2200 Oe, the point at which $H = \omega_k/\gamma$. At higher fields, modes at $\omega_k = \omega/2$ are not allowed and first order processes cannot occur.

Application of the theory to the portions of the butterfly curves below the minimum h_{crit} position gives ΔH_k versus $\mathbf{k}$. Recall that in the general theory, the $\mathbf{k}$ of the unstable mode is determined by minimizing $\Delta H_k/W$. The parameter W contains the $\mathbf{k}$ dependence of the coupling interaction. The k dependence of

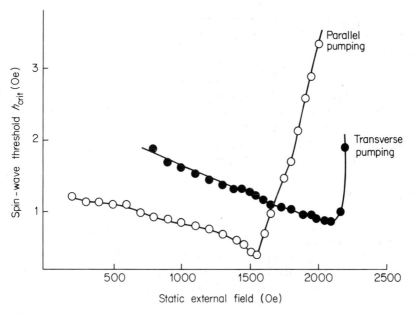

Figure 10.13 Butterfly curves for parallel and transverse pumping in single crystal YIG at 9·1 GHz.

ΔH_k, on the other hand, is not known. (ΔH_k versus **k** is the objective of the analysis in the first place.) The usual procedure is to ignore the **k** dependence of ΔH_k in the initial minimization, determine ΔH_k versus **k** from the data and then inspect the ΔH_k dependence for possible modifications in the previously determined k values for the unstable modes. In many situations ΔH_k is essentially **k**-independent, as for samples with a large impurity contribution to the relaxation, so that no difficulty arises. If ΔH_k does have a strong k dependence, the k for the unstable modes and the shape of the butterfly curves can be modified significantly (Patton, 1970a).

Another source of complication is the possible influence of two-magnon interactions on the threshold. Such interactions have been found to be extremely important for second order processes where the unstable spin-waves are at the driving frequency (Schlömann, 1959a; Schlömann et al., 1960b). Two-magnon interactions due to inhomogeneities give added coupling between the uniform precession and the spin-wave modes. Schlömann showed that the influence of inhomogeneities on resonance saturation for second order processes depends critically on the range of **k** for the modes which are strongly coupled to the uniform precession. If strong coupling to short wavelength modes exists, the susceptibility may exhibit an initial increase before saturation sets in at higher power levels.

The situation concerning two-magnon contributions to first order instability processes is more confused. Clearly, scattering involving the uniform precession at the driving frequency is not important because the potentially unstable modes at $\omega_k = \omega/2$ are not excited by the inhomogeneity interaction. The coupling between the spin-wave modes due to inhomogeneities, on the other hand, was initially believed to be the source of the large spin-wave linewidths observed in polycrystals. The argument was that two-magnon relaxation broadens the spin-wave linewidth in much the same way that it affects the resonance linewidth (see next two sections). This argument, however, is not consistent with the early work on single crystal YIG, in which ΔH_k was found to be extremely insensitive to surface roughness (Kasuya and LeCraw, 1961; LeCraw and Spencer, 1962). As will be discussed in the next section, surface roughness strongly contributes to the resonance linewidth.

A number of fairly quantitative arguments have been put forward to explain why ΔH, but not ΔH_k, is influenced by two-magnon interactions. One useful approach (Sparks, 1964) is to consider the difference between a resonance experiment and high power experiment. In an instability experiment, the threshold corresponds to the true normal mode with the lowest h_c. In the presence of inhomogeneities, this true mode is a mixture of the old spin-wave modes. Typical materials, however, contain only relatively long wavelength inhomogeneities (1 μm or larger), so that only modes with nearly the same k are mixed. As a result, the new modes are nearly identical with the old modes and the threshold is virtually unchanged. In a resonance experiment, the energy is coupled in to the uniform precession. Because of inhomogeneities, the true mode which most closely resembles the uniform mode contains some spin-wave components. The two-magnon resonance linewidth simply reflects the amount of spin-wave mixing. However, it should be realized that two-magnon interactions can contribute to ΔH_k if the coupling mixes modes with substantially different $\mathbf{k}$. In such a case, the scattering provides additional channels for energy to get out of the pumped mode and should result in an increased threshold. Sage (1968) has considered two-magnon effects on parallel pumping and discussed the various circumstances in which two-magnon interactions are important.

10.5 RELAXATION IN SINGLE CRYSTALS

The remaining three sections are concerned with applying the basic ideas developed in the previous sections. This and the following section consider saturated single crystal and polycrystalline materials respectively. The final section explores some of the effects of the partially magnetized state on microwave properties. In view of the excellent review of relaxation in single crystals by Sparks (1964), the emphasis here will be mainly on summarizing the theoretical and experimental situation. Two-magnon scattering due to inhomogeneities, temperature peak processes due to charge transfer or impurity relaxation, three-magnon processes and Kasuya–LeCraw magnon–phonon relaxation are discussed. A suitable combination of these mechanisms appears to adequately explain the observed linewidths for a large class of single crystal materials.

Two-magnon scattering is one very important source of relaxation in materials containing magnetic inhomogeneities. In single crystals, surface roughness, grain boundaries and atomic disorder are potentially important sources of the scattering. The basic idea, first pointed out by Clogston *et al.* (1956), is that such inhomogeneities result in a coupling between the otherwise orthogonal uniform precession and degenerate spin-wave modes and that energy transfer out of the uniform precession to the degenerate modes is important in the initial stages of resonance relaxation. The interaction is sensitive to the nature of the inhomogeneity. As a general rule, the coupling is large for spin-wave wavelengths greater than the dimensions of the inhomogeneity. The original calculation considered scattering due to atomic disorder. The wavelengths of spin-waves which are important at microwave frequencies are much bigger than lattice spacings, so that essentially all of the available degenerate modes are coupled to the uniform precession. Calculations by Callen and Pittelli (1960) as well as Hass and Callen (1961) and experimental results by Denton and Spencer (1962) on lithium ferrite single crystals indicate that such scattering can contribute to the linewidth.

The most convincing evidence for two-magnon relaxation in single crystals, however, is for materials with long wavelength inhomogeneities (generally 1 μm or larger). In this case, the important scattering occurs only to the relatively low k modes. The experimental evidence for a two-magnon relaxation contribution came from the early experiments of LeCraw *et al.* (1958) and Buffler (1959, 1960). They found that the linewidth increases with surface roughness and that the frequency dependence of ΔH reflected in a consistent way the shift of the spin-wave manifold with respect to the uniform precession frequency with biasing field (Figure 10.14). The peak in ΔH near 7 GHz is at $\omega = (2/3)\omega_m$, the

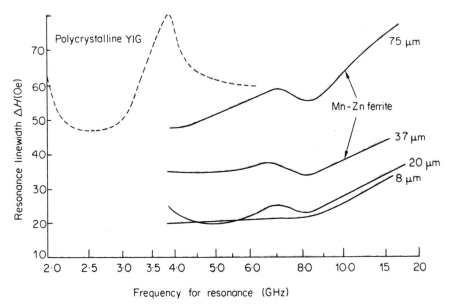

Figure 10.14 Resonance linewidth versus frequency for Mn–Zn ferrite single crystal sphere with different surface polish. The dotted curve is for polycrystalline YIG.

point at which the uniform precession resonance mode is coincident with the $\theta_k = \pi/2$ branch of the spin-wave manifold. At this point, the density of the degenerate low-k states is large. At lower frequencies, the uniform precession frequency is above the top of the manifold and the linewidth decreases. The effect in polycrystals, for which internal inhomogeneities also contribute to the two-magnon relaxation, is even more pronounced (dotted line in Figure 10.14). The peak for YIG is at lower frequency because $4\pi M_s$ is lower than for Mn–Zn ferrite.

Low-k scattering was first treated theoretically by Geschwind and Clogston (1957). They showed that the dipole coupling between the different regions serves to narrow the resonance line. The dipole narrowing effect for long wave-length inhomogeneity scattering is much the same as exchange narrowing in atomic disorder scattering (see also Clogston, 1958). The first detailed theory of scattering relaxation due to surface pits was done by Sparks (1961) and Sparks *et al.* (1961). In his theory, a single surface pit is represented by a spherical void in a saturated ferromagnetic medium of infinite extent. The model for the interaction is shown schematically in Figure 10.15. The pore produces a

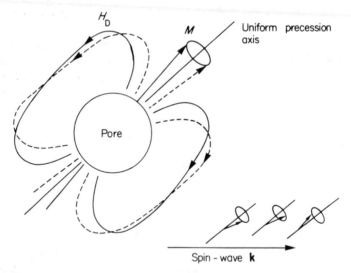

Figure 10.15 Schematic representation of interaction between spin-wave and the time varying demagnetizing field induced by the uniform precession around a spherical pore.

dipole field with an axis along the magnetization direction. If the magnetization precesses as in the case of uniform mode resonance, the dipole field $\mathbf{H_D}$ induced by the pore is modulated at the precession frequency. The wobble is indicated in the figure by showing $\mathbf{M}$ and $\mathbf{H_D}$ for two different points in the precession (solid and dotted lines for $\mathbf{M}$ and $\mathbf{H_D}$). The interaction between the dipole field $\mathbf{H_D}$ and the spin-wave with wave vector $\mathbf{k}$ couples energy from the uniform precession mode into the spin-wave mode. Sparks utilizes the transition probability method to calculate the relaxation rate, based on an interaction Hamiltonian of the form

$$\mathcal{H} = \sum_k F(\mathbf{k})(b_0^\dagger b_k + b_0^\dagger b_k) \tag{10.81}$$

with

$$F(\mathbf{k}) = 8\pi^2 \frac{a_0^3 \hbar}{V} \gamma (3 \cos^2 \theta_k - 1) \frac{j_1(ka_0)}{ka_0} \tag{10.82}$$

where a_0 is the pore radius, V is the volume of the sample and j_1 is the first spherical Bessel function. Equations (10.81) and (10.82) are obtained by evaluating the interaction energy $-\frac{1}{2}\int \mathbf{H_D} \cdot \mathbf{M}$ in terms of the operator formalism discussed in Section 10.2. Details of the analysis are given by Sparks (1961a,b, 1964) and Sparks et al. (1961). The coefficient $F(\mathbf{k})$ which couples the uniform precession amplitude b_0 with the b_k already exhibits several features of the relaxation expected from intuitive arguments. The interaction strength is large only for $ka_0 < 1$, as evident from the $j_1(ka_0)/ka_0$ term in $F(\mathbf{k})$. This dependence limits the important scattering to long wavelength modes with $\lambda_k > a_0$. In addition, the interaction has a strong θ_k dependence, varying as $(3 \cos^2 \theta_k - 1)$. Because of the axial symmetry of $\mathbf{H_D}$, the interaction should clearly depend on the direction of $\mathbf{k}$. The fact that $F(\mathbf{k})$ is zero at $\cos^2 \theta_k = 1/3$ is a direct result of the spherical pore assumption.

The transition probability analysis results in a relaxation rate expression of the form

$$\eta_k = (\pi/\hbar) \sum_k |F(\mathbf{k})|^2 \delta(\omega_0 - \omega_k) \tag{10.83}$$

Momentum is not conserved because the interaction is localized in the vicinity of the pore. In one sense, the pseudomomentum components of the pore which appear as terms in the Fourier expansion of $\mathbf{H_D}$ compensate for the momentum of the scattered spin-waves. The remainder of the analysis involves changing from a sum over $\mathbf{k}$ states to an integral over $\mathbf{k}$ space, and including a factor to account for surface pit scattering. The final linewidth result for a sphere completely covered with pits is (Sparks, 1964)

$$\Delta H_{\text{pits}} = \frac{\pi^3}{8} M \frac{a_0}{r_0} G(\theta_u) \tag{10.84}$$

$$G(\theta_u) = (3 \cos^2 \theta_u - 1)^2 / \cos \theta_u \tag{10.85}$$

where r_0 is the sphere radius and θ_u is the value of θ_k at $k = 0$ evaluated from the spin-wave dispersion relation for the particular biasing field used. Except for the $(3 \cos^2 \theta_u - 1)$ term, the basic features of the result is essentially model independent, even though a rather simple model was used. The M factor arises from the origin of the interaction in the demagnetizing field of the pore(s), the (a_0/r_0) factor simply reflects the volume fraction of pits in the sample, and the $\cos \theta_u^{-1}$ term comes from the density of degenerate states obtained from the approximate dispersion relation which results if the spin-wave ellipticity is neglected.

The $(3 \cos^2 \theta_u - 1)$ term, however, reflects the angular dependence of $\mathbf{H_D}$ for spherical pores. For a distribution of irregular shaped pits, the factor should be different. This difficulty was first realized when the theory was applied to single crystal data for FMR with $H_0 \gg 4\pi M$. In this region, (10.84) and (10.85) predict that ΔH_{pits} should vary as ω^{-2}. Data by Seiden and Sparks (1965) for several ferrites showed that ΔH was strongly frequency independent in the high field region. The predicted ω^{-2} dependence, however, is a direct consequence of the $(3 \cos^2 \theta_u - 1)$ term in the coupling coefficient. Seiden and Sparks argued that although the angular dependence should be approximately correct, the function should have no zero. They proposed a modified expression to account for pit non-sphericity.

$$\Delta H_{\text{pits}} = \frac{\pi^3}{32} M \frac{a_0}{r_0} \frac{\omega}{\omega_i} \frac{[(3 \cos^2 \theta_u - 1)^2 + 1 \cdot 6]}{\cos \theta_u} \tag{10.86}$$

where the 1·6 term is the angular average of $(3 \cos^2 \theta_u - 1)^2$. The ω/ω_i factor $(\omega_i = \gamma H)$ arises from using the exact dispersion relation rather than the approximation after the second Holstein–Primakoff transformation to derive the density of states. Recently Sage (1969) has re-examined the entire problem of scattering from non-spherical pores and justified the spirit of the Seiden–Sparks non-sphericity correction (but not the 1·6) in a quantitative manner. With this correction, the pit-scattering theory of Sparks gives qualitative agreement with the data of LeCraw et al. (1958). The pore calculation may be directly applied to porosity scattering in polycrystals and also gives agreement with data by a factor of two. Considering the rather crude model which was used in the theory, this agreement is quite good. The basic problem is that the data are for samples with highly irregular pits in close proximity while the theory is for isolated pits of specified symmetry. Recently, however, the effective linewidth approach has provided a way to extract the two-magnon contribution to FMR relaxation (Schlömann and Kohane, 1968; Vrehen, 1968; Patton, 1969a). Fairly close agreement between theory and data for polycrystals has been obtained (Patton, 1969a; Vrehen et al., 1969b) as discussed in the next section.

The two-magnon contribution to the single crystal linewidth can be all but eliminated by using polished samples or determining the spin-wave linewidth ΔH_k, which usually contains no two-magnon contribution, from high power data. However, FMR measurements on polished samples showed that, even with no two-magnon contribution, linewidths for some materials were quite large and, in many cases, exhibited a characteristic peak at low temperature. This behaviour has been attributed to various types of impurities in the ferrite and are usually explained by mechanisms which involve electronic excitations in the crystal. Three processes, valence exchange, slowly relaxing impurity and rapidly relaxing impurity, have been proposed to explain the linewidth peaks and associated behaviour in various materials. In the relaxation scheme of Section 10.3, they may be classified as direct magnon–phonon processes.

Valence exchange is based on an electronic transition between Fe^{2+} and Fe^{3+} ions on equivalent lattice sites in the ferrite, which is induced by the magnetization motion, as discussed in Chapters 5 and 8. Wijn and van der Heide (1953) first suggested this mechanism to explain the initial susceptibility of ferrites at low frequency, Galt (1954) applied the concept to wall motion in nickel ferrite and Clogston (1955) developed a thermodynamic theory for both wall motion and resonance. Consider a simplified example, due to Sparks (1964) and shown in Figure 10.16. The magnetization precession modulates the energy

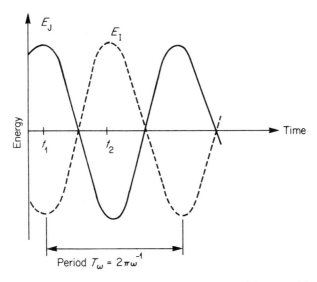

Figure 10.16 Energy level modulation for Fe^{2+} and Fe^{3+} ions on equivalent lattice sites (I and J) which gives rise to valence exchange relaxation.

levels for sites I and J at the precession frequency ω. At some time t_1, transitions from E_J to E_I are favoured and the energy change for each jump is absorbed by the lattice. At some later time t_2, the levels are reversed so that transitions from site I to site J occur. The average energy transfer to the lattice over a precession period determines the contribution to the uniform precession relaxation. The electronic relaxation time T_e plays a decisive role in determining this average. If the electronic relaxation time T_e is much longer than the modulation period T_ω, effectively no transitions occur and the mechanism does not contribute to the relaxation. At the other extreme, if T_e is short compared to the period the transitions, in essence, occur instantaneously and the sites

614

simply exchange electrons every half period, so there is also no net energy transfer to the lattice and no contribution to the linewidth. The largest contribution occurs when $\omega T_e \sim 1$, where a transition occurs once every cycle on the average. This is basically a thermal process in which the electron penetrates a potential barrier E_{ij}. Thus, T_e is strongly temperature dependent and behaves according to

$$1/T_e \sim \exp(-E_{ij}/k_b T)$$

where k_b is the Boltzmann constant and E_{ij} is on the order of 0·1–0·5 eV (Smit and Wijn, 1959). The linewidth peak occurs at the temperature where $\omega T_e \approx 1$ is satisfied. The peak is expected to shift to higher temperature (reduced T_e) with increasing frequency.

A similar process was proposed by Dillon and Nielsen (1959, 1962) to explain peaks in the linewidth temperature dependence for YIG doped with various rare earths. Essentially the same process was proposed by Teale and Tweedale (1962), Obata (van Vleck, 1962), and van Vleck and Orbach (1963). The mechanism involves electronic transitions between different levels associated with the 'impurity' atom. Consider another simple example from Sparks' treatment for ytterbium relaxation in YIG shown in Figure 10.17. The two lowest levels,

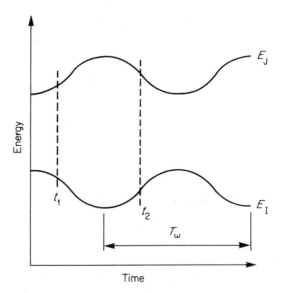

Figure 10.17 Energy level modulation scheme for impurity rare earth relaxation.

E_I and E_J, are degenerate in the crystalline field. The degeneracy is lifted by the exchange interaction with the Fe lattice, and the level separation is modulated by the magnetization precession. Because of the modulation, the thermal equilibrium population differential for the two levels changes with a period T_ω. The induced transitions give rise to a relaxation contribution as in the valence-exchange case. If the electronic transition time T_e is replaced by the rare earth electron transition time T_R, all of the remaining arguments carry over directly. The basic result in both cases is a linewidth of the form

$$\Delta H = C \frac{\omega T_{e,R}}{1 + (\omega T_{e,R})^2} \tag{10.87}$$

It must be realized that a number of different relaxation times may exist for any given material and orientation, so that the actual linewidth is more complicated than is indicated by (10.87). These points are adequately discussed in the cited references. This mechanism is often referred to as 'longitudinal relaxation' or 'slowly relaxing impurity' process. The term longitudinal refers to the z directed anisotropic exchange field which modulates the rare earth levels at the uniform precession frequency. The label 'slowly relaxing impurity' denotes that the relaxation time T_R is comparable to the uniform precession period T_ω in the range where the contribution is large.

The third mechanism was originally proposed by de Gennes et al. (1959) and Kittel (1960) to explain the linewidth temperature peak in rare earth substituted garnets. It also involves electronic transitions but in this case the level separation is much greater than the uniform precession magnon energy. For the uniform precession mode to excite transitions, energy conservation requires that the levels be broadened considerably so that the relaxation time T_R must be quite short. This process is usually termed the rapidly relaxing impurity mechanism. The largest contribution to the linewidth for this process occurs at the temperature where $\omega_{IJ} T_R \approx 1$, where $\hbar\omega_{IJ}$ is the energy level separation. Consequently, the peak in linewidth for this process occurs at the same temperature, independent of frequency. The linewidth varies according to

$$\Delta H = \frac{C\omega T_R}{1 + \omega_{IJ}^2 T_R^2} \tag{10.88}$$

in marked contrast with the dependences on temperature and frequency for valence exchange or longitudinal relaxation. From (10.88) the fast relaxation theory predicts that the peak in ΔH always occurs at the same temperature, independent of frequency, and the amplitude of the peak increases linearly with frequency. For slow relaxation, the amplitude is frequency independent and the position of the peak shifts to higher temperature with increasing frequency.

These differences in the linewidth behaviour in the slow and fast relaxation theories and the correlation between valence-exchange relaxation and resistivity due to electron hopping have provided a way to sort out, at least qualitatively, the various contributions to the linewidth. The processes are generally applicable to polycrystals as well as single crystals and to ΔH_k as well as ΔH. Most of the linewidth in spinel ferrites has been attributed to valence exchange. Fe^{2+} impurity in YIG has also been reported to give a small peak (Spencer et al.,

1961). The situation with regard to rare earth impurities in YIG and the choice between the slow or fast relaxation theories is not completely clarified.

In many cases, the slow relaxation model appears to be more applicable. The best evidence is the fit between the theory and data on ytterbium substituted YIG by Teale and Tweedale (1962): (Figure 10.18). Heeger *et al.* (1964) reported

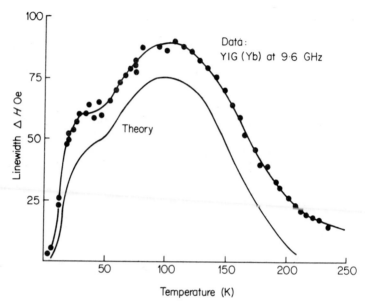

Figure 10.18 Temperature dependence of the resonance linewidth for YIG doped with Yb (after Teale and Tweedale, 1962).

two peaks for $MnFe_2O_4$, both of which they attribute to Fe^{2+} ions. One peak is due to a slow process (20 K) and the other due to fast relaxation (210 K). The possibility of Fe^{2+} acting as a rapid relaxer was first suggested by White (1959a). Other data (Spencer *et al.*, 1959) show a linewidth behaviour intermediate between the predictions of the slow and fast relaxation theories for YIG with unknown impurities. The peak shifts and the amplitude also increases with increasing frequency.

Seiden (1964) concluded that the details of the linewidth variation with temperature for the different rare earths doped into YIG cannot be accounted for by the slow-relaxation model. He did propose an empirical relation of the slow-relaxation type which provides a good fit to the data. Seiden also pointed out that the data exclude the fast-relaxation theory outright. The data show an inverse frequency dependence while the fast-relaxation theory results in a linear increase in ΔH with frequency. Other papers concerning rare earth relaxation include those of van Vleck (1964), Hartmann–Boutron (1964) and Dillon *et al.* (1967).

The two-magnon and temperature peak processes depend on some sort of inhomogeneity or impurity in the crystal. By eliminating such contributions by careful sample preparation and polishing (or alternatively by measuring ΔH_k), the intrinsic linewidth is obtained. LeCraw and Spencer (1962; see also Spencer et al., 1961) have obtained spin-wave linewidths of the order of 10–100 mOe at low temperature for YIG single crystals. The remainder of this section is concerned with the mechanisms which give rise to these small intrinsic linewidths, intrinsic in the sense that the losses are not due to inhomogeneities or impurities.

The spin-wave linewidth k-dependence at 300 K obtained by LeCraw and Spencer (1962) is shown in Figure 10.19. Except at very high k, the linewidth may be described by

$$\Delta H_k = A + Bk \tag{10.89}$$

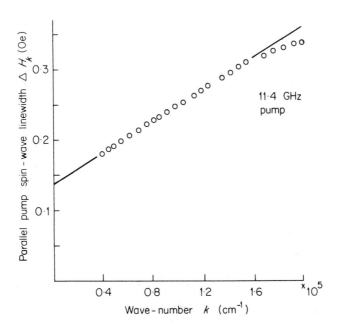

Figure 10.19 Wavenumber dependence of the parallel pump spin-wave linewidth for ultra pure YIG (after LeCraw and Spencer, 1962).

with $A = 136$ mOe and $B = 1.2 \times 10^{-6}$ Oe cm^{-1}. The temperature dependence of the linewidth extrapolated to $k = 0$ is shown in Figure 10.20. From about 150 K to 350 K, $\Delta H_{k \to 0}$ is linear in temperature. Above and below these limits the dependence is somewhat steeper. A small peak, presumably due to some residual impurity, shows up at 40 K. The linewidth extrapolates to approximately 15 mOe. The authors state that careful ΔH_k extrapolations to

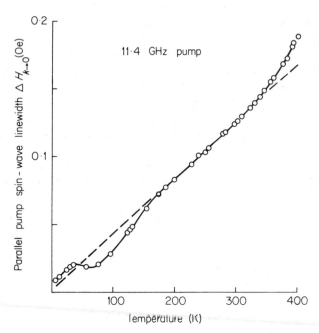

Figure 10.20 Temperature dependence of the parallel pump linewidth at $k = 0$ for ultra pure YIG (after LeCraw and Spencer, 1962).

$k = 0$ were not made for $T < 125$ K, so that the data in this region is only approximate. The frequency and magnetization dependence of $\Delta H_{k \to 0}$ at 300 K were also given. Above 3 GHz, the linewidth is linear in frequency and extrapolates to $\Delta H_{k \to 0} = 0$ at zero frequency. The response is approximately 23 mOe/GHz. Below 3 GHz, the linewidth increases with decreasing frequency (Figure 10.21). They also found that ΔH_k was an inverse function of M_s, described by

$$\Delta H_{k \to 0} \, (\text{Oe}) \sim 18 \cdot 5 / M_s(G) \qquad (10.90)$$

M_s was reduced from that for pure YIG by gallium substitution. These results have played a decisive role in clarifying the physical processes which contribute to the intrinsic relaxation in YIG and related low loss microwave materials. The mechanisms which appear to be important are the three magnon confluence and splitting processes (Schlömann, 1961; Sparks *et al.*, 1961) and the Kasuya–LeCraw two-magnon/one-phonon and three-magnon processes (Kasuya and LeCraw, 1961).

The details of three-magnon scattering calculations are given in the above references and discussed by Sparks (1964). From a theoretical viewpoint, three-magnon relaxation comes from a dipolar contribution to the Hamiltonian which is of third order in the spin-wave amplitudes. The interaction is long range, so that the scattering must conserve both energy and wave-vector. Two different processes must be considered, the confluence of two magnons to

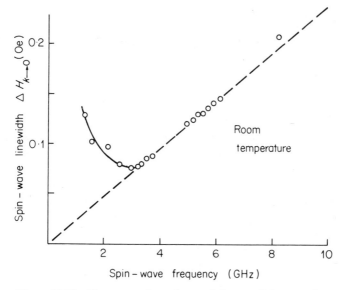

Figure 10.21 Frequency dependence of the parallel pump, $k = 0$ linewidth of ultra pure YIG (after LeCraw and Spencer, 1962).

produce a third and the splitting of a single incident magnon to produce two final magnons.

The confluence process appears to explain the linear k dependence of ΔH_k. The Schlömann theory gives

$$\Delta H_{k\text{-confl}} = \frac{k_b T \omega_m k}{8 D \omega_k} \tag{10.91}$$

for small k and $\omega_m/\omega_k \gtrsim 1$. The result must be multiplied by a correction factor

$$F = [1 + \theta(\omega_m/\omega^2/4] - (\omega_m/\omega)/6$$

at low frequency. The Sparks et al. (1961) result is essentially the same. (10.91) indicates that the B coefficient of (10.89) for the LeCraw–Spencer data should be $1\cdot04 \times 10^{-6}$ Oe cm^{-1}, in good agreement with the observed $1\cdot2 \times 10^{-6}$ Oe cm^{-1} value. Relatively minor deviations from the theory, however, should be noted. The weaker than linear k-dependence at high-k is opposite to the next order k^3 term predicted by Schlömann. In addition, the theory predicts a linear temperature dependence for the B coefficient, while the observed dependence is somewhat weaker. A more sophisticated theory has been reported by Sparks (1967) which satisfactorily accounts for, at least qualitatively, these apparent discrepancies in the earlier results. Other and perhaps more serious discrepancies with data (Comstock et al., 1964, 1966; Nilsen et al., 1965; Comstock, 1965) were discussed but not completely explained in Sparks's 1967 paper.

Three-magnon splitting appears to explain the sharp rise in the parallel pump linewidth below 3 GHz reported by Green and Schlömann (1961) and LeCraw and Spencer (1962) (Figure 10.21), as pointed out by Schlömann (1961). From energy and momentum considerations, three-magnon splitting can only contribute to the parallel pump linewidth $\Delta H_{k \to 0}$, attributed to $k = 0$ modes at $\theta_k = \pi/2$, below some critical frequency. Since the initial magnon is at $\omega_k = \omega/2$, the two scattered magnons must be at $\omega_k = \omega/4$. At the bias field required for parallel pumping, modes with $\omega_k = \omega/4$ are allowed only for $\omega < 4\omega_m/3$. This result follows directly from the dispersion relation when the relaxing (parallel pumped) mode has $k = 0$ and $\theta_k = \pi/2$. For YIG, this corresponds to a spin-wave frequency of 3·3 GHz for the pumped mode. In theory, the splitting process should also contribute to the relaxation of $k \neq 0$ modes, provided k is sufficiently large so that energy and momentum conservation is possible (Sparks, 1964) but no experimental evidence for such a contribution has been reported.

The only part of the YIG linewidth which remains unexplained is the $\Delta H_{k \to 0}$ which varies linearly with frequency above 3 GHz (Figure 10.21). A splitting process is not allowed above 3 GHz. A confluence process which strictly conserves energy and momentum is also forbidden when one incident magnon has $k = 0$. The other incident magnon and the scattered magnon must have the same k and hence the same energy, so that energy conservation is violated. After enumerating twenty odd possible scattering mechanisms, Kasuya and LeCraw (1961) singled out two processes which give temperature, frequency and magnetization dependences consistent with the YIG data. They proposed that two-magnon/one-phonon scattering was the origin of the linewidth temperature dependence below 300 K (apart from the small impurity peak at 40 K), the linear increase with frequency above 3 GHz and the inverse dependence on M_s. The more rapid increase in $\Delta H_{k \to 0}$ above 300 K was attributed to three magnon scattering. Both processes were attributed to a uniaxial term in the Hamiltonian. Unfortunately, except for a review in Sparks's book (1964), a full account of the complete theory has never been published.

Recently, Schlömann has presented an alternative explanation based on a three-magnon confluence process in which energy need not be conserved (Schlömann and Green, 1967). The finite relaxation time of the scattered magnons allows energy conservation to be violated by an amount determined by the uncertainty principle and in such a case the confluence contribution to ΔH_k does not vanish for $k = 0$. The theory appears to explain the observed $\sin^2 2\theta_k$ dependence of ΔH_k for $k = 0$, obtained from analysing the high field side of the parallel pump butterfly curve for YIG at 10 GHz and 300 K, and also predicts the linear temperature and frequency dependence discussed above. The calculated linewidth, however, is linear in M_s, in contrast with the observed M_s^{-1} dependence of LeCraw and Spencer (1962). Schlömann suggests that this apparent discrepancy is removed if the optical branches of the magnon spectrum are important in the interaction. The most serious discrepancy lies in the theoretical prediction of a background absorption due to high-k spin-waves which results in a field independent longitudinal susceptibility of approximately 10^{-4} which is not observed experimentally (Kohane, 1967).

In discussing the linewidth data in terms of the initial scattering interactions involving the driven mode, it has tacitly been assumed that subsequent relaxation processes, considered together, relax the scattered modes so quickly that they play essentially no role in determining the linewidths. The large degree of success in explaining data in terms of the initial processes indicates that this is indeed the case. Sparks and Kittel (1960) proposed that the scattered magnons thermalize by additional three magnon or higher-order interactions. The thermal magnons, at a temperature slightly higher than the lattice, relax by magnon–phonon interactions. Various calculations for these higher order processes have been reported (see Sparks and Kittel, 1960) but the details appear to have little direct bearing on the present discussion. Of course, some mechanisms, such as the temperature peak processes, serve to relax the driven mode directly to the lattice. These channels parallel the magnon channels and the linewidths are simply additive.

10.6 RELAXATION IN POLYCRYSTALLINE MATERIALS

Linewidths in polycrystals are generally much larger than in single crystals. The increase is generally attributed to some aspect of the microstructure, such as porosity, grain size, grain size distribution, or anisotropy in the randomly oriented crystallites. This section is concerned exclusively with the loss properties which are related to microstructure. The resonance losses are reasonably well understood in terms of simple inhomogeneous broadening or spin-wave scattering. Both linewidth and near resonance effective linewidth data will be considered here. The off-resonance losses appear to be related to microstructure but the physical mechanisms are not clearly understood. A few processes which appear promising will be presented, but main emphasis is on recent experimental results.

One of the obvious sources of line-broadening in polycrystals is the distribution of resonance frequencies or fields for the randomly oriented crystallites, due to crystalline anisotropy. For simple inhomogeneous broadening, the linewidth should be on the order of the spread in the distribution of internal fields from crystallite to crystallite. Kittel and Abrahams (1953) originally proposed that the linewidth due to anisotropy broadening should be approximately one quarter of H_A ($= 2|K_1|/M_s$) provided H_A is much smaller than the field required for resonance. In his classic article on line-broadening in polycrystals, Schlömann (1956) estimated the anisotropy linewidth to be equal to the anisotropy field.

$$\Delta H_{anis} = H_A \qquad (10.92)$$

The situation has, however, proved to be somewhat more complicated than these early papers would indicate. In the limit of very large anisotropy fields, with $H_A \gg 4\pi M_s$, the dipolar coupling between the crystallites is very weak compared to the anisotropy within grains and the crystallites tend to resonate independently with a resonance position determined by the magnitude of the anisotropy and the crystallite orientation. Consequently, the observed absorption curve is simply a superposition of the individual grain resonances. Using a free energy formulation to determine the resonance condition (Smit and Beljers, 1955), Schlömann (1958b) evaluated the distribution in resonance positions for polycrystals comprised of grains of cubic symmetry with negative K_1. The calculated lineshapes for this independent grain model were in good agreement with data for nickel aluminum ferrite (Schlömann and Zeender, 1958; Schlömann, 1959b) and for hexagonal ferrites (Schlömann and Jones, 1959) (Figure 10.22). The experimental curve is essentially a smeared out version

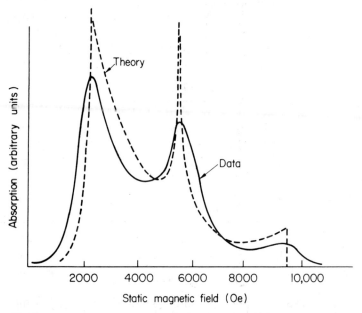

Figure 10.22 Resonance absorption curve for hexagonal Ba–Mg ferrite at 16 GHz (after Schlömann and Jones, 1959).

of the theoretical curve. This difference is presumably due to the neglect of the small dipolar coupling between grains in the theory. The separation of the peaks can be used to determine the K_i. Each peak can be related to resonance in crystallites with a particular orientation. The details depend on the crystal symmetry as well as the signs and relative sizes of the anisotropy constants. Recently, the technique has been utilized by van Hook and coworkers (van Hook et al., 1968; Euler and van Hook, 1970) to study the crystalline anisotropy of calcium–vanadium–indium substituted yttrium iron garnet materials.

As H_A is reduced in comparison to $4\pi M_s$, the independent grain model, becomes less and less valid because of the dipolar coupling between grains leading to a tendency for the crystallites to resonate collectively, so that the individual peaks due to crystallographic resonances (like those in Figure 10.22) no longer appear and the absorption curve is narrowed. The amount of narrowing should be of the order of $H_A/4\pi M_s$ when $H_A \ll 4\pi M_s$ is satisfied, so that the dipolar narrowed anisotropy linewidth is approximately

$$\Delta H_{\text{anis}}^{(\text{dipolar})} \approx H_A^2/4\pi M_s \qquad (10.93)$$

Physically, this is quite different from the large anisotropy case. Here, the strong dipolar forces cause the grains to resonate collectively, in a uniform precession mode, and the anisotropy field fluctuations are only a small perturbation. In this limit, then, the collective or spin-wave approach originally developed by Clogston et al. (1956) for short wavelength disorder scattering in single crystals is a convenient starting point for a quantitative theory of relaxation in polycrystals. The collective theory was first applied to polycrystals by Geschwind and Clogston (1957). Almost all of the qualitative arguments concerning two-magnon scattering discussed in Section 10.5 carry over directly.

For polycrystals, however, the situation is somewhat less complicated because scattering to very high k modes need not be considered. Grain sizes are usually of the order of one micron or greater and the coupling is limited to relatively long wavelength modes. The basic approach (Clogston, 1958; Schlömann, 1958a) is similar to that for pit scattering except that the inhomogeneity is the fluctuation in H_A from grain to grain, rather than the demagnetizing field of a pore. In one sense, anisotropy scattering is isotropic. Spin waves 'see' the same inhomogeneity distribution in the form of randomly oriented crystallites, regardless of the spin-wave propagation direction. The troublesome point of the θ_k dependence of the coupling strength which depends on the shape of the pore is avoided. Schlömann obtained a spherical sample linewidth given by

$$\Delta H = (H_A^2/4\pi M)(8\pi\sqrt{3}/21)G(\omega/\omega_m) \qquad (10.94)$$

where

$$G(x) = \frac{x^2 - x/3 + 19/360}{\sqrt{(x - 1/3)^3(x - 2/3)}} \qquad (10.95)$$

for $\omega > (2/3)\omega_m$. The initial term is the same as (10.94). $G(\omega/\omega_m)$ reflects the change in the density of low k states degenerate with the uniform mode (at resonance) with frequency. In the high frequency limit ($\omega \gg \omega_m$), G is equal to unity.

Van Hook and Euler (1970) has measured the resonance linewidth in polycrystalline Ca–V–In substituted garnets as a function of anisotropy field H_A by varying temperature and found good agreement with the theory in the high H_A range, 50 Oe $< H_A <$ 200 Oe. For larger H_A, the collective theory is no longer applicable. For lower H_A, other residual broadening effects obscure the somewhat smaller anisotropy contribution (Patton and Van Hook, 1971).

Porosity is another very important source of line-broadening in polycrystals. Rather loosely defined, porosity can mean anything from actual voids in the

624

material to regions containing non-magnetic or magnetic second phases. The early Schlömann (1956) theory gave

$$\Delta H_{\text{porosity}} = 1 \cdot 5 (4\pi M_s) v / V \qquad (10.96)$$

where v is the total pore volume and V is the sample volume. This was obtained by considering the inhomogeneous broadening due to the demagnetizing field of a spherical pore at the centre of a spherical sample. The more elaborate pit scattering calculation of Sparks was actually a pore calculation, so that his result may be applied directly to porosity in polycrystals.

$$\Delta H_{\text{porosity}} = \frac{\pi^2}{2} \frac{v}{V} \frac{\omega}{\gamma H_0} M_s \left[\frac{(3 \cos^2 \theta_u - 1)^2 + 1 \cdot 6}{\cos \theta_u} \right] \qquad (10.97)$$

Both theories give essentially the same result. For YIG ($4\pi M_s \approx 1750\,\text{G}$) at 10 GHz, (10.96) gives $\Delta H = 26$ Oe/per cent porosity, while (10.97) gives 27 Oe/per cent; Seiden and Grunberg (1963) give an experimental value of 23 Oe/per cent (Figure 10.23).

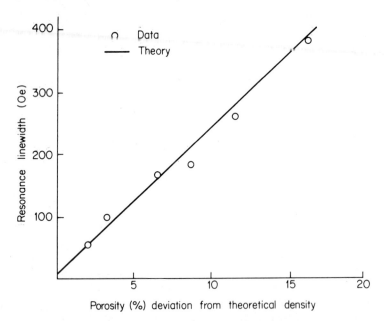

Figure 10.23   Resonance linewidth versus porosity for YIG at 10 GHz (after Seiden and Grunberg, 1963).

These early results clearly showed that the dominant contributions to the microwave losses in polycrystals are due to inhomogeneities. Recently, workers have begun to examine the field dependences expected for the two-magnon contribution to the relaxation. The basic idea is that two-magnon relaxation depends on the presence of spin-wave modes which are degenerate with the uniform precession at the driving frequency. By modifying the density of degenerate states, the relaxation can be changed and detailed information concerning the two-magnon scattering can be obtained.

One approach is to drive the uniform precession off resonance and determine the relaxation rate from susceptibility data. The density of degenerate states can be varied over wide ranges simply by changing the static magnetic field. The results are limited only by the accuracy with which the susceptibility components can be measured. The data appear in the form of an effective linewidth parameter ΔH_{eff} as a function of applied field discussed in Section 10.3. The off-resonance susceptibility technique, first suggested by Motizuki et al. (1965) was used by Liu (1963) and by Kohane and Schlömann (1968) to determine ΔH_{eff} from measurements of the imaginary part of the microwave susceptibility. This procedure is valid far from resonance. Most of the interesting changes in ΔH_{eff} due to two-magnon effects, however, occur near resonance. In this region, it is necessary to measure the real part of the susceptibility as well, in order to accurately determine ΔH_{eff}. This procedure has been applied by Vrehen (1968, 1969), Vrehen et al. (1969a, 1969b, 1970) and Patton (1969a, 1970b, 1970c).

A more limited technique used by several workers is to operate at resonance, but measure linewidths at different frequencies (Buffler, 1959, 1960), for different sample orientations (Risley and Bussey, 1959), or for different magnetostatic modes (White, 1959b; Nemarich, 1964). The variation in the density of degenerate states which can be obtained with such approaches is usually smaller and harder to obtain experimentally than with the off-resonance ΔH_{eff} approach. The rest of this discussion of resonance losses will be concerned with ΔH_{eff} in the vicinity of resonance.

First consider the two factors which determine the two-magnon relaxation contribution, the density of degenerate modes and the coupling between the uniform precession and the degenerate modes. Both factors can be strongly field dependent because the position of the manifold shifts with respect to the driving frequency. Consider the spin-wave manifold diagrams in Figure 10.24 for different values of the applied field. $H_{\parallel}$ and $H_{\perp}$ are the field values where $\omega_k = \omega$ is satisfied for $(k = 0, \theta_k = 0)$ and $(k = 0, \theta_k = \pi/2)$ modes, respectively. The density of low-k modes degenerate with ω is large only for a field range

$$H_{\parallel} = \omega/\gamma + \tfrac{4}{3}\pi M_s > H_0 > [(2\pi M_s)^2 + (\omega/\gamma)^2]^{\frac{1}{2}} - \tfrac{2}{3}\pi M_s = H_{\perp} \qquad (10.98)$$

(10.98) is for spherical samples. Only the low-k modes (shaded region of the manifold in each diagram) need be considered. The coupling to high k modes is very weak because most of the inhomogeneities in polycrystals are long wavelength (one micron or larger). In the intermediate field or manifold region, the density of low-k states varies approximately as $\cos \theta_u^{-1}$. (Recall that the $\cos \theta_u^{-1}$ term in the pit scattering linewidth of Section 10.5 was due to the density-of-states variation.) θ_u is the spin-wave angle at $\omega_k = \omega$ in the $k \to 0$ limit. The resultant field dependence of the density of states is shown

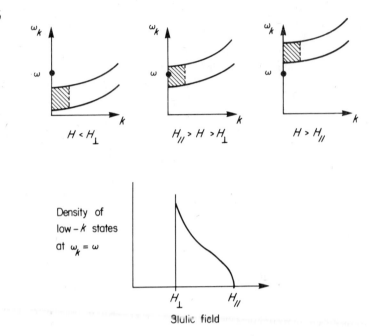

Figure 10.24 Shift of spin-wave manifold with bias field and the resultant variation in density of states at $\omega_k = \omega$.

schematically in Figure 10.24. It is zero at very low field, exhibits a discontinuous jump at $H_0 = H_\perp$, and falls to zero again as H_0 approaches $H_\parallel$.

While the density of states variation is fairly general, the field dependence of the coupling to the degenerate modes depends on the source of the scattering. The two important cases, porosity and anisotropy, are sketched in Figure 10.25. For anisotropy scattering, all k-values see essentially the same inhomogeneity distribution, independent of the propagation direction, because the crystallite orientation distribution is random. For porosity scattering, the coupling comes from the dipolar field generated by the magnetization divergence at the pore surfaces. The spin-wave coupling due to this field depends on θ_k. For a spherical pore, the dependence is given by $(3\cos^2\theta_u - 1)$, where coupling has been restricted to low-k modes so that $\theta_u \approx \theta_k$ is justified. This dependence is just the angular factor in Sparks's theory. For non-spherical pores, the coupling need not vanish at $\cos^2\theta_u = \frac{1}{3}$ but is still expected to have a minimum at the corresponding field value. The two dependences are shown in Figure 10.25.

Experimental data (Patton, 1969a) exhibit the general features expected from the above discussion. Figure 10.26 shows the effective linewidth field dependence at 10 GHz for dense Ca–V substituted YIG with $4\pi M_s = 650$ G and $H_A = 178$ Oe. Anisotropy scattering is the primary source of the relaxation in this material. The experimental curve has essentially the same shape as the density of states curve, except for some rounding and a small shoulder at 3700 Oe applied field. Data for dense YIG with a small porosity, of the order

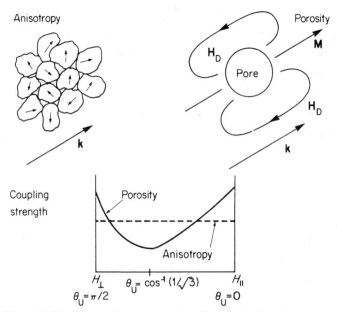

Anisotropy

Porosity

H_D

Pore

M

H_D

k

k

Coupling strength

Porosity

Anisotropy

$H_\perp$ $\theta_U = \cos^{-1}(1/\sqrt{3})$ $H_\parallel$

$\theta_U = \pi/2$ $\theta_U = 0$

Figure 10.25 Schematic representation of interaction for anisotropy and porosity scattering and the variation in coupling with bias field in each case.

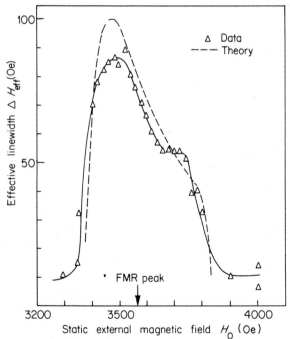

Figure 10.26 Effective linewidth versus bias field for Ca–V substituted YIG at 10 GHz and room temperature (after Patton, 1969a).

of 0·5 per cent (Figure 10.27) exhibit sharp increases in ΔH_{eff} at 3350 Oe and 4100.Oe, as well as a local minimum near 3600 Oe. This curve has roughly the same shape as that given by the coupling strength field dependence for pore scattering in Figure 10.25. These results show the distinct differences in ΔH_{eff} versus applied field for anisotropy and porosity scattering. The differences should readily enable one to sort out the two contributions to the scattering relaxation in polycrystals.

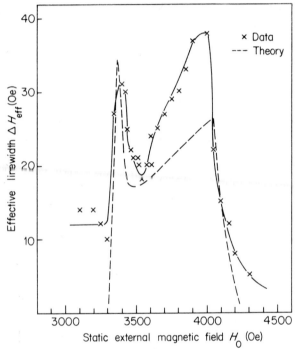

Figure 10.27 Effective linewidth versus bias field for YIG
with a small porosity (after Patton, 1969a).

One disturbing feature of both sets of data is the non-zero value of ΔH_{eff} outside the manifold region. The residual ΔH_{eff} very far from resonance will be discussed shortly. Of more immediate concern in the two-magnon context, is the fact that the decrease in ΔH_{eff} as H_0 falls below $H_\perp$ or goes above $H_\parallel$ is not sharp, but gradual. Significant tails extend outside this region of allowed degeneracy (Kohane and Schlömann, 1968). These tails result because the spin-wave levels are broadened because of their own relaxation times. Consequently, some scattering is allowed even for H_0 somewhat less than $H_\perp$ or greater than $H_\parallel$. Quantitative results can be obtained by including spin-wave relaxation in the two-magnon theory (Motizuki et al., 1965; Schlömann, 1969a). Schlömann treated the problem by considering secondary scattering among the spin-waves as the main source of the spin-wave broadening. He considered the special case of isotropic anisotropy scattering. A computed theoretical curve for the vanadate garnet with no adjustable parameters is shown in Figure 10.26. The agreement is excellent. Vrehen *et al.* (1970) have also considered the scattering relaxation outside the manifold region in some detail. Their theory includes additional broadening of the spin-wave band because of the anisotropy field and fair agreement with data is obtained. Sparks also attempted to introduce spin-wave relaxation into the porosity linewidth theory by substituting a gaussian function for the delta function $\delta(\omega - \omega_k)$ in the original theory (Motizuki *et al.*, 1965). A theoretical curve obtained assuming that η_k for the spin-waves is the same as η_0 at resonance is shown for 0·5 per cent porous YIG in Figure 10.27. The curve shows qualitative agreement with the data.

In contrast with the fairly detailed understanding of resonance losses in polycrystals, present day knowledge concerning the spin-wave linewidth and the effective linewidth far from resonance is very incomplete. Two-magnon contributions can be almost completely eliminated at the outset. The effect of two-magnon relaxation on ΔH_k for typical polycrystalline ferrites should be negligible for reasons covered in Section 10.4. ΔH_{eff} for $H_0 \ll H_\perp$ or $H_0 \gg H_\parallel$ should not be affected because there are no degenerate low-k modes available for scattering. The remainder of this section will summarize some recent data for polycrystals, primarily YIG, and discuss some possible (but not established) origins for the observed behaviour.

First consider the effect of grain size and porosity on the spin-wave linewidth ΔH_k (Patton, 1970b, 1970c). Experimental results for YIG at 9·1 GHz and room temperature are shown in Figures 10.28 and 10.29 (solid lines). In Figure 10.28, ΔH_k versus k is plotted with average grain diameter a_0 as a parameter. The 30 μm grain size sample shows a linear k dependence, similar to that observed in single crystals except for the somewhat larger ΔH_k. As the grain size is reduced, ΔH_k increases and for $a_0 = 1$ μm exhibits a strong inverse k dependence. The spin-wave linewidth extrapolated to $k = 0$ is proportional to a_0^{-1}, as shown in Figure 10.30. This result can be qualitatively explained by a model, recently proposed by Vrehen *et al.* (1970), in which the lifetime of the excited spin-wave is limited by the transit time for the mode across an individual grain, $\tau_k \approx a_0/|v_g|$, where v_g is the spin-wave group velocity. The transit time limited linewidth $\Delta H_k^{(t)}$ is given by

$$\Delta H_k^{(t)} = 1/\gamma \tau_k = |v_g|/\gamma a_0 \qquad (10.99)$$

The observed a_0^{-1} dependence of $\Delta H_{k \to 0}$ indicates that the model is qualitatively correct and that $|v_g|$ for the minimum k mode which can be excited does not change appreciably with grain size. Assuming that the $\Delta H_{k \to 0}$ in Figure 10.30 arises from a transit time mechanism, (10.99) gives values of $|v_g|$ of the order of 10^4 cm/s.

Since $v_g = \nabla_k \omega_k$, where ω_k is the spin-wave frequency, the transit time model can also be used to calculate the k-dependence of $\Delta H_k^{(t)}$ through the usual dispersion relation for ω_k. The group velocity is given by

$$v_g = 2\gamma Dk\hat{k} + \gamma 4\pi M k^{-1} \sin\theta_k \cos\theta_k \hat{\theta} \qquad (10.100)$$

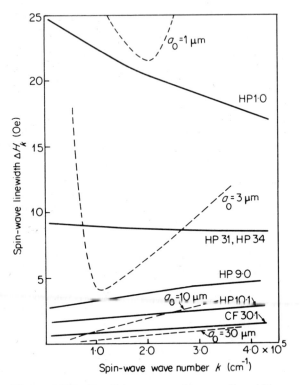

Figure 10.28 Parallel pump spin-wave linewidth versus wave-number for hot pressed YIG with various grain sizes. The dotted lines show theoretical values calculated from the transit time model (after Patton, 1970b).

where $\hat{k}$ and $\hat{\theta}$ are unit vectors in the $\mathbf{k}$ and $\mathbf{k} \times (\mathbf{k} \times \mathbf{H}_0)$ directions, respectively. Further details of the calculation are given in Patton (1970b). Calculated curves are shown in Figure 10.28 (dotted lines). The agreement is rather poor. The theory for $a_0 = 1$ μm and $a_0 = 3$ μm gives distinct minima in the 1–2 × 10^5 cm^{-1} range and the ΔH_k increase sharply at higher k-values. The theory for $a_0 = 10$ μm and $a_0 = 30$ μm give the right kind of k-dependence, but the increase is more rapid than indicated by the data.

One possible reason for the discrepancy is that the transit time mechanism is based on an independent grain model in which spin-waves are completely destroyed at the grain boundaries. In dense materials, some propagation across the grain boundaries probably occurs. The model should be more applicable to porous material in which the grains are separated to some degree. The ΔH_k versus k data for porous YIG (Patton, 1970c) shown in Figure 10.24 support this expectation. The dense material, with less than one per cent porosity, has a large (30 μm) grain size. The weak linear k-dependence is comparable to that expected from theory. The more porous materials, on the

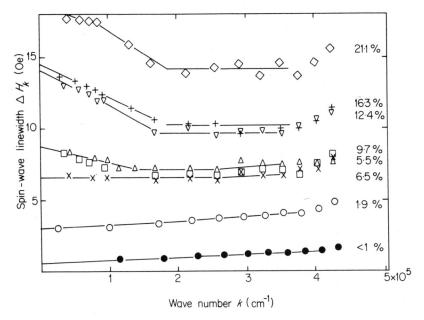

Figure 10.29 Parallel pump spin-wave linewidth versus wave-number for porous YIG (after Patton, 1970c).

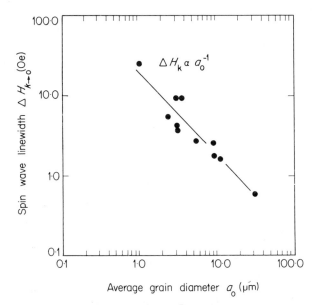

Figure 10.30 Parallel pump $k = 0$ spin-wave linewidth versus average grain diameter (after Patton, 1970b).

other hand, generally have a much smaller grain size, on the order of a few microns. The ΔH_k for these materials show a distinct minimum in the $k = 1\text{–}2 \times 10^5\ \text{cm}^{-1}$ range, and increase rapidly at lower k. This is precisely the behaviour expected from the transit time theory (Figure 10.28). From two pieces of experimental evidence, the inverse grain size dependence and the k-dependence, it appears that transit-time effects may play a significant role in determining ΔH_k for polycrystals.

Problems still exist, however. If (10.99) and (10.100) are applied to the $\Delta H_{k \to 0}$ data to determine the actual lower limit on $k\ (k_{\min})$ for the unstable modes, $k_{\min}$ is about $2 \times 10^5\ \text{cm}^{-1}$, independent of grain size. This means that the portions of the experimental curves for $k < 2 \times 10^5\ \text{cm}^{-1}$ are meaningless. These values of $k_{\min}$, moreover, correspond to a wavelength λ_k in the 0·5 micron range. Thus, if the transit-time explanation is correct, it appears that a submicron scale inhomogeneity operates to limit the wavelength of the unstable modes. It should also be noted that the instability theory (Section 10.4) was developed under the assumption that the dimensions of the region supporting the mode were much larger than the wavelengths of the relevant spin-waves. If the modes are confined to single micron size grains, this assumption is no longer satisfied.

Finally, consider the off-resonance effective linewidth in polycrystals. The effective linewidth as a function of static field with grain size as a parameter are shown in Figure 10.31 for YIG at 10 GHz. Far from resonance ΔH_{eff} is quite small compared to both ΔH and the near-resonance ΔH_{eff} because of the absence of two-magnon scattering to low k modes. In addition, ΔH_{eff} is field

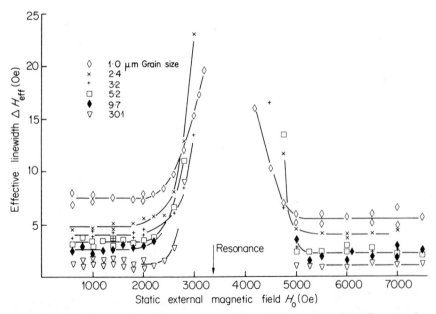

Figure 10.31 Effective linewidth versus bias field for YIG samples with different grain sizes (after Patton, 1970b).

independent below 2000 Oe and above 5000 Oe. This suggests that multimagnon processes do not contribute. If they did, the contribution should change at least slightly as the position of the manifold is shifted by changing the applied field. The values of ΔH_{eff} in the low field region ($H_0 < 2000$ Oe) are slightly larger than at high field ($H_0 > 5000$ Oe). This increase is probably due to a small two-magnon contribution because of the high k modes which are degenerate with the uniform precession at low field. For $H_0 < 2000$ Oe, the wavelengths of the degenerate modes are less than 0·1 μm, much smaller than the grain size, so that the coupling should be quite small. The data confirms this expectation since the largest increase in ΔH_{eff} from high field to low field is only 2 Oe for $a_0 = 1$ μm and is generally much smaller for the larger grain size materials.

The origin of the relatively large ΔH_{eff} values in Figure 10.31, compared to the 0·5 Oe (or less) linewidths for single crystals is at present unexplained (Patton and van Hook, 1971). Since two-magnon contributions are negligible, the high field ΔH_{eff} should be comparable to single crystal linewidths. The data indicates a $\sqrt{a_0}$ dependence for the high field ΔH_{eff}. This dependence excludes the transit time explanation. Schlömann (1969b) has recently proposed that magnetoelastic scattering due to the random orientation of the crystallites may be important. His theory does predict a weak inverse grain size dependence which is consistent with the observed $\sqrt{a_0}$ relationship, but the calculated values are several orders of magnitude smaller than indicated by the data.

Slightly different results have been obtained for the off-resonance ΔH_{eff} in porous YIG. First, the low and high field values are practically the same. Presumably, there is no high-k two-magnon contribution at low field because of a larger inhomogeneity wavelength for the pores. The ΔH_{eff} is linear in porosity, increasing from 1·5 Oe in the dense material to 13·5 Oe in the 21 per cent porous sample. In addition the ΔH_{eff} were linearly related to the spin-wave linewidth values extrapolated to $k = 0$, with $\Delta H_k / \Delta H_{\text{eff}} \approx 1·7$. This result strongly suggests that the two linewidths have a common origin. The case for a common origin is weakened somewhat by the transit time explanation of ΔH_k. The transit time model, in the form discussed above, does not appear to be directly applicable to uniform precession (off-resonance) relaxation.

In summary, the resonance losses in polycrystals can be explained almost completely by two-magnon processes or inhomogeneous anisotropy broadening. The non-resonant losses are not completely understood. The characteristic dependences of ΔH_k and the off-resonance ΔH_{eff} on grain size and porosity should provide important clues to the explanation of these properties.

10.7 PARTIALLY MAGNETIZED MATERIALS

So far, only phenomena associated with saturated materials have been considered. Partially magnetized materials played an important role in many of the early studies which followed Griffith's (1946) initial FMR discovery. In the last decade or so, moreover, interest in this area has increased substantially because of the increased technological importance of partially magnetized materials.

634

This section will touch briefly on some of the important aspects of the subject, beginning with a qualitative description of gyroscopic resonance processes in the presence of domain structure and a few remarks on the early work which separated the gyroscopic phenomena from domain wall motion effects. The wall motion effects are generally important at lower frequencies and are mentioned only briefly. After these qualitative considerations, the actual tensor susceptibility is examined and the influence of domain structure on the high power properties is discussed.

First, consider the possibility of gyroscopic resonance of the type discussed in Section 10.3 for materials magnetized below saturation. Because of the domain structure, the internal demagnetizing fields are modified so that substantial losses and dispersion can occur. Essentially, these fields are such that the sample 'thinks' that it is biased at (or near) ferromagnetic resonance, even though a saturated sample in the same static field would be biased very far from resonance. The basic theory is due to Polder and Smit (1953). See also Rado's comment following this same article, an earlier paper by Polder (1951) and the discussion of LeCraw and Spencer (1957).

Consider an ellipsoidal sample divided into thin layer domains separated by 180° walls normal to one principal axis (x axis), as shown in Figure 10.32, with the magnetization vectors along the z axis in the antiparallel domains. For this hypothetical domain structure, the demagnetizing factors which determine, in part, the natural precession frequency from the Kittel formula are not simply specified by the sample shape. First, the static demagnetizing field is essentially zero if the layers are sufficiently thin. This field arises from the free poles connected with the discontinuity of M_z at the sample surface. These poles alternate

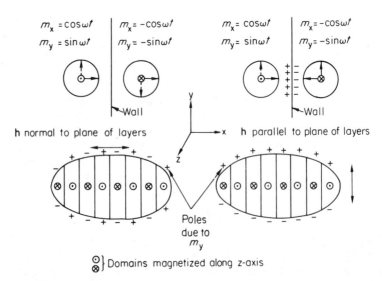

Figure 10.32 Polder–Smit domain model for two different microwave field directions.

in sign from one domain to the next, so that the field averages to zero. The demagnetizing fields connected with the microwave magnetization components, m_x and m_y in this example, depend strongly on the relative phase of the precessional motion in adjacent domains. If the precession is driven by a microwave field normal to the layers (Figure 10.32a) the phase is such that m_x is always continuous across the domain boundaries and m_y changes sign from one layer to the next. Hence, the x component of the demagnetizing field is simply that given by the sample shape, $-4\pi m_x N_x$, because there are no free poles at the walls to contribute. Since m_y gives rise to poles only at the sample surface, and the sign alternates from domain to domain, that component of the microwave demagnetizing field is zero. From these considerations, the natural resonance frequency for this microwave field orientation is given by the Kittel condition with $N_{x,y,z}$ replaced by effective demagnetizing factors,

$$N_x^{\text{eff}} = N_x \qquad (10.101a)$$

$$N_y^{\text{eff}} = 0 \qquad (10.101b)$$

and the internal field $H = H_0 - N_z^{\text{eff}} 4\pi M_s$ set to zero. It is assumed that the domain configuration adjusts itself so that $H = 0$ is satisfied. For $\mathbf{h}$ normal to the layers, the above considerations result in a frequency $\omega_\perp = 0$. Small effective fields due to anisotropy etc. have been neglected. These fields result in a non-zero, but still small, resonance frequency.

If the microwave field is parallel to the plane of the layers (y-directed as in Figure 10.32b), the precession phase is such that m_x is discontinuous across the domain boundary and m_y is continuous. As a result, the x component of the demagnetizing field alternates from one domain to the next with magnitude $4\pi m_x$. The surface free-poles associated with m_y do *not* alternate in sign. There is a net y-directed microwave demagnetizing field of magnitude $4\pi N_y m_y$. The appropriate demagnetizing factors

$$N_x^{\text{eff}} = 1 \qquad (10.102a)$$

$$N_y^{\text{eff}} = N_y \qquad (10.102b)$$

result in

$$(\omega/\gamma)_\| = 4\pi M_s \sqrt{N_y} \qquad (10.103)$$

The maximum natural frequency occurs for the parallel configuration and $N_y = 1$.

$$\omega_{\text{max}} = \gamma \cdot 4\pi M_s = \omega_m \qquad (10.104)$$

For YIG, ω_{max} is approximately 5 GHz. This simple analysis shows that, depending on the particulars of the domain structure and the microwave field orientation, ferromagnetic materials in a partially magnetized or demagnetized state may exhibit gyroscopic resonance at anywhere from very low frequencies up to frequencies in the microwave range. This result is often used as a qualitative criterion for remanent device design. In order to simply avoid resonance losses of the type described above, care is usually taken to ensure that $\omega/\omega_m > 1$ is satisfied.

Curiously enough, the early data that provided the major impetus for the Polder–Smit model showed large dispersion in the radio-frequency range, from 5 MHz to 500 MHz (Snoek, 1947, 1948; Beljers and Snoek, 1950), which were later shown to be primarily due to an entirely different phenomena, a type of domain wall oscillation resonance (Rado et al., 1950). The work of Rado and coworkers (Johnson et al., 1947; Johnson and Rado, 1949; Rado et al., 1950, 1952; Rado, 1953a, 1953b) concerning the initial permeability spectra of ferrites provides one of the most interesting sequels in the field of high frequency properties. In their 1950 paper, they reported the observation of two loss peaks and dispersion regions for a commercial iron-magnesium ferrite, one in the R.F. range (~ 50 MHz) and one at microwave frequencies (~ 4 GHz). By grinding up the solid material to obtain a fine powder and making identical measurements on a paraffin power mixture, they showed that the low frequency resonance disappears while the high frequency mode remains and is essentially unchanged. Their results are shown in Figure 10.33. The low frequency dispersion and loss is clearly absent in the powder-paraffin mixture. Rado proposed

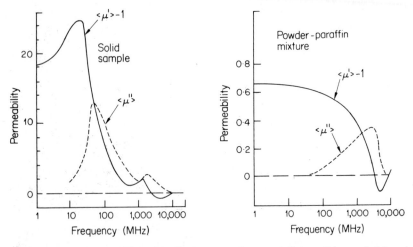

Figure 10.33 Permeability components versus frequency for a solid sample (a) and paraffin–powder mixture (b) of Mg-ferrite in zero static field (after Rado et al., 1950).

that the low frequency peak was due to domain wall oscillation resonances and the high frequency loss peak due to domain rotation (gyroscopic) resonance. The wall-resonance presumably vanishes in the powder sample because the particles are too small to support a multi-domain structure. Much of the important early literature is reviewed in Rado's (1953a) paper. More recently, Pippin and Hogan (1958) have discussed similar effects. Since domain wall oscillation resonance is essentially a low frequency (R.F. or lower) phenomenon and was discussed at length in Chapter 9, it will not be considered further here.

Following these early results which established the basic effect of gyroscopic resonance in partially magnetized materials, more and more attention has been given to the actual behaviour of the susceptibility in such materials. The first problem which arises, however, is the definition of susceptibility for a multi-domain sample. Any uniform external field, microwave or static, gives rise to position dependent internal fields. The response in individual domains is different and depends on such things as the domain orientation and domain shape. Rado (1956) has carefully examined the problem of characterizing such media in terms of 'average' field vectors and a susceptibility tensor defined in terms of such average quantities. Analogous with the analysis of Section 10.3, the average internal susceptibility is defined by

$$\langle \mathbf{b} \rangle = \begin{vmatrix} \langle \mu \rangle & -i\langle \kappa \rangle & 0 \\ i\langle \kappa \rangle & \langle \mu \rangle & 0 \\ 0 & 0 & \langle \mu_z \rangle \end{vmatrix} \langle \mathbf{h}_i \rangle \tag{10.105}$$

One important difference from Section 10.3 is that μ_z (or $\langle \mu_z \rangle$) is no longer unity. $\langle M_z \rangle$ is usually less than M_s and may change with field. Because of this difference, the discussion here will deal with the permeability μ, rather than the susceptibility χ ($\mu = 1 + 4\pi\chi$) which was considered in Section 10.3. (Note that κ in this section is actually $4\pi\kappa$ of Section 10.3.)

The first actual calculation of the permeability was done by Rado (1953b) using a basic torque equation analysis with damping neglected. Let the position dependent quantities $\mathbf{H(r)}$ and $\mathbf{M(r)}$ be written as

$$\mathbf{H} = \mathbf{h}(\mathbf{r}, t) \tag{10.106a}$$

and

$$\mathbf{M} = \mathbf{M_s} + \mathbf{m}(\mathbf{r}, t) = \mathbf{M_s}\mathbf{u} + \mathbf{m}(\mathbf{r}, t) \tag{10.106b}$$

The static part of the internal field is assumed to be zero (or negligible) for a non-saturated sample. The unit vector $\mathbf{u}$ designates the saturation direction in the individual domains. If the cross-product $[\mathbf{m} \times \mathbf{h}]$ is neglected to first order (low power and small excitations are assumed), for an $\exp(i\omega t)$ time dependence the local torque equation reduces to

$$\mathbf{b} = \mathbf{h} - i(\gamma 4\pi M_s/\omega)\mathbf{h} \times \mathbf{u} \tag{10.107}$$

In order to calculate the average quantities required to define the permeability, Rado makes the crucial assumption that

$$\langle \mathbf{h} \times \mathbf{u} \rangle_{av} = \langle \mathbf{h} \rangle_{av} \times \langle \mathbf{u} \rangle_{av} \tag{10.108}$$

(10.108) is equivalent to saying that the fluctuations in $\mathbf{h}$ and $\mathbf{M}$ from domain to domain are not correlated. Sandy (1969) has considered the validity of this assumption in some detail and we will return to this point shortly. Assume, for the moment, that (10.108) is correct. If the direction of the average net magnetization is along the z axis, $\langle \mathbf{u}_x \rangle = \langle \mathbf{u}_y \rangle = 0$ and $\langle \mathbf{u}_z \rangle = \langle \mathbf{m}_z \rangle / M_s$ are satisfied and the final result is given by (10.105) with

$$\langle \mu \rangle = 1 \qquad (10.109\text{a})$$

$$\langle \kappa \rangle = \gamma 4\pi \langle M_z \rangle / \omega \qquad (10.109\text{b})$$

$$\langle \mu_z \rangle = 1 \qquad (10.109\text{c})$$

All of these quantities are real because damping was excluded from the torque analysis. Others (Soohoo, 1956; Allen, 1966) have obtained essentially the same result by similar methods.

Experimental measurements of the permeability tensor components for various materials in a partially magnetized state have been reported by LeCraw and Spencer (1956, 1957), Sandy and Green (1967), Green et al. (1968) and Green and Sandy (1969). LeCraw and Spencer's early data on $\langle \mu' \rangle$, $\langle \kappa' \rangle$, and $\langle \mu_z' \rangle$ for Mg–Mn ferrite are shown in Figure 10.34. Primes denote the real parts

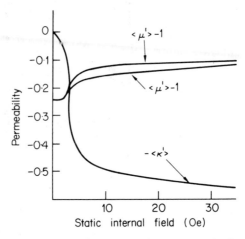

Figure 10.34 Permeability components (real parts) versus field for Mg–Mn ferrite (after LeCraw and Spencer, 1956).

of the measured quantities. These data show that Rado's prediction of $\langle \kappa' \rangle$ is essentially correct. As the internal field increases, $\langle M_z \rangle$ increases toward the saturation value M_s, causing $\langle \kappa' \rangle$ to increase as predicted by (10.109b). The actual dependence of $\langle \kappa' \rangle$ on $\gamma 4\pi \langle M_z \rangle / \omega$ is shown for polycrystalline YIG at 5·5 GHz and 9·2 GHz in Figure 10.35 (Green and Sandy, 1969). The solid line corresponds to (10.109b). The agreement with Rado's simple theory is quite good. Of significance is the fact that $\langle \kappa' \rangle$ exhibits no hysteresis. It is a unique function of $\gamma 4\pi \langle M_z \rangle / \omega$. This, however, is not a general result. Significant hysteresis (but qualitative agreement with Rado's theory) was reported for Mg–Mn ferrites.

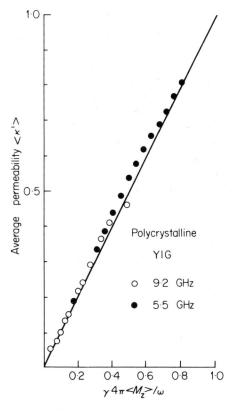

Figure 10.35 Permeability component $\langle \kappa' \rangle$ versus average magnetization (after Green and Sandy, 1969). The solid line is from Rado's theory.

Figure 10.34 shows, however, that the simple theory fails miserably in explaining $\langle \mu' \rangle$ and $\langle \mu_z' \rangle$. Both quantities are significantly less than unity and have strong dependences on $\langle M_z \rangle$ (through the internal field in this case). As LeCraw and Spencer pointed out, this result is related to gyroscopic resonance in line with the Polder–Smit model. Sandy and Green (1967) later extended this model to explain initial permeability data (zero static field) for a number of garnet and ferrite materials. They rendered the Polder–Smit model quantitative by assuming that each region of the sample undergoes resonance at some frequency between zero and $\gamma 4\pi M_s$, from the analysis presented earlier and calculating the initial permeability by assuming various distribution functions for the resonance frequencies. The permeability was obtained from

$$\langle \mu \rangle = \tfrac{2}{3} \int_0^{\gamma 4\pi M_s} \mu_{\text{local}} D(\omega_{\text{res}}) \, d\omega_{\text{res}} \tag{10.110}$$

640

In the demagnetized state, μ_{local} is given by the appropriate expression in Section 10.3 with the internal field set equal to zero. The $\frac{2}{3}$ factor in (10.110) comes from the assumption that the domain orientations are at random with respect to the microwave field. They found that $\langle \mu \rangle$ was not strongly dependent on the choice of the distribution function $D(\omega_{res})$ and were able to obtain fair agreement between the calculated curves and experimental data.

An important question still remains, however. What is wrong with the Rado calculation which causes it to fail for $\langle \mu' \rangle$ and $\langle \mu'_z \rangle$, but give good agreement for $\langle \kappa' \rangle$? It appears that the answer involves the validity of the assumption in (10.108), namely, that $\mathbf{h}$ and $\mathbf{M}$ are not correlated. Sandy (1969) has shown that this condition is not satisfied in general. Part of $\mathbf{h}$ is due to demagnetization effects in individual domains which are related to the direction of $\mathbf{M}$ in that domain. Sandy argued that the correlation makes a significant contribution to the reduction in $\langle \mu \rangle$ and $\langle \mu_z \rangle$ predicted from Rado's theory and also contributes to the hysteresis in $\langle \kappa' \rangle$ for some materials. Unfortunately, this important commentary on and extension of Rado's widely used theory has not been published. Nor has any comprehensive comparison with data been presented. One available comparison for the dependence of the initial permeability component $\langle \mu' \rangle$ on $\gamma 4\pi M_s / \omega$ is shown in Figure 10.36. The theory predicts that

$$\langle \mu' \rangle = 1 - \frac{2N\omega_d^2/3\omega}{1 - N^2\omega_d^2/3} \tag{10.111}$$

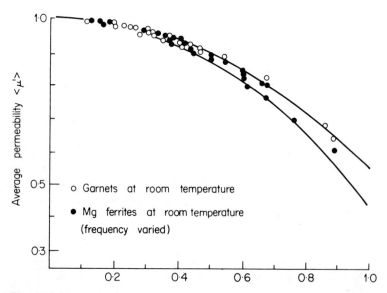

Figure 10.36 Dependence of $\langle \mu' \rangle$ on $\gamma 4\pi \langle M_z \rangle / \omega$ for several materials. The solid curves are for $N = 0.6$ and $H_a = 0$ (upper curve) or $H_a = 4\pi M_s/10$ (lower curve) (after Sandy, 1969).

where

$$\omega_d = \omega_m + \gamma H_a/N \qquad (10.112)$$

Here H_a, the average internal field and N is a demagnetizing factor which characterizes, in some average way, the domain shapes. The data are for several different garnets and spinel ferrites.

The experimental work by Green and coworkers (1968, 1969) shows that $\langle\mu'\rangle$, while similar in behaviour to that shown in LeCraw and Spencer's early paper, may also exhibit significant hysteresis as a function of $\gamma4\pi M_z$. Data on a Mg–Mn ferrite (Trans Tech TT1-390) at 5·5 GHz, taken from their study, is shown in Figure 10.37. The size of the hysteresis was reported to vary greatly

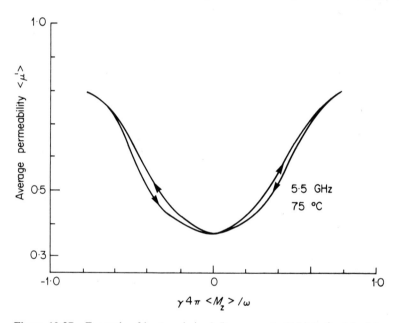

Figure 10.37 Example of hysteresis in $\langle\mu'\rangle$ versus $\gamma4\pi\langle M_z\rangle/\omega$ for Mg–Mn ferrite at 5·5 GHz (after Green and Sandy, 1969).

from material to material. The behaviour of $\langle\mu'_z\rangle$ is qualitatively similar (Figure 10.34). Green and coworkers report that no hysteresis in $\langle\mu'_z\rangle$ could be observed, although they point out that this may be due to the relative insensitivity of the measurement for this particular case.

Turn now to the imaginary parts of the tensor components. Rado's statistical theory has nothing to say about $\langle\mu''\rangle$, $\langle x''\rangle$, or $\langle\mu''_z\rangle$. The data of LeCraw and Spencer are shown in Figure 10.38. Curves of $\langle\mu''_\pm\rangle = \langle\mu''\rangle \pm \langle k''\rangle$ are shown, rather than $\langle\mu''\rangle$ and $\langle k''\rangle$ in order to show explicitly the anti-Larmor resonance peak for $\langle\mu''_-\rangle$ at about 15 Oe internal field. $\langle\mu''_\pm\rangle$ correspond to the magnetization response to circularly polarized fields rotating in the usual sense to

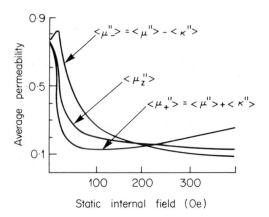

Figure 10.38 Permeability components (imaginary parts) versus field for Mg–Mn ferrite (after LeCraw and Spencer, 1956).

excite gyroscopic resonance (+ sign) or in the opposite sense (− sign). The peak in $\langle \mu''_- \rangle$ is somewhat surprising, in that the so-called 'anti-Larmor' resonance is excited by the wrong sense of circular polarization. They explain this resonance by a qualitative extension of the Polder–Smit domain model to include an applied static magnetic field and take the sense of the microwave field polarization into account (LeCraw and Spencer, 1957).

Green and coworkers (1968, 1969) also report extensive data for $\langle \mu'' \rangle$, $\langle k'' \rangle$, and $\langle \mu''_z \rangle$. Data for $\langle \mu'' \rangle$ on annealed samples of yttrium iron garnet doped with one per cent Dy are shown in Figure 10.39. The rare earth doping ensures that the relaxation rate remains essentially the same so that the observed dependences reflect changes due to the state of magnetization only. $4\pi M_s(\omega_m)$ was changed by heating the sample. Note that as M_s is decreased (reduction in ω_m/ω), the peak in $\langle \mu'' \rangle$ at $\langle M_z \rangle = 0$ then gradually changes to a broad minimum. They propose that the change is due to the shift of the natural resonance frequencies with ω_m. When ω_m/ω is near unity (upper curve), resonances close to ω_m predominate for the demagnetized state and cause the peak in $\langle \mu'' \rangle$. At remanence ($|\langle M_z \rangle| > 0$) the natural resonance frequencies are shifted to lower values. The domain regions 'think' that they are biased further from resonance and $\langle \mu'' \rangle$ is decreased. When ω_m/ω becomes very small, however, the natural resonance frequencies are quite low in any case. The effect of demagnetization is simply the occurrence of more domains which are aligned parallel to the microwave field. These regions do not contribute substantially to $\langle \mu'' \rangle$ so that a minimum occurs at $\langle M_z \rangle = 0$. The peak in $\langle \mu''_z \rangle$ at $\langle M_z \rangle = 0$ (see Figure 10.38) is caused by the same effect. In the demagnetized state there are more domains aligned perpendicular to the z axis which can contribute to gyroscopic resonance.

Before closing this discussion of permeability, it is important to mention a somewhat different approach from the statistical theories of Rado and other

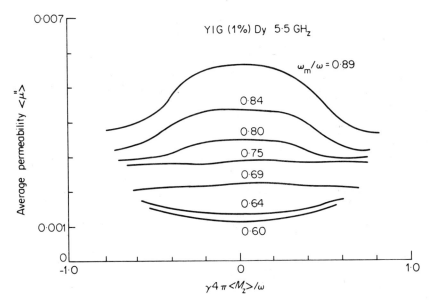

Figure 10.39 Variation of $\langle\mu''\rangle$ with $\gamma 4\pi\langle M_z\rangle/\omega$ in rare earth doped YIG at 5·5 GHz for different values of ω_m/ω (after Green *et al.*, 1968 and Green and Sandy, 1969).

workers. If it is possible to make assumptions about the domain structure which are simultaneously plausible and simple, analytic rigorous calculations are possible. Schlömann (1970) has recently demonstrated the usefulness of such an approach and shown that relatively simple (and perhaps implausible) models can be used to predict rather accurately the permeability in real materials. He considered a structure consisting of concentric 'up' and 'down' domains (M along z axis), calculated the local field quantities of interest and proceeded to obtain general expressions for the average permeability. Among other things, the theory accounts qualitatively for many of the phenomena described above. These include:

(1) Losses for $\omega_m/\omega > 1$.
(2) The variation in $\langle k'\rangle$ and $\langle\mu'\rangle$ with $\gamma 4\pi\langle M_z\rangle/\omega$, and in particular the hysteresis which is observed in some cases.
(3) The variation of the initial permeability (both $\langle\mu'\rangle$ and $\langle\mu''\rangle$) with ω_m/ω.
(4) The anti-Larmor resonance peak.

Finally, consider the effect of a domain structure on the behaviour of microwave materials at high power levels. How are spin-wave instability processes discussed in Section 10.4 affected? The answer turns out to be rather simple. The spin-wave processes which lead to instability are about the same, but the domain shape replaces the sample shape in determining the demagnetizing factors which are important. These basic conclusions were reached by a study of spin-wave thresholds in spheres, rods and discs of polycrystalline YIG by Patton and

Green (1969). They found that disc data for a variety of field orientations (both microwave and static) could be consistently related by a simple assumption that the discs demagnetized by the formation of elongated rod shaped domains with axes normal to the plane of the disc. The results were also consistent with data on rods. The results are summarized in Figure 10.40. The thresholds in the

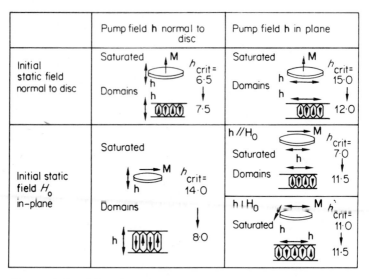

Figure 10·40   Tabulation of high power results for disc samples of polycrystalline YIG at 9·1 GHz (after Patton and Green, 1969).

demagnetized state (next to each domain picture) depend only on the microwave field polarization direction. The two configurations with **h** normal to the disc have $h_{crit} \approx 7\cdot5\text{--}8$ Oe. The three for in-plane microwave excitation have $h_{crit} \approx 11\text{--}12$ Oe. This indicates that the domain configuration does not depend on the static field direction before demagnetization. The next point to notice is that the threshold when **h** is normal to the disc is the same as for the rod sample. This suggests a similar type of domain shape for the demagnetized discs, i.e., rod-shaped domains with axes normal to the disc and oriented with **M** along that axis. For such rod-shaped, axially magnetized regions, the parallel pump threshold should be lower than the transverse pump threshold, and this result is also true experimentally. When **h** is normal to the disc (parallel to rod-domain axis), $h_{crit} \approx 8$ Oe from the first column of Figure 10.40. When **h** is in the disc plane (normal to domain axis), $h_{crit} \approx 12$ Oe. Our hypothetical domain model also explains the large changes in h_{crit} which occur for some configurations, but not for others, as the sample demagnetizes. For example, the biggest change occurs when H_0 is initially in the disc plane and **h** is normal to the disc (first column, case 2). This is because the pump configuration changes from transverse pumping for an in-plane magnetized disc to parallel pumping in axially magnetized rod domains. The smallest change occurs when H_0 and h are both normal to the disc plane. In this case, the pump field h remains parallel to M as demagnetization proceeds so that parallel pump excitation is preserved. The change in the transverse demagnetizing factor from zero (normally magnetized disc) to one-half (axially magnetized rod) does not affect the parallel pump threshold. The above arguments are only qualitative. As is evident from the discussion of the Polder–Smit model, there are important differences between demagnetizing fields which arise because of domain structure and those which are due to sample shape. Nevertheless, the results show that substantial progress is possible on the basis of some rather simple models.

REFERENCES

Akhieser, A., 1946, *J. Phys. U.S.S.R.*, **10**, 217.
Allen, J. L., 1966, Thesis, Georgia Institute of Technology.
Anderson, P. W. and H. Suhl, 1955, *Phys. Rev.*, **100**, 1788.
Beljers, H. G. and J. L. Snoek, 1950, *Philips Tech. Review*, **11**, 313.
Bertaut, F. and F. Forrat, 1956, *Compt. Rend.*, **242**, 382.
Bloch, F., 1946, *Phys. Rev.*, **70**, 460.
Bloembergen, N., 1956, *Proc. I.R.E.*, **44**, 1250.
Bloembergen, N. and R. W. Damon, 1952, *Phys. Rev.*, **85**, 699.
Bloembergen, N. and S. Wang, 1954, *Phys. Rev.*, **93**, 72.
Buffler, C. R., 1959, *J. Appl. Phys.*, **30**, 172S.
Buffler, C. R., 1960, *J. Appl. Phys.*, **31**, 222S.
Callen, H. B., 1958, *J. Phys. Chem. Solids*, **4**, 256.
Callen, H. B. and E. Pittelli, 1960, *Phys. Rev.*, **118**, 1523.
Churkin, V. I. and N. N. Chelishchev, 1967, *Fiz. Tverdogo Tela*, **9**, 1814 (*Sov. Phys. Solid State*, **9**, 1423).
Clogston, A. M., 1955, *Bell Syst. Tech. J.*, **34**, 739.
Clogston, A. M., 1958, *J. Appl. Phys.*, **29**, 334.
Clogston, A. M., H. Suhl, L. R. Walker and P. W. Anderson, 1956, *J. Phys. Chem. Solids*, **1**, 129.
Comstock, R. L., 1965, *Appl. Phys. Lett.*, **6**, 29.
Comstock, R. L. and W. G. Nilsen, 1964, *Phys. Rev.*, **136**, A442.
Comstock, R. L., J. J. Raymond, W. G. Nilsen and J. P. Remeika, 1966, *Appl. Phys. Lett.*, **9**, 274.
Damon, R. W., 1953, *Rev. Mod. Phys.*, **25**, 239.
De Gennes, P. G., C. Kittel and A. M. Portis, 1959, *Phys. Rev.*, **116**, 323.
Denton, R. T. and E. G. Spencer, 1962, *J. Appl. Phys. Suppl.*, **33**, 1300.
Desormiere, B. and E. Milot, 1967, *I.E.E.E. Trans. Mag.*, **MAG-3**, 402.
Desormiere, B., E. Milot and H. LeGall, 1969, *J. Phys. Chem. Sol.*, **30**, 1135.
Dillon, J. F., 1962, *J. Phys. Soc. Japan*, **17**, sup. B1, 376.
Dillon, J. F. and J. W. Nielsen, 1959, *Phys. Rev. Lett.*, **3**, 30.
Dillon, J. F., J. P. Remeika and L. R. Walker, 1967, *J. Appl. Phys.*, **38**, 2235.
Euler, F. and H. J. Van Hook, 1970, *J. Appl. Phys.*, **41**, 3325.
Fletcher, R. C., R. C. LeCraw and E. G. Spencer, 1960, *Phys. Rev.*, **117**, 955.
Galt, J. K., 1954, *Bell Syst. Tech. J.*, **33**, 1023.
Geller, S. and M. A. Gilleo, 1957a, *Acta Cryst.*, **10**, 239, 797.
Geller, S. and M. A. Gilleo, 1957b, *J. Phys. Chem. Solids*, **3**, 30.
Geschwind, S. and A. M. Clogston, 1957, *Phys. Rev.*, **108**, 49.
Gilbert, T. A., 1955, *Armour Research Rept.*, No. 11, January 25 (unpublished).

Green, J. J. and T. Kohane, 1964, *Semicond. Prod.*, **7**, No. 8, 46.

Green, J. J. and F. Sandy, 1969, *Final Technical Report, RADC-TR-6938*, Rome Air Dev. Center, Griffiss AFB, New York.

Green, J. J. and E. Schlömann, 1961, *J. Appl. Phys.*, **32**, 168S.

Green, J. J. and E. Schlömann, 1962, *J. Appl. Phys.*, **33**, 535.

Green, J. J., C. E. Patton and F. Sandy, 1968, *Final Report RADC-TR-68-312*, Rome Air Dev. Center, Griffiss AFB, New York.

Griffiths, J. H. E., 1946, *Nature*, **158**, 670.

Gurevich, A. G., 1960, *Ferrites at Microwave Frequencies*, English translation by A. Tybulewicz, Consultants Bureau, New York.

Hartmann-Boutron, F., 1964, *J. Appl. Phys.*, **35**, 889.

Hass, C. W. and H. B. Callen, 1961, *Phys. Rev.*, **122**, 59.

Heeger, A. T., G. Blocker and S. K. Ghosh, 1964, *Phys. Rev.*, **134**, A399.

Herring, C. and C. Kittel, 1951, *Phys. Rev.*, **81**, 869.

Holstein, P. and H. Primakoff, 1940, *Phys. Rev.*, **58**, 1098.

Iida, S., 1963, *J. Phys. Chem. Solids*, **24**, 625.

Johnson, M. H. and G. T. Rado, 1949, *Phys. Rev.*, **75**, 841.

Johnson, M. H., G. T. Rado and M. Maloof, 1947, *Phys. Rev.*, **71**, 322.

Joseph, R. I., C. P. Hartwig, T. Kohane and E. Schlömann, 1966, *J. Appl. Phys.*, **37**, 1069.

Kasuya, T. and R. C. LeCraw, 1961, *Phys. Rev. Lett.*, **6**, 223.

Kitaev, L. V. and I. G. Fedoseeva, 1968, *Fiz. Tverdogo Tela*, **10**, 1198 (*Sov. Phys. Solid State*, **10**, 951).

Kittel, C., 1948, *Phys. Rev.*, **73**, 155.

Kittel, C., 1949, *Phys. Rev.*, **76**, 743.

Kittel, C., 1960, *J. Appl. Phys.*, **31**, 11S.

Kittel, C. and E. Abrahams, 1953, *Rev. Mod. Phys.*, **25**, 233.

Kohane, T., 1967, private communication.

Kohane, T. and E. Schlömann, 1963, *J. Appl. Phys.*, **34**, 1544.

Kohane, T. and E. Schlömann, 1968, *J. Appl. Phys.*, **39**, 720.

Landau, L. and E. Lifshitz, 1935, *Phys. Z. Sowjet Union*, **8**, 153.

Lax, B. and K. J. Button, 1962, *Microwave Ferrites and Ferrimagnetics* (McGraw-Hill: New York).

LeCraw, R. C. and E. G. Spencer, 1956, *I.R.E. Conv. Rec. Part.*, V, 66.

LeCraw, R. C. and E. G. Spencer, 1957, *J. Appl. Phys.*, **28**, 399.

LeCraw, R. C. and E. G. Spencer, 1962, *J. Phys. Soc. Japan*, **17**, sup. B-1, 401.

LeCraw, R. C., E. G. Spencer and C. S. Porter, 1958, *Phys. Rev.*, **110**, 1311.

LeCraw, R. C., W. G. Nilsen, J. P. Remeika and J. H. van Vleck, 1963, *Phys. Rev. Lett.*, **11**, 490.

LeGall, H., 1964, *Solid State Comm.*, **2**, 377.

LeGall, H., 1967, *Solid State Comm.*, **5**, 637.

LeGall, H., 1968, *Phys. Stat. Sol.*, **28**, 495.

LeGall, H., B. Desormiere and E. Milot, 1969, *J. Phys. Chem. Solids*, **30**, 979.

Liu, S., 1963, Stanford Univ., Microwave Lab. Rept. SUML-1092.

Monosov, Y. A., 1969, *I.E.E.E. Trans. Mag.*, **MAG-5**, 61.

Monosov, Y. A. and V. V. Surin, 1967, *Fiz. Tverdogo Tela*, **9**, 1246 (*Sov. Phys. Solid State*, **9**, 970).

Monosov, Y. A., F. V. Lisovsky and V. V. Surin, 1968, *J. Appl. Phys.*, **39**, 1080.

Morgenthaler, F. R., 1960, *J. Appl. Phys.*, **31**, 95S.

Morgenthaler, F. R., 1964, *J. Appl. Phys.*, **35**, 900.

Motizuki, K., M. Sparks and P. E. Seiden, 1965, *Phys. Rev.*, **140**, A972.

Nemarich, J., 1964, *Phys. Rev.*, **136**, A1657.

Nilsen, W. G., R. L. Comstock and L. R. Walker, 1965, *Phys. Rev.*, **139**, A472.

Patton, C. E., 1968, *J. Appl. Phys.*, **39**, 3060.

Patton, C. E., 1969a, *Phys. Rev.*, **179**, 352 (see also *J. Appl. Phys.*, **40**, 1427).

Patton, C. E., 1969b, *J. Appl. Phys.*, **40**, 2837.

Patton, C. E., 1970a, *J. Appl. Phys.*, **41**, 431.

Patton, C. E., 1970b, *J. Appl. Phys.*, **41**, 1637 (see also *J. Appl. Phys.*, **41**, 1355).

Patton, C. E., 1970c, *Proc. Intern. Conf. Ferrites*, July, Kyoto, Japan, 524.

Patton, C. E., 1971, *Czech. J. Phys.* (Proceedings of IVth International Colloquium on Magnetic Thin Films, Prague, Sept. 1970), **B21**, 490.

Patton, C. E. and J. J. Green, 1969, *I.E.E.E. Trans. Mag.*, **MAG-5**, 626.

→ Patton, C. E. and T. Kohane, 1972, *Rev. Sci. Instrum.*, **43**, 76.

Patton, C. E. and H. J. van Hook, 1972, *J. Appl. Phys.*, **43**, 2872.

Pippin, J. E. and C. L. Hogan, 1958, *Sol. State Phys. in Elec. and Telecomm.*, **3**, 462.

Polder, D., 1948, *Phys. Rev.*, **73**, 1116.

Polder, D., 1949, *Phil. Mag.*, **40**, 99.

Polder, D., 1951, *J. Phys. et Radium*, **12**, 337.

Polder, D. and J. Smit, 1953, *Rev. Mod. Phys.*, **25**, 89.

Rado, G. T., 1953a, *Rev. Mod. Phys.*, **25**, 81.

Rado, G. T., 1953b, *Phys. Rev.*, **89**, 529.

Rado, G. T., 1956, *I.R.E. Trans.*, **AP-4**, 512.

Rado, G. T., R. W. Wright and W. H. Emerson, 1950, *Phys. Rev.*, **80**, 273.

Rado, G. T., R. W. Wright, W. H. Emerson and A. Terris, 1952, *Phys. Rev.*, **88**, 909.

Risley, A. and H. Bussey, 1959, *J. Appl. Phys.*, **35**, 896.

Sage, J. P., 1968, *J. Phys. Chem. Solids*, **29**, 2199.

Sage, J. P., 1969, *Phys. Rev.*, **185**, 859.

Sandy, F., 1969, *Raytheon Technical Report T-815*, March 5 (unpublished).

Sandy, F. and J. J. Green, 1967, *J. Appl. Phys.*, **38**, 1413.

Schlömann, E., 1956, *A.I.E.E. Spec. Pub. T-91*, 600 (Proc. Conf. Magnetism and Magnetic Materials).

Schlömann, E., 1958a, *J. Phys. Chem. Solids*, **6**, 242.

Schlömann, E., 1958b, *J. Phys. Chem. Solids*, **6**, 257.

Schlömann, E., 1959a, *Raytheon Technical Report R-48*, 'Ferromagnetic resonance at high power levels'.

Schlömann, E., 1959b, *J. Phys. Rad.*, **20**, 327.

Schlömann, E., 1961, *Phys. Rev.*, **121**, 1312.

Schlömann, E., 1962a, *J. Appl. Phys.*, **33**, 527.

Schlömann, E., 1962b, *J. Appl. Phys.*, **33**, 2822.

Schlömann, E., 1963, *J. Appl. Phys.*, **34**, 1998.

Schlömann, E., 1969a, *J. Appl. Phys.*, **40**, 1422.

Schlömann, E., 1969b, described at a workshop on microwave materials, Intermag Conference, Amsterdam, April.

Schlömann, E., 1970, *J. Appl. Phys.*, **41**, 204.

Schlömann, E. and J. J. Green, 1959, *Phys. Rev. Lett.*, **3**, 129.

Schlömann, E. and J. J. Green, 1963, *J. Appl. Phys.*, **34**, 1291.

Schlömann, E. and J. J. Green, 1967, *Raytheon Tech. Memorandum T-760*, Dec. 7.

Schlömann, E. and R. V. Jones, 1959, *J. Appl. Phys.*, **30**, 177S.

Schlömann, E. and R. I. Joseph, 1961, *J. Appl. Phys.*, **32**, 165S.

Schlömann, E. and R. I. Joseph, 1968, *Raytheon Technical Report R-68*, March 15, 'Spin waves and magnetoelastic waves'.

Schlömann, E. and T. Kohane, 1968, *J. Appl. Phys.*, **39**, 720.

Schlömann, E. and J. Zeender, 1958, *J. Appl. Phys.*, **29**, 341.

Schlömann, E., J. J. Green and U. Milano, 1960a, *J. Appl. Phys.*, **31**, 386S.

Schlömann, E., J. Saunders and M. Sirvetz, 1960b, *I.R.E. Trans. Microwave Theory and Techniques*, **MTT-8**, 96.

Seiden, P. E. and J. G. Grunberg, 1963, *J. Appl. Phys.*, **34**, 1696.

Seiden, P. E., 1964, *Phys. Rev.*, **133**, A728.

Seiden, P. E. and M. Sparks, 1965, *Phys. Rev.*, **137A**, 1278.

648

Smit, J. and H. G. Beljers, 1955, *Philips Res. Rep.*, **10**, 113.
Smit, J. and H. P. J. Wijn, 1959, *Ferrites—Physical Properties of Ferrimagnetic Oxides in Relation to Their Technical Applications* (Wiley: New York).
Snoek, J. L., 1947, *Nature*, **160**, 90.
Snoek, J. L., 1948, *Physica*, **14**, 207.
Soohoo, R. F., 1956, *I.R.E. Conv. Rec.*, Part V, 84.
Soohoo, R. F., 1960, *Theory and Application of Ferrites* (Prentice-Hall: New Jersey).
Sparks, M., 1961, Thesis, University of California, Berkeley.
Sparks, M., 1964, *Ferromagnetic Relaxation Theory* (McGraw-Hill: New York).
Sparks, M., 1967, *Phys. Rev.*, **160**, 364.
Sparks, M. and C. Kittel, 1960, *Phys. Rev. Lett.*, **4**, 232, 320.
Sparks, M., R. Loudon, and C. Kittel, 1961, *Phys. Rev.*, **122**, 791.
Spencer, E. G., R. C. LeCraw and A. M. Clogston, 1959, *Phys. Rev. Lett.*, **3**, 32.
Spencer, E. G., R. C. LeCraw and R. C. Linares, 1961, *Phys. Rev.*, **123**, 1937.
Starobinets, A. G. and A. G. Gurevich, 1968, *J. Appl. Phys.*, **39**, 1075.
Suhl, H. J., 1956a, *Proc. I.R.E.*, **44**, 1270.
Suhl, H. J., 1956b, *Phys. Rev.*, **101**, 1437.
Suhl, H. J., 1957, *J. Phys. Chem. Solids*, **1**, 209.
Sohl, H. J., 1958, *J. Appl. Phys.*, **29**, 416.
Teale, R. W. and K. Tweedale, 1962, *Phys. Lett.*, **1**, 298.
van Hook, H. J. and F. Euler, 1970, *J. Appl. Phys.*, **40**, 4001.
van Hook, H. J., T. Kohane and F. Euler, 1968, *Final Report No. 68-0467*, Air Force Cambridge Research Laboratories, Bedford, Mass.
van Vleck, J. H., 1949, *Physica*, **15**, 197.
van Vleck, J. H., 1962, *J. Phys. Soc. Japan*, **17**, Sup. B1, 352.
van Vleck, J. H., 1964, *J. Appl. Phys.*, **35**, 882.
van Vleck, J. H. and R. Orbach, 1963, *Phys. Rev. Lett.*, **11**, 65.
von Aulock, W. H., 1965, *Handbook of Microwave Ferrite Materials* (Academic Press: New York).
Vrehen, Q. H. F., 1968, *I.E.E.E. Trans. Mag.*, **MAG-4**, 479.
Vrehen, Q. H. F., 1969, *J. Appl. Phys.*, **40**, 1849.
Vrehen, Q. H. F., A. Brose van Groenou and J. G. M. deLau, 1969a, *J. Appl. Phys.*, **40**, 1426.
Vrehen, Q. H. F., A. Brose van Groenou and J. G. M. deLau, 1969b, *Sol. State Comm.*, **7**, 117.
Vrehen, Q. H. F., A. Brose van Groenou and J. G. M. deLau, 1970, *Phys. Rev.*, **B1**, 2332.
Wang, S., G. Thomas and T. Hsu, 1968, *J. Appl. Phys.*, **39**, 2719.
Weiss, M. T., 1958, *Phys. Rev. Lett.*, **1**, 239.
White, R. L., 1959a, *Phys. Rev. Lett.*, **2**, 465.
White, R. L., 1959b, *J. Appl. Phys.*, **30**, 182S.
Wijn, H. P. J. and H. van der Heide, 1953, *Rev. Mod. Phys.*, **27**, 98.
Winkler, G. and P. Hansen, 1969, *Mat. Res. Bull.*, **4**, 825.

ACKNOWLEDGEMENTS

The author is indebted to Professor S. Chikazumi, University of Tokyo, the Japan Society for the Promotion of Science and Raytheon Company for support during the preparation of this report. He is grateful to E. Schlömann, J. J. Green, F. Sandy, J. Sage, N. Ogasawara and H. J. van Hook for many stimulating discussions on the subject matter of this report. R. Kaelberer is thanked for help in proofing the final text.

11 *Magnetic oxides in geomagnetism*

K. M. CREER, I. G. HEDLEY and W. O'REILLY

11.1 GENERAL INTRODUCTION

11.1.1 Role of magnetism of oxides in geophysics

The field of earth sciences comprises studies not only of the physical properties and internal constitution of the earth as it is at the present time but also of its evolution throughout geological time.

Many physical parameters are now measured with great precision both on the surface of the earth and from satellites. Of these the most important are the acceleration due to gravity, heat flux, travel times of seismic waves originating from earthquakes and the geomagnetic field. However, we are able to measure past variations of only one of these quantities, viz. the geomagnetic field vector, and we are able to do this because many rocks became magnetized in the ambient geomagnetic field when they were formed. Although the intensities of magnetization are weak, from 10^{-2} gauss for igneous rocks to 10^{-7} gauss for sediments (intensities lower than this cannot be accurately measured), the stability is very high compared with most commercial permanent magnetic materials, the coercivity of remanence of some red sediments being of the order of 10,000 oersteds while that of the stronger igneous rocks may be of the order of a few hundreds of oersteds. This remanence is carried by impure iron oxides.

The conclusions drawn from studies of the natural remanent magnetization (NRM) of rocks have made a highly significant impact in the field of earth science in the years since the war. The earth has had a magnetic field of about its present intensity for about 3000 million years. For the past few hundred million years, i.e. during Phanerozoic time, the field has been, on average, axial and dipolar, and this has enabled geophysicists to reconstruct the position of the landmasses throughout this time and helped establish the validity of the much discussed hypothesis of continental drift. Studies of the time scale of geomagnetic field reversals as evidenced by magnetic anomalies over the ocean floors have resulted in estimates of the rates for continental drift during the past 50 million years or so.

Workers in geomagnetism were reluctant at first to accept that the geomagnetic field had reversed in the past and alternative explanations on the basis of a self reversal of the remanent magnetization were sought. This led Néel to put forward four mechanisms for self reversal. However, self reversal has only been found in minerals of unusual and atypical composition. Furthermore the observed correlation of reversed magnetization with K-Ar absolute ages and the consequent establishment of a geomagnetic polarity time scale has led nearly all geophysicists to accept the reality of geomagnetic field reversals.

Impetus was also given to the development of the theory for the behaviour of fine grained antiferromagnetic materials (Néel, 1962) by the discovery of Curie law behaviour of red sandstones containing finely divided haematite.

11.1.2 Chemical composition of crust

The most common rocks such as granites, basalts, sandstones, shales and limestones are assemblages of about ten different mineral series of which the most abundant are quartz and the feldspars, the latter being a solid solution series between albite, $NaAlSi_3O_8$ and anorthite $CaAl_2Si_2O_8$. The most abundant elements in crustal rocks are thus (1) oxygen 46·6%, (2) silica 27·7%, (3) aluminium 8·1%, (5) calcium 3·6%, (6) sodium 2·8% and (7) potassium 2·6%. The fourth most abundant element is iron, 5%, and while it occurs in some silicates such as olivines and pyroxenes, the chief occurrences are in the carbonate (siderite) and the iron oxides. In the last form, iron is responsible for nearly all the magnetic remanence encountered in rocks and, because of this geophysical and geological interest, a great deal of effort has been applied to the study of iron oxides in recent years. There are basically two types of rocks, primary and secondary. Primary rocks are igneous in origin; secondary rocks have been formed by the deposition of the erosion products of primary rocks. At the present point in geological time many mineral grains have been through several cycles of sedimentation.

11.1.3 Igneous rocks

Igneous rocks, such as basalt lavas, diabase dikes and sills, cool from the melt when they are ejected through or injected into the earth's crust. They acquire a stable remanent magnetization when they cool through the Curie temperatures of their constituent magnetic minerals (Section 11.4.2).

The particular kind of iron oxide (Figure 11.1) formed from the melt depends on the conditions of cooling. We will consider three main cases.

(1) In the case of a lava, the melt is quenched from about 1000 °C–1300 °C and if the cooling is fast enough, homogeneous but impure titanomagnetites $Fe_{3-x}Ti_xO_4$ ($0 < x < 1$) are formed. The usual compositions are $0.75 > x > 0.45$ and the major cation impurities are Mg^{2+} and Al^{3+} together with smaller quantities of Cr^{3+} and Mn^{2+}. The corresponding Curie points are between 0 °C and 250 °C. The solvus of the Fe_3O_4–Fe_2TiO_4 binary series indicates that

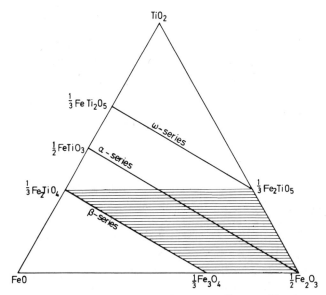

Figure 11.1 The TiO_2–FeO–Fe_2O_3 ternary diagram. The three principal solid solution series are shown (i) the rhombohedral haemo-ilmenite (α series), (ii) the cubic titanomagnetite (β series) with spinel structure and (iii) the orthorhombic pseudo-brookite (ω series). The shaded region represents the stability field of non-stoichiometric, cation deficient (oxidized) titanomagnetites below about 350 °C. Lines of constant Fe:Ti ratio (oxidation lines) are parallel to the base of the triangle.

such compositions are not stable, although unmixing is extremely slow at ambient temperatures, if it occurs at all.

(2) If the rock cools rather more slowly, e.g. the inner parts of a lava flow, and is self buffering, the iron–titanium oxide system equilibriates with the oxygen pressure corresponding to each temperature. This results in the formation of an intergrowth of oxides of spinel and rhombohedral structure. As the temperature falls, the spinel phase is progressively enriched in iron and depleted in titanium: it approaches magnetite in composition. The rhombohedral phase (a haemo-ilmenite $Fe_{2-y}Ti_yO_3$, $0 < y < 1$) becomes enriched in titanium, eventually approaching a composition given by $y \sim 0.9$ (Buddington and Lindsley, 1964). Sills and dykes are masses of solidified magma which have been injected either between beds of sedimentary rocks or along cracks in rock strata, usually at an angle to the bedding. Because of the slower rate of cooling the diabase rocks formed under these conditions are much coarser grained than basalt. If the cooling rate is slow, on reaching the solvus curve, (~ 600 °C) for the $Fe_{3-x}Ti_xO_4$ solid solution, further exsolution into two spinel phases takes place. Single phase titanomagnetites are hardly ever found in dikes and sills: the iron titanium oxide grains nearly always contain intergrowths, the

phase carrying the remanent magnetization having its Curie point between 560 °C and 575 °C.

(3) For lavas the oxygen pressure is no longer determined by the basalt alone but by the availability of atmospheric oxygen. The iron-titanium oxides may become excessively oxidized especially if the cooling rate is slow. The oxidation of titanomagnetites is a complex process and oxidation products may be either (a) a cation deficient material of spinel structure, (b) an intergrowth between phases of spinel and corundum structures, or (c) a combination of (a) and (b). The ultimate oxidation products are haematite together with TiO_2 (as either anatase or rutile) or pseudo-brookite Fe_2TiO_5.

The coercivities of many igneous rocks are notably high and this would seem to indicate that the magnetic unit particles are single domain (see Sections 1.5 and 12.3.3.1). The lowest coercivities are measured for basalts containing homogeneous titanomagnetite grains of a few tens of microns in diameter in which domain structure may be observed. The configuration usually consists of 180° walls, the domain thickness depending on the square root of the grain diameter as expected from theory. The coercivity and stability of the remanence of such rocks to demagnetization in alternating fields increases when oxidation has occurred. This is accompanied by an increase in Curie point from less than 300 °C to more than 500 °C and is attributed to unmixing on a submicroscopic scale. Similarly, in many diabases, the grains responsible for the remanence are about 100 microns in diameter and contain ilmenite and magnetite lamellae which typically may be 10 microns thick. The critical single domain size is about 0·7 microns and the high stability of these rocks is usually due to fine scale exsolution within the magnetite lamellae as illustrated (Figure 11.2).

11.1.4 Sedimentary rocks

Sedimentary rocks vary in texture from coarse sandstones to fine mudstones and shales and their magnetic mineral content ranges from abundant in sedimentary iron ores to nearly zero in non-magnetic limestones.

Between these two extremes, red sandstones have been most frequently chosen for palaeomagnetic studies, partly because they occur abundantly throughout the geological column, but mainly because of their high magnetic stability. These rocks typically contain a few per cent of haematite, α-Fe_2O_3 (Section 11.3) and while the remanence is extremely weak, of the order 10^{-5} to 10^{-6} gauss, its coercive force is very high, of the order of kilo-oersteds. Grains of specular haematite sometimes carry the remanence which may thus possibly have originated when they settled in the water on deposition. In some rocks fine grained haematite is present in the pigment which gives the sandstones their characteristic red colour and carries the remanence (Collinson, 1966).

Some of the specular haematite has a characteristic triangular texture when viewed under polarized light indicating that it was formed from the oxidation of magnetite. Indeed in some instances remnants of the original magnetite

Figure 11.2 (a) Ore microscope photograph of polished section of an Fe–Ti oxide grain containing 'ilmenite' (white) and 'magnetite' (darker) lamellae. (b) Replica of a small area of an etched surface of the above grain. Note the flat topography of the ilmenite areas and the broken topography of the magnetite areas, indicating submicroscopic exsolution in the latter.

contribute to the remanence (Chamalaun, 1964). Partly because of the conversion to haematite, magnetite occurs less frequently in ancient red sediments (van Houten, 1968). The haematite can also come directly from haematite bearing igneous rocks (Phillips, 1964).

Maghemite, γ-Fe_2O_3, occurs chiefly as the oxidation product of titanomagnetites and as such is thought to be responsible for the strong magnetization of 'lodestones' (Nagata and Kobayashi, 1961). Although it has been rarely reported in red sandstones, maghemite may occur on a world-wide scale in laterites (Frankel, 1966). Strictly, these are not sediments as they are produced by intense weathering of igneous rocks when the silica is removed, leaving a residue rich in alumina and ferric oxide. The remanent direction of the original lava tends to be preserved in laterites but the magnetization is unstable with additional components present (Wilson, 1961). This may be due to the fine grained nature of the ferric oxide.

Ferric oxyhydroxides such as goethite, α-FeOOH, occur in red beds and the latter is thought to be responsible for the unstable magnetization of some rocks (Strangway et al., 1968).

Naturally occurring material from the rhombohedral, α, solid solution series (see Figure 11.1) with Curie temperatures above the ambient occur rarely. Pure ilmenite often occurs in igneous rocks, either as a separate phase or intergrown with an iron-rich spinel phase.

Two processes of primary magnetization are possible in sedimentary rocks which were formed at ambient temperatures. These are (a) depositional magnetization (DRM) where the magnetic particles are aligned in the ambient field during deposition of the water-borne sediment and (b) chemical or crystallization magnetization (CRM) where a magnetic material is produced by a chemical change in the rock and so becomes magnetized in the direction of the external field. Some of the possible reactions are (i) the oxidation of magnetite, Fe_3O_4 to haematite, α-Fe_2O_3, (ii) the dehydration of the oxyhydroxides of iron and (iii) direct precipitation from iron-rich solutions. The presence of very finely divided material in red mudstones and sandstones is revealed by the Curie–Weiss law behaviour, and Creer (1961) has estimated the grain size of the haematite in the pigment of these rocks to be about 20 Å.

To summarize, igneous rocks usually have a thermo-remanent magnetization (TRM) carried by titanomagnetite, oxidation during geological time subsequent to the formation of a rock may cause chemical remanent magnetization (CRM) carried by secondary mineral phases such as titanomaghemite to be impressed. In sedimentary rocks, the ferric oxides and hydroxides usually carry the remanent magnetization which is commonly a CRM. In some sedimentary rocks such as inland lake deposits, the remanence was impressed by alignment of magnetic grains such as magnetite by the ambient geomagnetic field during deposition (DRM).

We shall now discuss these materials carrying natural remanent magnetization (NRM) in detail. Sections 11.2 and 11.3 describe the properties of minerals most commonly found respectively in igneous rocks and in sediments. In the final section (11.4), we briefly describe the physical basis of the acquisition of TRM and CRM and also some of the mechanisms by which the remanence measured in the laboratory may be found to be reversed with respect to the geomagnetic field at the time of formation of the rock.

11.2 OXIDES CARRYING THE REMANENT MAGNETIZATION OF IGNEOUS ROCKS: MAGNETITE, THE TITANOMAGNETITES AND HAEMO-ILMENITES

11.2.1 Introduction

The titanomagnetites ($Fe_{3-x}Ti_xO_4$) are ferrimagnetic and have the spinel structure but, having appreciable electrical conductivity, they have no technical interest and it has been left to researchers in rock magnetism to work out the properties of the system in detail. Owing to the presence of impurities and possible non-stoichiometry in naturally occurring samples, the greater part of the work on this series has been done using synthetic materials. Polycrystalline titanomagnetites may be made by quenching mixtures of the constituent oxides fired at about 1100 °C in evacuated, sealed, quartz ampoules. Large single crystals have been produced by Syono (1965) using the Bridgman method. The melting point of the series decreases with increasing titanium content (from 1594 °C for $x = 0$ to 1395 °C for $x = 1$) and the equilibrium oxygen pressure at the melting point falls from about 10^{-8} torr for $x = 0$ to 10^{-12} torr for $x = 1$. Small single crystals have recently been made by Hauptman and Stephenson (1968) using a strip furnace. The addition of Ti^{4+} to magnetite produces extra Fe^{2+} ion and the unit cell size therefore increases steadily, but not linearly (Figure 11.3). The distortion of the oxygen lattice of the structure from pure cubic increases as the larger radius Fe^{2+} ions enter tetrahedral sites

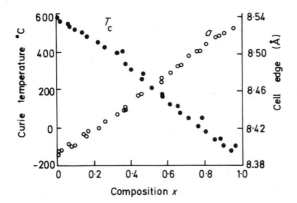

Figure 11.3 Variation of unit cell size (Å) and Curie temperature (°C) as a function of composition parameter x in the titanomagnetite solid solution $Fe_{3-x}Ti_xO_4$ (after Akimoto, 1962).

of the structure, the u parameter rising from about 0·379 for magnetite to 0·386 for ulvöspinel.

11.2.2 Cation distribution

The cation distribution $Fe^{3+}[Fe^{3+}Fe^{2+}]O_4$ (see Section 1.4.3.1) indicates that the high electrostatic and anion polarization contributions to the total lattice energy are offset by the energy reduction due to the formation of covalent bonds by Fe^{3+} in tetrahedral sites. Ulvöspinel has the distribution $Fe^{2+}[Fe^{2+}Ti^{4+}]O_4$ which is favoured by electrostatic and anion polarization energies. Ti^{4+} also has a strong preference for the octahedral sites of the spinel structure, being found in tetrahedral sites only when ions with an even stronger preference (e.g. Ni^{2+}) are also present (Blasse, 1964).

The cation distribution for intermediate compositions has been studied by Banerjee et al. (1967) who conclude that tetrahedral sites contain both Fe^{2+} and Fe^{3+} ions throughout most of the series, i.e. a compromise between the competing energy contributions. Stephenson (1969) and Bleil (1971) have discussed the temperature dependence of cation distribution in the series.

11.2.3 Anisotropy and magnetostriction

Contributions to the first cubic anisotropy constant $K_1(T)$ in magnetite come from single ion terms resulting from the various combinations of ions and symmetries:

$$K_1(T) = K_1(T)[Fe_A^{3+}] + K_1(T)[Fe_B^{3+}] + K_1(T)[Fe_B^{2+}]$$

The three terms depend on temperature in different ways and do not all have the same sign. Thus at 130 K for pure magnetite, K_1 vanishes as contributions from the various sources exactly cancel. Below 130 K, K_1 is positive ($\sim 10^5$ ergs/cm^3 at 80 K, see also Chapter 5).

The zero point of K_1 may be shifted to higher or lower temperatures depending on doping. It is interesting to compare the effect of Ti^{4+} substitution (producing more octahedrally sited Fe^{2+}) and Co^{2+} replacement of octahedral Fe^{2+}. Small amounts of titanium depress the zero point in K_1 (Figure 11.4) indicating that octahedral Fe^{2+} gives a negative contribution to K_1 (Syono and Ishikawa, 1963a). This is also evident from the reduced value of K_1 at room temperature in non-stoichiometric samples (Gyorgy and O'Bryan, 1966). As K_1 is negative in the higher temperature range, the Fe^{2+} contribution must decrease more slowly than other terms as temperature rises. This is consistent with the description of Fe^{2+} in magnetite by Slonczewski (1961) described in Chapter 5. $Fe^{2+}(3d^6)$ is non-degenerate in the ground state but as temperature rises progressive occupation of the first excited, degenerate, state gives a relatively large contribution to the anisotropy which tends to offset the fall in anisotropy with rising temperature. $Co^{2+}(3d^7)$ is degenerate in octahedral sites in magnetite and the isotropic point is shifted upwards in temperature (Bickford et al., 1957) indicating a large positive contribution to K_1 from Co^{2+}.

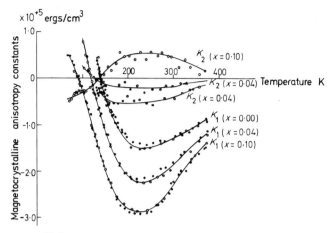

Figure 11.4 Variation of magnetocrystalline anisotropy constants K_1 and K_2 for magnetite and the titanomagnetites between 80 °C and 300 K (after Syono and Ishikawa, 1963a).

Fletcher (1971) has measured K_1 of magnetite and titanium poor titanomagnetites between room temperature and T_c (Figure 11.5). In this range it is found that the anisotropy is related to M_s by

$$\frac{K_1(T)}{K_1(0)} = \left[\frac{M_s(T)}{M_s(0)}\right]^8$$

in contrast to the tenth power law deduced from a phenomenological pair model in cubic symmetry (Zener, 1954). As shape anisotropy depends on M_s^2 only, the acquisition of thermoremanent magnetization in titanomagnetites, which is important in rock and palaeomagnetism, will be determined by shape and not by crystal anisotropy.

At room temperature K_1 falls with increasing hydrostatic pressure with $K_1^{-1}(dK_1/dP) = -13\cdot5\,\text{kbar}^{-1}$ (Sawaoka and Kawai, 1968) and it is suggested that the Fe^{2+} contribution is becoming less negative due to an increasing u parameter.

The second anisotropy constant K_2 is negative at all temperatures for most spinel oxides. In magnetite, however, K_2 changes sign at about 150 K with a value of about $-3 \times 10^4\,\text{ergs/cm}^3$ at room temperature (Bickford et al., 1957). Samples containing small amounts of titanium and lithium also show this effect (Syono, 1965) but small quantities of cobalt result in a large and negative K_2. Nagata and Kinoshita (1967) found that, for pure magnetite, K_2 at room temperature is reduced by about 10 % on application of a hydrostatic pressure of 2 kbar.

In pure magnetite or samples with only a small quantity of impurities which still show the Verwey transition (Section 11.2.5), the anisotropy below the transition is correctly described by uniaxial anisotropy constants. Palmer (1963) has

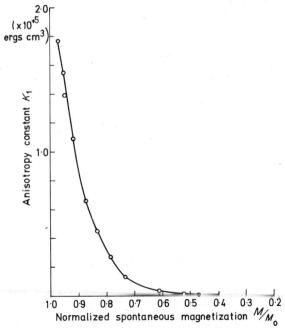

Figure 11.5 Variation of K_1 for titanomagnetite ($x = 0.1$) as a function of reduced saturation magnetization M/M_0. The temperature range covered is 300–900 K (after Fletcher, 1971).

obtained values of 20.5×10^5 ergs/cm^3 and 2×10^5 ergs/cm^3 for K_a and K_b which are almost constant between 4.2 K and the transition temperature.

Titanium rich titanomagnetites show a large anisotropy ($\sim 10^7$ ergs/cm^3 at 80 K) and coercive force (Syono and Ishikawa, 1964; Banerjee and O'Reilly, 1967; Ishikawa, 1967) with a strong temperature dependence. This may result from a degenerate ground state for octahedrally sited Fe^{2+} giving a large positive contribution to K_1 like that found in Co^{2+} doped magnetite (Readman et al., 1968).

Figure 11.6 shows magnetostriction data for magnetite and the titano-magnetites (Bickford et al., 1957; Syono and Ishikawa, 1963b, 1964). For small x, λ_{100} is independent of temperature from 77 K to 293 K and ranges from -2×10^{-5} ($x = 0$) changing sign with composition to $\sim +10^{-5}$ ($x = 0.10$). For higher x, λ_{100} becomes large and positive and strongly temperature dependent having values for $x = 0.56$ of 2×10^{-4} at 300 °C and 10^{-3} at 77 K. For low x values, λ_{111} is positive and increases with increasing temperature but after about $x = 0.2$ decreases with increasing temperature, being 10^{-4} at 300 °C and about 3×10^{-4} at 77 K for $x = 0.56$. For a single crystal of composition $Fe_{2.05}Ti_{0.95}O_4$, $\lambda_{100} = (4.7 \pm 0.3) \times 10^{-3}$ at 77 K, attributed to Jahn-Teller distortions of tetrahedral sites containing Fe^{2+} ions (Ishikawa and Syono, 1971).

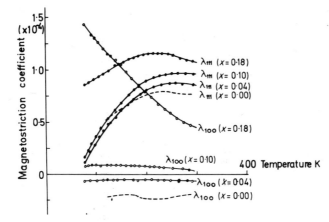

Figure 11.6 Temperature dependence of magnetostriction constants for magnetite and the titanomagnetites (after Syono and Ishikawa, 1963b).

For magnetite λ_{111} and λ_{100} increase by about 30% on application of a hydrostatic pressure of 2 kbar at room temperature (Nagata and Kinoshita, 1967).

11.2.4 Superexchange interactions

The steady decrease in Curie temperature (Figure 11.3) with increasing titanium content ($T_c = 850$ K for Fe_3O_4, 120 K for Fe_2TiO_4) indicates a progressive weakening of the AB interaction as the Fe^{3+}–Fe^{3+} interactions are replaced by weak Fe^{2+}–Fe^{2+} interactions (Geller et al., 1964). For magnetite T_c increases with increasing hydrostatic pressure with a 'magnetic Gruneisen parameter', $d(ln\ T_c)/d(ln\ V) = -4$ (Schult, 1970). This parameter increases with increasing x value suggesting an increasingly positive BB interaction (Schult, 1970) which may also account for the apparent Néel P-type character of the temperature dependence of M_s in titanium-rich samples. Alternative explanations for this behaviour, especially for the end member Fe_2TiO_4 have been given by Readman et al. (1967) and Ono et al. (1968).

11.2.5 Electrical resistivity of magnetite

As described in Section 8.5.2 there is a conductivity minimum of about $10^{-15}\ \Omega^{-1}\ cm^{-1}$ at about 5 K, also observed in impure and non-stoichiometric samples, and up to about 120 K the conductivity rises by about 9 orders of magnitude according to $\log \sigma \propto T^{-\frac{1}{4}}$ (see also Drabble et al., 1971). The Verwey transition (119 K in pure magnetite) is depressed by impurities or non-stoichiometry, or possibly suppressed altogether if the impurities are present in sufficient concentration (Verwey and Haayman, 1941; Epstein, 1953; Fujisawa, 1965). The usual description of the Verwey transition is that above the transition temperature the high conductivity results from electron hopping between iron ions of mixed valency in octahedral sites. Below the transition, electronic ordering takes place, with alternate (001) layers of Fe^{2+} and Fe^{3+} in octahedral sites, accompanied by a cubic or orthorhombic distortion. This picture, however, is not consistent with Mössbauer effect spectra observed below the transition (Sawatzky et al., 1969; Hargrove and Kundig, 1970) in which 4 spectra are found to describe octahedral site ions. Cullen and Callen (1971) taking a band-model approach show that additional, transverse, orderings are necessary to produce insulating behaviour and thus explain the extra spectra.

Above the transition the conductivity rises slowly in the region of 10^2 $\Omega^{-1}\,cm^{-1}$ and passes through a maximum at 80 °C (Domenicali, 1950) above which magnetite shows metallic behaviour.

11.2.6 Mössbauer effect spectra

Above the Verwey transition the spectra consist of two superimposed six line patterns due to Fe^{3+} in tetrahedral sites and an averaged spectrum for Fe^{3+} and Fe^{2+} in octahedral sites, arising from the rapid electron exchange which takes place faster than the nuclear precession frequency. The temperature dependence of the resultant line broadening of the octahedral site spectrum between 100 K and 200 K yields an activation energy of the electron hopping process of 0·065 eV (Sawatzky et al., 1969) in agreement with electrical conductivity data (Verwey and Haayman, 1941). The hyperfine fields of ^{57}Fe nuclei on the two sublattices as determined from the Mössbauer spectra (van der Woude et al., 1968) have the same temperature dependence between room temperature and T_c as the sublattice magnetizations determined by neutron diffraction (Riste and Tenzer, 1961).

Below the transition Fe^{2+} and Fe^{3+} in octahedral sites show their individual spectra although these are not consistent with simple Verwey ordering as described in Section 11.2.5. Evans and Hafner (1969) have shown that for a highly non-stoichiometric sample, $Fe_{2.83}\square_{0.17}O_4$, the Verwey temperature lay between 4·2 K and 77 K. For titanomagnetites (Figure 11.7) the randomness of the distribution of Fe^{2+} and Ti^{4+} within the octahedral sublattice gives a hyperfine field which varies from site to site and the resultant averaged spectra show a temperature independent broadening (Banerjee et al., 1967).

11.2.7 Cation deficient titanomagnetites

The most interesting oxidation process of titanomagnetites is the production at very low temperatures (~ 300 °C) of oxidized non-stoichiometric titanomagnetites of general composition $Fe_a Ti_b \square_c O_4$ where $a + b + c = 3$ and $\square$ indicates a vacant lattice site normally occupied in stoichiometric spinels. The oxidation of magnetite $Fe_3 O_4$ to maghemite, $\gamma\text{-}Fe_2 O_3$, is a special case of this oxidation process, simplified by the fact that Fe^{2+} only occurs in the octahedral sites of magnetite and vacancies occur only in the octahedral sites of

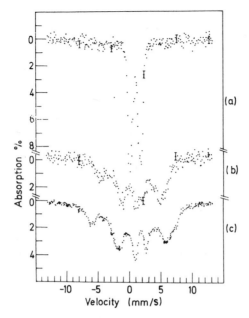

Figure 11.7 Mössbauer effect spectra of Fe_2TiO_4 at (a) 290 K, (b) 77 K and (c) 4·2 K. The line broadening is due to the random distribution of Fe^{2+} and Ti^{4+} within the octahedral sublattice (after Banerjee *et al.*, 1967).

maghemite. In the titanomagnetites the oxidation process is complicated because Fe^{2+} is located on both sites of the structure for most of the series.

The conditions determining whether magnetite oxidized to maghemite, γ-Fe_2O_3 or haematite α-Fe_2O_3 in the laboratory have been studied notably by Colombo *et al.* (1964) and Gallaher *et al.* (1968). It appears that a fast reaction rate at relatively low temperature is needed to maintain the spinel structure. This requires a large specific surface and a fast ionic diffusion rate promoted by a high degree of crystal imperfection (Lepp, 1957). Such conditions are usually to be found in magnetite made by precipitation from aqueous solution (David and Welch, 1956) and water absorbed on such samples may also play an important role (Elder, 1965).

Titanomagnetites cannot be made in this way and precipitated samples have been simulated by ball milling sintered titanomagnetites for prolonged periods in a water slurry (Sakamoto *et al.*, 1968; Sakamoto and Ozima, 1969; Readman and O'Reilly, 1970). Preground samples can then be oxidized in air at $\sim 300\,°C$ to produce single phase deficient spinel oxidation products.

The magnetic properties of non-stoichiometric titanomagnetites have been studied by Readman and O'Reilly (1971) who conclude, from the observed cation distributions, that Fe^{2+} in tetrahedral sites have a limited availability

662

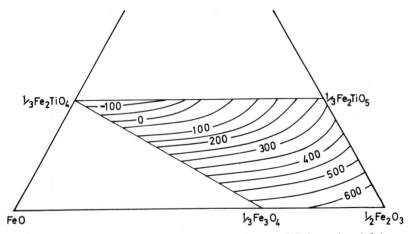

Figure 11.8 Contours of constant Curie temperatures (°C) for cation deficient spinel oxides in the Fe_3O_4–Fe_2TiO_4–Fe_2TiO_5–Fe_2O_3 quadrilateral of the FeO–TiO_2–Fe_2O_3 ternary diagram (after Readman and O'Reilly, 1971).

for oxidation—about 20% of the availability of Fe^{2+} in octahedral sites. Curie temperatures and unit cell edges of oxidized titanomagnetites are shown in Figures 11.8 and 11.9 as contours on the FeO–Fe_2O_3–TiO_2 ternary diagram.

T_c for non-stoichiometric titanomagnetitel rises with increasing oxidation indicating a strengthened AB interaction due to increasing Fe^{3+} concentration. This offsets the increasing vacancy concentration as expected (O'Reilly, 1968). Curie temperatures greater than about 400 °C for samples with moderate to high degrees of non-stoichiometry are difficult to measure because they tend

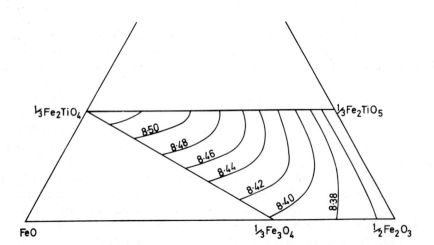

Figure 11.9 Contours of constant extrapolated cell edge (Å) of cation deficient spinels in the Fe_3O_4–Fe_2TiO_4–Fe_2TiO_5–Fe_2O_3 quadrilateral (after Readman and O'Reilly, 1971).

to invert to a multiphase intergrowth of the same chemical composition on heating in a non-oxidizing atmosphere (Readman and O'Reilly, 1970). This is exactly equivalent to the well-known inversion of maghemite to haematite. Extrapolation of measured Curie temperatures for non-stoichiometric titano-magnetites to the case of zero titanium concentration yields a T_c for maghemite of 695 °C (see also Section 11.3.3).

11.2.8 The haemo-ilmenite series

These rhombohedral oxides comprise a solid solution series represented as $(1 - y)Fe_2O_3 yFeTiO_3$. T_N increases from 50 K ($y = 1$) to 950 K ($y = 0$) and and the cell parameters also vary smoothly from $\alpha = 53° 46'$, $a = 5.48$ Å for $y = 1$ to $\alpha = 55° 17'$, $a = 5.413$ Å for $y = 0$.

Members of this series occur naturally in oxidized or metamorphosed igneous rocks and exhibit several interesting properties such as the variation of M_s with y. Ilmenite, $y = 1$, is antiferromagnetic. The hexagonally close-packed oxygen lattice contains alternate layers of Fe^{2+} and Ti^{4+} in planes normal to the trigonal axis [111]. A weak antiferromagnetic superexchange interaction exists between the nearest layers of Fe^{2+} which are separated by a layer of Ti^{4+}.

When ilmenite is doped with small quantities of Fe^{3+}, some Fe^{3+} are introduced into the Ti^{4+} layers and become coupled antiferromagnetically with adjacent layers of Fe^{2+} forming ferrimagnetic clusters of superparamagnetic size (Ishikawa, 1962). But when $y = 0.9$ approximately, there are sufficient Fe^{3+} in the Ti^{4+} layers to convert the antiferromagnetic coupling between Fe^{2+} layers into a parallel alignment throughout the whole crystal. Antiferromagnetic coupling now exists between the Fe^{2+} layers and the Fe^{3+} residing in the Ti^{4+} layers. Within the range $y = 0.9$ to $y = 0.45$ approximately, members of this solid solution series are ferrimagnetic with spontaneous magnetization $4y\beta$ per formula unit.

When ilmenite is substituted into haematite however, the Ti^{4+} and Fe^{2+} are randomly introduced into the layers of Fe^{3+} and members between $y = 0$ and $y = 0.4$ approximately are antiferromagnetic, apart from the weak parasitic ferromagnetism due to canting of spins in the basal plane for values of y close to zero.

From $y = 0.45$ to $y = 0.55$ approximately, both ordered and disordered cation distributions occur. The activation energy of the order–disorder process is about 0.3 eV (Ishikawa and Syono, 1963) and this process can be completed in laboratory times at 600–800 °C in naturally quenched crystals.

11.3 OXIDES AND OXYHYDROXIDES CARRYING THE REMANENT MAGNETIZATION OF SEDIMENTARY ROCKS

11.3.1 Forms of ferric oxide and oxyhydroxide

Of the three forms of ferric oxide only haematite, α-Fe_2O_3, which is the most stable, is found in any abundance as a mineral. Maghemite, γ-Fe_2O_3, is much less common whilst the third oxide ε-Fe_2O_3 exists only as a synthetic (Schrader and Buttner, 1963). Haematite crystallizes in the rhombohedral system, space group R3C. The structure is made up of distorted hexagonally close-packed layers of oxygen ions. It has a canted antiferromagnetic structure and so carries a weak parasitic ferromagnetism amounting to 0.1 gauss cm^3 g^{-1}. Although maghemite is normally written as γFe_2O_3, it has the spinel crystal structure. Because all the iron is ferric, the structure is cation deficient and it may therefore be written $Fe_{\frac{8}{3}}\square_{\frac{1}{3}}O_4$. Like magnetite it has a considerable M_s exceeding 70 gauss cm^3 g^{-1} at room temperature (Aharoni et al., 1962). It is relatively unstable and inverts to α-Fe_2O_3 in the laboratory in the range 400–600 °C.

Of the iron oxyhydroxides (oxide hydrates), goethite, α-FeOOH, is the most common and occurs naturally. It is antiferromagnetic with $T_N \sim 110$ °C, but also is weakly ferromagnetic and this makes it the most important form of oxyhydroxide in rocks. Next in importance as a mineral is the γ-form, lepidocrocite, which is antiferromagnetic below 65 K. The β isomer is extremely rare and also is magnetically ordered only below room temperature. δ-FeOOH is highly unstable and it is therefore hardly surprising that it has never been found as a mineral. However, magnetically, it is the most interesting of the oxyhydroxides.

11.3.2 Haematite

11.3.2.1 Structure

Ferric ions fill two-thirds of the octahedral interstices within the hexagonally close-packed oxygen lattice (Pauling and Hendricks, 1925). A more refined description of the structure (Blake et al., 1966) maintains the essential features described by the earlier investigators. Shull et al. (1951) showed by neutron diffraction that, above room temperature, the spins lie in (111) planes with the relative orientation $+ - - +$ as one goes along the trigonal axis (Figure 11.10). Below about -10 °C in pure crystals the spins are directed along the [111] axis and the temperature at which the transition occurs is known as the Morin temperature T_M.

11.3.2.2 The Morin transition

This spin-flip transition in haematite is named after Morin (1950) although Honda and Sone (1914) had earlier noticed a decrease in the susceptibility just below room temperature whilst Charlesworth and Long (1939) had observed thermal hysteresis in cycling through the transition temperature. Above T_M the anisotropy within the basal plane is small and highly variable but the anisotropy field confining the spins to this plane is several orders of magnitude greater: ~ 30 kOe (Anderson et al., 1954). Below T_M the spins are pinned along the [111] axis by a strong field determined to be 65 kOe by the Mössbauer effect (Blum et al., 1965).

In the transitional region, haematite behaves as a metamagnet (Kaczer and Shalnikova, 1964) and only moderate fields of a few kilo-oersteds are required

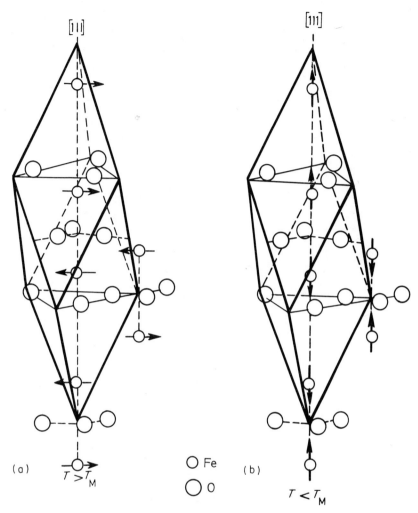

Figure 11.10 Crystallographic and magnetic structure of α-Fe_2O_3 (a) above the Morin temperature and (b) below the Morin temperature. The arrows illustrate the directions of the magnetic moments of the Fe^{3+} (after Shull *et al.*, 1951).

to flip the spins against the anisotropy forces (Figure 11.11). The spin-flip produces a spurious peak in the susceptibility curves at the Morin temperature due to the retention of the weakly ferromagnetic state in high fields (Flanders, 1966).

Imbert and Gerard (1963) interpreted their Mössbauer measurements on a natural crystal by a model in which the spins flip discontinuously from the basal plane (111) to the rhombohedral axis [111]. The width of the transition observed for natural crystals is thus thought to be due to a wide range of transition temperatures throughout the crystal rather than a gradual rotation of all the

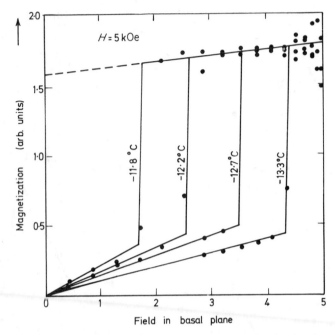

Figure 11.11 Metamagnetic behaviour of haematite in the region of the Morin temperature. Magnetization (measured by torque method) plotted against magnetizing field in basal plane (after Kaczer and Shalnikova, 1964).

spins through 90°. However, the detailed magnetic study by Flanders (1969) on small synthetic crystals which have a sharp transition shows that spin rotation takes place in at least three stages, due to the different anisotropy terms, and it is unlikely that the small initial and final rotations would have been observed by the Mössbauer effect. Discrete steps in the transition were also present in the resonance observations of Besser *et al.* (1967) but they could be associated with inhomogeneous doping.

The temperature of the transition is raised by increasing the pressure but the coefficient has not been firmly established. Whorlton *et al.* (1966) found a coefficient of 3·8°/kbar for pressures up to 6 kbar where it changed to 1°/kbar for pressures up to 26 kbar (Figure 11.12). However, Kawai and Ono (1966) raised the transition temperature of their sample to above room temperature by applying only 3 kbar. The discrepancy may lie in the nature of the sample used. Searle (1967) has explained the pressure sensitivity in terms of a magneto-elastic model. The effective shift of Morin temperature with field is small, amounting to 0·2°/kOe for (111) and 0·5°/kOe for [111]. Tasaki and Iida (1961) and later Artman *et al.* (1965) interpreted the Morin transition in terms of two competing anisotropies: (a) a fine structure (single ion) anisotropy energy constant K_{FS} due to spin–orbit effects which keeps the spins in [111] direction

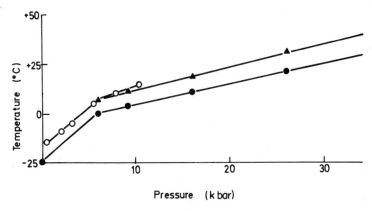

Figure 11.12 Morin temperature as a function of hydrostatic pressure (after Whorlton *et al.*, 1966).

and (b) an anisotropy, K_{MD} of dipolar origin which constrains the spins into the (111) plane.

For $T > T_M$, $K_{FS} < K_{MD}$ whilst for $T < T_M$, $K_{FS} > K_{MD}$. There is thus a change of sign of the resultant anisotropy constant, K, as T is increased through T_M with the spins flipping from the c axis [111] to the (111) plane. The predominance of the single-ion anisotropy below T_M fits in with the absence of a weak ferromagnetism in pure single crystals at these temperatures for Moriya (1960) requires an anisotropic dipolar interaction to explain the presence of a weak ferromagnetic moment.

11.3.2.3   Weak ferromagnetism

Two types of weak ferromagnetism have been recognized in haematite:

(1) An anisotropic type which is now known to be an intrinsic property of the haematite structure. It is only observed above the Morin temperature when it is confined to the basal plane.

(2) An isotropic type, considered to be due to structural defects. This is observed both above and below the Morin temperature.

The anisotropic type is carried by pure synthetic single crystals (Figure 11.13) in which its intensity is invariably 0.5 gauss cm^3 gm^{-1}. Below the Morin temperature such crystals are purely antiferromagnetic (Lin, 1961). Li (1956) attributed it to magnetic defects in domain walls, domain patterns being subsequently observed by Williams *et al.* (1958) using the Faraday effect. However, this idea was soon discarded in favour of an explanation put forward by Dzyaloshinsky (1958). By expanding the thermodynamic potential of the crystal as a function of spin density, Dzyaloshinsky showed that a stable state exists in which the spins are slightly canted within the basal plane so producing a weak net ferromagnetism whose strength is 10^{-2} to 10^{-5} of the nominal moment. This should be compared with the experimental value of 2×10^{-4}.

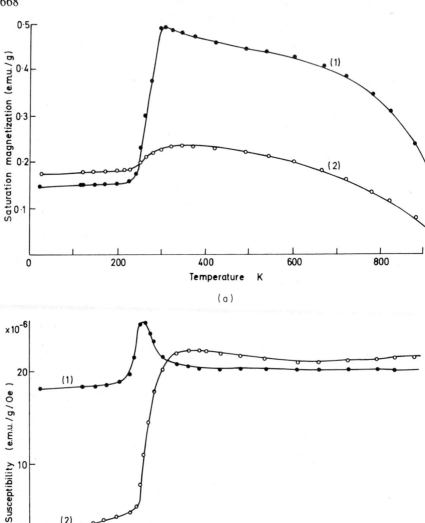

Figure 11.13 Temperature variation of saturation magnetization (a) and susceptibility (b) of a single crystal of haematite in the basal plane (1) and along the trigonal axis (2) (after Néel, 1953).

These symmetry requirements do not allow a weak ferromagnetism in pure stoichiometric synthetic crystals below the transition temperature. Moriya (1960) has expressed Dzyaloshinsky's spin canting term in the antisymmetric form $\mathbf{D} \cdot \mathbf{S}_1 \times \mathbf{S}_2$ and he calculated $\mathbf{D}$ which is a constant vector along the trigonal axis by developing Anderson's (1959) theory of anisotropic super-exchange to include spin–orbit coupling. His estimated magnitude of the parasitic ferromagnetism is in good agreement with the experimental value. Other materials which are normally regarded as purely antiferromagnetic such as CoF_2 (Borovik–Romanov, 1960) frequently have a small ferromagnetic moment.

This weak ferromagnetic moment due to the canting of the spins in the basal plane is not located at the sites of the Fe^{3+} ions but has been shown, by neutron diffraction, to be displaced slightly towards a nearest neighbour cation in the same (111) plane (Pickert et al., 1964).

In most natural crystals the remanence in basal plane decreases rapidly at the Morin temperature but does not completely disappear below it (Flanders and Remeika, 1965).

In one of the first attempts to explain the weak ferromagnetism of haematite, Néel (1953) attributed it to ferrous ion impurities. While their presence has since been disproved in pure synthetic samples exhibiting only the anisotropic ferro-magnetism, ferrous ions and other structural defects provide a good explana-tion of the isotropic ferromagnetism as suggested by the following evidence: (a) the absence of a ferromagnetic moment below T_M in pure stoichiometric synthetic crystals, (b) the decrease in this remanence in natural crystals on annealing (Gallon, 1968) and (c) an increase observed in an Elba crystal after fast neutron irradiation (Ogilvie, 1963). However, as regards (c), it should be pointed out that Gallon observed only a small change after similar irradiation of a synthetic crystal whilst no change was produced in natural crystals. Clearly further work is needed to determine the nature of the defects responsible. Pos-sible sources of defects include those produced by internal stress (producing a piezomagnetic moment) which would be at least partly removed by annealing. A defect moment parallel to the spin axis would be produced by an inequality in the number of cations on each sublattice or by a difference in the magnitude of the moments associated with each sublattice site. There is also the possi-bility of a contribution to the isotropic moment from regions in the crystal which for some reason do not undergo the spin-flip transition and so retain the canted basal plane configuration.

The properties of the small moment which persists in some crystals below T_M are very variable. Thus while the crystal from Elba was ferromagnetic in both the (111) plane and the [111] direction (Néel and Pauthenet, 1952), the remanence of Lin's (1961) sample, also from Elba, was solely parallel to [111]. Another crystal from Elba (Gallon, 1968) possessed remanence which lay almost in the basal plane (111) above T_M and which turned to make an angle of 30° with [111] below the transition. Bitter patterns led Gustard (1967) to suggest that a low temperature moment in the basal plane of a synthetic crystal (Gallon, 1968) was localized at certain regions in the crystal and Gallon suggests that the ferromagnetism is held by small regions which have not undergone the Morin transition. It is more usual, however, for the direction of the remanence in some natural crystals to lie parallel to [111] below T_M.

11.3.2.4 Anisotropy

A variability of the basal plane anisotropy amongst different samples is seen in studies of rotational hysteresis loss (i.e. the net energy loss per cycle, (W_R) on rotating the sample in a magnetic field). For coherent rotation, W_R is maximum for $H = K/M_s$ and drops to zero for $H = 2K/M_s$ where K is the effective anisotropy. (Although single crystals of haematite are undoubtedly multi-domain they can be considered as monodomain in the fields used.)

The peak rotational hysteresis loss occurs at fields of several kOe, indicating considerable anisotropy even in the basal plane (Vlasov et al., 1967). The loss curves are similar to those of typical hard magnetic materials (Wohlfarth, 1963). Extreme magnetic hardness is indicated by the almost constant W_R in a natural crystal, even in large fields (Flanders and Schuele, 1964). Porath (1968a) found similar behaviour in haematite powder obtained by heating maghemite and there is other evidence that energy loss in high fields is characteristic of powdered haematite. It is therefore not necessary to call upon exchange anisotropy coupling between α- and γ-Fe_2O_3 to explain the observed non-vanishing W_R in partially inverted maghemite as supposed by Banerjee (1966).

Crystalline perfection affects the in-plane anisotropy and natural crystals which contain few imperfections have a low K_3 (Sunagawa and Flanders, 1965). Flanders and Remeika (1965) found a correlation in some crystals between a high tin content and the magnitude of the triaxial anisotropy energy constant K_3 from torque measurements. An effective triaxial anisotropy can also be produced by twinning under applied stress (Porath and Raleigh, 1967) but the stress induced anisotropy of natural crystals is largely uniaxial (Porath, 1968b). It is particularly marked for compression within the basal plane and is therefore of a magnetoelastic origin.

Torque measurements reveal a basically uniaxial behaviour (Tasaki and Iida, 1963), but with additional smaller 4θ and 6θ components in the torque curves of some natural crystals (Banerjee, 1963; Porath, 1968b). This anisotropy may be observed more easily by antiferromagnetic resonance and the sixfold term is remarkably small, 10^{-2} Oe, in good single crystals (Morrish and Searle, 1964; Iida and Tasaki, 1965). Mizushima and Iida (1966) have shown both experimentally and theoretically that magnetoelastic coupling is responsible for the observed uniaxial anisotropy.

The coercivities of synthetic haematite powders are several kilo-oersteds, much larger than for most single crystals and much greater than expected either from a shape effect (in view of the small M_s) or from the intrinsic magnetocrystalline anisotropy.

Urquhart and Goldman (1956) found a magnetostriction coefficient of 8×10^{-6} linked with the weak ferromagnetism within the basal plane. This is appreciable for such a weakly magnetic material and this led Porath (1968b) to suggest that the origin of the magnetic hardness of small haematite particles lies in a stress induced anisotropy.

11.3.2.5   Effect of doping

The transition temperature is highly sensitive to the presence of foreign cations which usually lower it (Haigh, 1957; Kaye, 1961; Tasaki and Iida, 1961; Besser et al., 1967). Titanium has the largest effect, only 0·3% completely suppressing the Morin transition, and tin also has a strong influence (Flanders and Remeika, 1965). Rhodium appears to be unique in raising T_M (Krens et al., 1965).

The high sensitivity of T_M to titanium substitution has been explained by Besser in terms of the lattice distortion produced by the large Ti^{4+} and Fe^{2+} cations, as compared with Fe^{3+}, combined with the spin–orbit coupling of Fe^{2+}. Care is needed in interpreting the results obtained with flux grown single crystals which may have retained some flux; and inhomogeneous doping is thought to be responsible for the observed broadening of the transition in some titanium doped crystals (Besser et al., 1967).

The competing anisotropy model (Section 11.3.2.2) was used by Tasaki and Iida (1961) to estimate the effect of impurities on the transition temperature. By assuming that the dipolar term K_{MD} decreases quadratically and the single-ion anisotropy K_{FS} decreases linearly with doping they concluded that T_M should actually increase with the doping level in contradiction to experiment. Artman et al. (1965) consider that K_{MD} most probably varies as $(1 - f)^2$, where f is the doping level, and if K_{FS} does not decrease as rapidly as this, T_M should increase with f for mixed $(Fe_{1-f}Al_f)_2O_3$ crystals. As this was not found experimentally, K_{MD} must be less sensitive to impurities than supposed. Using the above approach, Besser et al. (1967) have successfully predicted T_M for $f = 0.01$ in the above system by assuming that the decrease in effective anisotropy field H_K observed on doping is solely due to a change in the single-ion term H_{FS} where $H_{FS} = H_K - H_{MD} : H_K = K/M$ where M is the sublattice magnetization. H_{MD} can be calculated for given lattice parameters (Artman et al., 1965) and the experimental H_K from spin-flip measurements at low temperature (Besser et al., 1967) gives an estimate of H_{FS}. In terms of Besser's model (Figure 11.14) $H_{MD} > H_{FS}$ at all temperatures below the Curie point for high enough doping. Thus the spin-flip transition is quenched.

11.3.2.6   Memory effect

Haigh (1957) observed that when both natural crystals and powders were magnetized and then cycled through the transition temperature in zero field, some of the initial remanence was recovered. Gallon (1968) reported a weak memory in certain synthetic crystals. The latter was much improved on crushing and then reduced again on annealing and so could have been due to internal stresses. The memory has also been attributed to impurities.

Assuming that the isotropic type of weak ferromagnetism in haematite is caused by lattice defects or impurities, any isotropic magnetization present in powders and natural crystals would preserve its direction and intensity below the T_M where the intrinsic remanence is lost, and on warming to room temperature

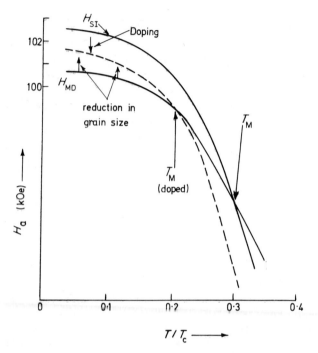

Figure 11.14 The competing anisotropy model for the Morin transition. Above the Morin temperature, T_M, dipolar anisotropy dominates and holds the spins on the basal plane. Below T_M, single ion anisotropy is dominant and the spins lie along the [111] axis. Doping effects only the single ion term (curve H_{SI}) lowering T_M as shown. Similarly, reduction of crystallite sizes increases the dipolar curve (H_{MD}) without affecting the single ion term (H_{SI}) again lowering the transition temperature (after Besser *et al.*, 1967).

again the latter might be recovered. The persistence of magnetic regions below the transition temperature in natural crystals and even in some synthetic crystals has been revealed by studies of domain patterns.

A remarkable alternation in the quality of the memory was observed by Nagata *et al.* (1961) on repeated cycling through the transition. The remanence was larger after an even number of cycles than after an odd number (Figure 11.15) by up to 15% of the initial remanence depending on the initial magnetizing process. Yamamoto and Iwata (1964) proposed a spin-fanning model to explain this oscillation in memory.

11.3.2.7 *Dependence of properties on crystal size*

The red sedimentary rocks encountered in rock magnetism contain small crystallites with a distribution of sizes. Thus it is important to consider the effect on the magnetic properties of haematite of reducing the crystallite dimensions.

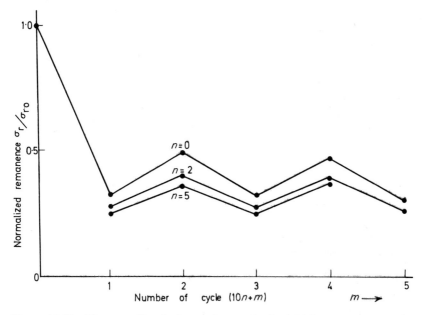

Figure 11.15 Memory effect in haematite. σ_{r0} is the initial remanence, σ_r, the remanence after cycling through the Morin temperature. The recovery is represented by σ_r/σ_{r0} (after Nagata *et al.*, 1961).

Chevallier and Mathieu (1943) prepared crystallites of various sizes by crushing a natural crystal. The weak ferromagnetism decreased with decreasing grain size and was effectively zero in grains of diameter 1 micron. The coercive force, which is 300 Oe for grains of 300 micron, increased to 1500 Oe at 15 micron and then decreased to 400 Oe for 1 µm (Figure 11.16). We would expect the strain produced by grinding to increase the coercive force and this behaviour has not been satisfactorily explained. In fact synthetic powders of the order of a micron in size obtained by low temperature sintering have large coercivities reaching several kOe. These values are much greater than can be attributed to shape anisotropy and so must have some microscopic origin. The magnetic properties of synthetic crystallites are also sensitive to the method of preparation (Takada *et al.*, 1965).

The weak M_s decreases for crystallites smaller than 1000 Å and disappears at about 200 Å (Hedley, 1968); T_M also decreases, and becomes less pronounced and is absent in particles less than 300 Å in diameter. This depression of T_M may be attributed to its known sensitivity to pressure. Schroer and Nininger (1967) measured an abnormally large cell size in synthetic fine grained material supported on microporous silica and equated this to a negative equivalent pressure which would depress the transition temperature. Another explanation (Takada *et al.*, 1965) is based on the competing anisotropy model (Section 11.5.2). The dipolar anisotropy is greatly enhanced in fine particles due to the large

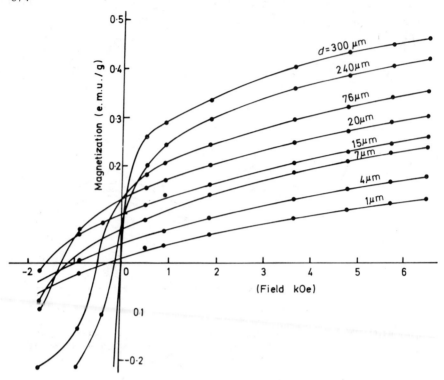

Figure 11.16 Effect of grain size on magnetization of haematite (after Chevallier and Mathieu, 1943).

number of surface spins and the temperature at which the single-ion and dipolar terms are equal is lowered. Hence the observed T_M is lowered (Figure 11.14).

The susceptibility of haematite increases in grains smaller than 200 Å and is temperature dependent. Particles smaller than 100 Å also have a weak but measurable ferromagnetism. These phenomena are common to other antiferromagnets such as NiO and Cr_2O_3 (Cohen et al., 1962). Néel (1961, 1962) has explained how such fine antiferromagnetic particles can possess a permanent moment due to an imbalance in the number of spins occupying the two magnetic sublattices caused by internal defects and the dominance of surface spins. In sufficiently small particles the direction of this moment can be mobile as in the case of fine ferromagnetic particles, i.e. they are also superparamagnetic. The maximum in the susceptibility–temperature curves is interpreted by Néel as a blocking temperature below which an appreciable thermo-remanence can be acquired. The TRM predicted by Néel's theory is in reasonable agreement with the experimental value.

11.3.2.8 Domain patterns

Domains in haematite were first observed on a (111) face by Williams et al. (1958) using the Faraday effect. Since then, Eaton and Morrish (1969) have noted that domain patterns are usually seen on faces perpendicular to the basal plane. These are in the form of parallel slabs or cylinders of the order of 100 microns broad. The fact (Gallon, 1968) that the developed face of pure crystals generally tends to be (110) whilst that for impure crystals tends to be (111) leads us to question the purity of the sample used by Williams and coworkers.

As one might expect, the domain walls are more regular in the synthetic than in the natural crystals, where they tend to be ill-defined and convoluted (Gustard, 1967). Only a few oersteds are necessary to move the domain walls in good synthetic crystals whilst much larger fields are needed in natural haematites. The room temperature domain pattern persists in Elba crystals below the Morin temperature in agreement with the observed memory (Gustard, 1967). Even in synthetic crystals the domains do not completely disappear on cooling in zero magnetic field (Blackman and Gustard, 1962) but are replaced by irregular knots of colloid which are magnetically quite hard (Gallon, 1968). Domain patterns in thin sections have been described by Eaton and Morrish (1969). In no study of domains has any large movement of walls been reported in passing through the transition temperature.

Sunagawa and Flanders (1965) found a correlation between the magnetic properties and surface features of crystals such as twinning and misorientation, both of which can cause internal strain. Gallon (1968) attempted to explain the behaviour of natural single crystals by the pinning of domain walls caused by local strain.

11.3.2.9 Piezomagnetism

Magnetic symmetry considerations show that haematite should be piezomagnetic (Dzyaloshinski, 1957) but the only experimental study is that of Andratskii and Borovik-Romanov (1967) who found a measurable piezomagnetic effect in two natural crystals only for stresses σ_{yy} and σ_{xz}* with the induced moments in the basal plane in the x and y directions respectively at low temperatures while at room temperature, i.e. above T_M the only piezomagnetic effect was for stress σ_{yy} with the associated moment along the trigonal axis [111] in agreement with symmetry requirements (Birss, 1966). The observed piezomagnetic coefficients were $\Lambda_1 = 3.6 \times 10^{-5}$ and $\Lambda_2 = 6.4 \times 10^{-5}$ below T_M and $\Lambda_6 = 5 \times 10^{-5}$ gauss cm^3 g^{-1}/kg m^{-1} above it.

The considerable piezomagnetism of haematite, 1 kbar producing a 50% increase in magnetization, is of considerable interest in rock magnetism. In particular the internal stresses associated with imperfections produced during crystal growth could be responsible for producing a significant ferrimagnetic moment. The perturbing effect of pressure on the acquisition of a remanence or on an already magnetized rock has received considerable attention (Stacey, 1963; Nagata and Carleton, 1969).

Haematite also exhibits a pressure demagnetizing effect (Kume, 1962) common to other magnetic minerals where the application of a pressure tends

* The directions refer to an orthogonal set of axes with z along the trigonal axis and x in the basal plane parallel to an axis of twofold symmetry.

e the remanent magnetization. This property has been explained by
ρ *et al.* (1969) in terms of the theory of Néel (1949) for single-domain
blages.

3.2.10 *Curie temperatures*

Haematite has a higher Curie point than any other oxide material. The value
normally quoted, 675 °C (Chevallier, 1951) is slightly lower than the many later
measurements and there is a remarkable range of temperatures in the literature even up to 710 °C (Vlasov and Fedoseyeva, 1968). Some of this variation
is clearly due to experimental errors as well as the presence of impurities. Strictly
speaking, magnetostatic experiments measure the Curie point of the weak
ferromagnetism as the low anisotropy within the basal plane makes it difficult
to observe the antiferromagnetic Néel point.

11.3.3 Maghemite

11.3.3.1 *Cation models*

The formula for the contents of the unit cell of maghemite, γ-Fe_2O_3 is
$Fe_8[Fe_{40/3}\square_{8/3}]O_{32}$. The presence of octahedral vacancies $\square$ has been confirmed by neutron diffraction (Ferguson and Haas, 1958) and Mössbauer
measurements (Armstrong *et al.*, 1966), but there is some doubt as to their exact
location. Because of the extra reflexions in the X-ray powder pattern and its
similarity to that of ordered lithium ferrite, Braun (1952) proposed a lithium
ferrite structure $Fe_8[Li_4Fe_{12}]O_{32}$ with vacancies replacing some of the lithium
ions, viz. $Fe_8[(Fe_{4/3}\square_{8/3})Fe_{12}]O_{32}$. The ordering of the vacancies within the
lithium B sites means that the structure is no longer spinel but needs a tetragonal
unit cell with $c/a = 3$ (van Oosterhout and Rooijmans, 1958; Schrader and
Buttner, 1963).

The chemical preparation of maghemite requires hydrogen ions and is usually
carried out either by the oxidation of wet-ground Fe_3O_4 or by the decomposition of γ-FeOOH. (The formation of maghemite, rather than α-Fe_2O_3, from
magnetite also requires the presence of water as discussed in Section 12.3.1.6.2.)
This led Braun to suggest that maghemite is a hydrogen ferrite which can range
in composition from $Fe_8[H_4Fe_{12}]O_{32}$ to $Fe_8[(Fe_{4/3}\square_{8/3})Fe_{12}]O_{32}$. From a
thermomagnetic and chemical study, Aharoni *et al.* (1962) concluded that
maghemite is actually a mixture of the cation-deficient spinel and hydrogen
ferrite types, the latter being stable on heating in a helium atmosphere. However
the existence of a hydrogen ferrite is in considerable doubt following the
neutron diffraction study of Uyeda and Hasegawa (1962) which pointed to an
ordered octahedral vacancy model. Similarly the chemical study of Strickler
and Roy (1961) has shown that the hydrogen content plays no part in determining the properties of maghemite and is most probably held as water. Furthermore the estimated moment of a hydrogen ferrite model is 1·08 β per Fe^{3+}
compared to experimental values of 1·18 β per Fe^{3+} (Henry and Boehm, 1956)
and 1·19 β per Fe^{3+} (Weiss and Forrer, 1929). The octahedral vacancy model

corresponds to $1{\cdot}25\ \beta$ per Fe^{3+}. Care should be taken in evaluating the experimental data as the samples can contain up to several per cent of Fe^{2+} and the magnetization of these powders is known to be slightly dependent on crystallite size (Berkowitz *et al.*, 1968).

The only single crystal study is that of Takei and Chiba (1966) on an epitaxially grown thin film which, unlike the powder samples, does not show any vacancy ordering. Resonance data gives $1{\cdot}45\ \beta$ per Fe^{3+} for the saturation moment and to explain this large figure they postulate A site vacancies, viz. $Fe_{7.5}\square_{0.5}[Fe_{2.9}\square_{2.1}]O_{32}$.

11.3.3.2   Curie temperature

Because of the inversion of maghemite to haematite between 400 °C and 600 °C, T_c cannot be measured directly. However, Curie points of maghemite doped with varying amounts of sodium (Michel and Chaudron, 1935) and aluminium (Michel *et al.*, 1951) have been measured. The presence of foreign cations tends to stabilize the structure and reduce T_c. Michel *et al.* (1951) claim that numerous determinations based on extrapolation to zero impurity cation in such stabilized substances indicate a Curie temperature of 675 °C. This technique has been criticized by Aharoni *et al.* (1962), because of the distortions produced by the added cations, and these workers obtained a Curie point of 590 °C in a sample of maghemite containing 1 % FeO which stabilized it to heating in a helium atmosphere.

The stabilizing influence of their thin film substrate permitted Takei and Chiba (1966) to directly observe the Curie point. Their vacancy disordered sample gave a Curie temperature of 470 °C. The poor agreement amongst the various estimates of the Curie point of maghemite is a reflexion of the considerable errors involved in the extrapolation procedures not to mention the assumption involved in their employment.

Banerjee (1965) used a thermodynamic approach based on the theory of Belov and obtained a Curie temperature of 545 °C which is very close to the high inversion temperature of 548 °C. On superexchange grounds Readman and O'Reilly (1972) estimate the Curie temperature to be 712 °C.

11.3.3.3   Other properties

The magnetic properties of maghemite powders have received considerable attention because of their use as a magnetic recording medium as described at length in Chapter 12. Such powders are essentially single domain with a coercive force of 200–300 Oe which can be attributed to a 'chain of spheres' reversal model (Section 12.3.3.1.1).

There is X-ray evidence for an order–disorder transformation in alkali-stabilized maghemite at 630 °C (Behar and Gollonques, 1957) similar to that seen in lithium ferrite.

The temperature of the transition to haematite is lowered by increased pressure with a coefficient of 3 °C per bar (Kushiro, 1960).

11.4 PHYSICAL BASIS OF ROCK MAGNETISM

11.4.1 Time dependence of magnetization—superparamagnetism

The magnetic material carrying the most stable remanence found in rocks is invariably so finely divided that we may apply single domain theory. Although, in many igneous rocks, the stably magnetized grains may be optically visible (larger than a few microns), replicates of etched surfaces exhibit evidence of subdivision into fine exsolution lamellae under the electron microscope (Figure 11.2). Hence we will first briefly consider single domain theory, the essential aspects of which were developed by Néel (1951) and Brown (1959).

If we consider an assemblage of particles with easy axes parallel, initially magnetized to saturation along the easy axes by an external field which is removed at time $t = 0$, then the remanence M_r varies with time as

$$M_r = M_s \exp(-t/\tau)$$

where τ is the relaxation time where $1/\tau$ is proportional to the probability of the spontaneous magnetization changing from one easy direction to another. For a particle of volume v this probability depends on the energy barrier, Cv, between easy directions and the available thermal energy kT. It may be shown by arguments of varying degrees of rigour (see, for example, Morrish, 1965, p. 361).

$$\frac{1}{\tau} = 10^9 \exp(-Cv/kT)$$

where C describes the anisotropy energy per unit volume arising from shape, crystal or strain anisotropies. For a given value of C, if v is small enough or T high enough τ may be small enough for the particles to behave like a paramagnetic material with no permanent bulk magnetization (for $\tau = 100$ s, $Cv \simeq 50kT$). Such a situation is described by the term superparamagnetism: the magnetization follows the applied field reversibly and Curie's Law, $1/\chi \propto T$ holds (see Section 1.3.2). Saturation is obtained in smaller fields than for the same material in the true paramagnetic state and the volumes of an assemblage of uniform grains may be determined simply from measurements of the Curie constant.

Superparamagnetism (which is referred to again in Section 12.3.3.1) is rarely observed in igneous rocks in which the carrier of magnetization is usually a titanomagnetite, because the critical volume V_B below which magnetite behaves superparamagnetically at ambient temperatures is very small. Superparamagnetism in fine grain samples of magnetite has been studied by McNab et al. (1968) using the Mössbauer effect. Here in contrast to magnetostatic measurements, the relaxation time necessary to obtain a collapsed spectrum is of the order of 10^{-8} s. However, for a given C and T this only reduces the critical radius by a factor of about 2. They found that in a sample with uniaxial shape anisotropy ($K_u \sim 8 \times 10^4$ ergs/cm^3) an almost completely collapsed spectrum was obtained at 295 K for a mean grain diameter of 121 Å. For magnetite spheres having only magnetocrystalline anisotropy the variation of τ at 300 K with grain diameter is:

d (Å)	440	485	505	565
τ	2·45 seconds	9 minutes	28 days	19 years

Hence it is sensible to define a critical volume V_B, the blocking volume, above which the grain has stable magnetization for each temperature T. Similarly for fixed volume, τ changes very rapidly with temperature and a blocking temperature exists which may be well below the Curie temperature of the material. In the calculation of blocking volumes and temperatures the variation of K_u or K_1 with temperature and possibly with volume (as in the case of α-Fe_2O_3 with dipolar anisotropy) must be taken into account. Hedley (1968) has shown that the blocking diameter for α-Fe_2O_3 at room temperature is about 200 Å.

Creer (1961) studied the superparamagnetism of α-Fe_2O_3 occurring naturally in red sedimentary rocks and this led Néel (1961) to develop a theory of superparamagnetism in antiferromagnetic materials. If the cation sites in a grain were filled at random, we should expect to find that the number of effective carriers of magnetic moment $p = n^{\frac{1}{2}}$ where n is the total number of cations. In this case the Curie constant will be the same as that of truely paramagnetic material formed of the same cations. If, however, the grains are formed by stacking up successive lattice planes, there will be two kinds of grain, viz. 'neutral' in which there are an even number of layers and which therefore have zero resultant magnetic moment and 'active' grains which contain an odd number of layers and thus have a magnetic moment proportional to the surface area so that $p = n^{\frac{2}{3}}$. In the latter case, the specific susceptibility will be proportional to $n^{\frac{1}{3}}$, i.e. to the grain diameter. Classical antiferromagnetics such as NiO and Cr_2O_3 have been shown to possess superparamagnetism (Cohen et al., 1962) so verifying Néel's theory. In addition, very fine grained antiferromagnetics exhibit an enhanced antiferromagnetic (temperature independent) susceptibility because the spins at and near the surface are not so firmly bound by exchange forces as in the bulk material. This property of superantiferromagnetism has also been described by Néel (1961) but is not of interest in rock magnetism.

11.4.2 Thermoremanent and chemical remanent magnetization

The stable remanence of igneous rocks is acquired when the minerals cool through their blocking temperatures T_B in the presence of the geomagnetic field, i.e. thermo-remanence, TRM. For an assemblage of superparamagnetic particles with uniaxial anisotropy and parallel easy axes the intensity M induced by an external field H at temperature T is:

$$M(T) = M_s(T) \tanh\left[\frac{M_s(T)Hv}{kT}\right]$$

As T falls through T_B further spontaneous redistribution of the moments between directions along the easy axis is negligible and if $H \ll K_u/M_s$, the remanence at room temperature, $M_r(T_0)$ is

$$M_r(T_0) = M_s(T_0) \tanh\left[\frac{M_s(T_B)Hv}{kT_B}\right]$$

It can be seen that for small H/T,

$$M_r(T_0) = \frac{M_s(T_0)M_s(T_B)v}{kT_B} \cdot H$$

The description of chemical remanent magnetization (CRM) in which grains of magnetic material grow from small nuclei through the blocking volume in the presence of the geomagnetic field is exactly analogous to that of TRM. Thus:

$$M_r(T_0) = \frac{V_B[M_s(T_0)]^2}{kT_0} \cdot H$$

We note that the intensity of both TRM and CRM are proportional to the external field strength for weak fields of the order of the geomagnetic field intensity.

In multidomain particles a blocking temperature may also exist above which thermal energy is equal to or greater than the barriers preventing domain wall motions, thus domain wall translation is relatively easy (Néel, 1955). For weak fields no work is done against anisotropy so that in a simple picture it is only necessary to minimize the remaining energy terms—the magnetostatic self-energy and the energy due to the external field—to determine the moment at temperature $T(> T_B)$ in a field H for a particle with demagnetizing factor N. The thermo-remanence at room temperature is given by

$$M(T_0) = [HM_s(T_0)]/[NM_s(T_B)]$$

11.4.3 Self-reversal of remanent magnetization

The mechanisms by which the magnetic minerals in a rock may acquire a remanent magnetization in the opposite direction to that of the geomagnetic field at the time of the formation of the rock or by which the remanence may reverse at a later time independent of any polarity changes of the geomagnetic field may be divided broadly into two groups—those involving a single phase system or those in which there are interactions between more than one phase (Néel, 1951).

11.4.3.1 Single phase systems

Self-reversal in a ferrimagnetic material is obtained when the sublattice which initially dominates magnetically, changes with temperature or time. An obvious example is the Néel N-type ferrite (see Section 1.4.3.1, 2) in which a TRM acquired just below the Curie temperature spontaneously reverses as the temperature falls through the compensation temperature. To produce self-reversal of natural remanent magnetization in rocks both Curie and compensation temperatures of the magnetic mineral must be above room temperature. Such materials are not found to occur naturally.

The magnetically dominant sublattice may change with time if there is a change in the cation populations of the two sublattices due to (a) the substitution of a diamagnetic ion into the sublattice which initially dominates; (b) oxidation, that is, the introduction of vacancies together with change in the Fe^{3+}/Fe^{2+} ratio; (c) ordering of an initially random cation distribution due to the annealing of quenched samples (Verhoogen, 1956). None of these mechanisms have yet been shown to produce self-reversal in geophysically relevant materials but mechanism (b) may be the most promising.

11.4.3.2 Two phase systems

We consider two cases here, one in which the two phases are linked magnetostatically and the second, more important, case in which the phases are coupled by exchange interactions (exchange anisotropy).

(1) *Magnetostatic coupling.* If an intergrowth or mixture of particles of magnetic materials A (spontaneous magnetization M_A, Curie or blocking temperature T_{CA}, demagnetizing factor N_A, volume fraction f_A) and B(M_B, T_{CB}, N_B, f_B) are cooled in an external field from above the higher Curie temperature (say T_{CB}) to below the lower Curie temperature, it is possible that the net remanent magnetization of the system may be opposed to the external field. To obtain self-reversal the geometry of the system must be such that the magnetic field due to the first magnetized phase (B) experienced by phase A exceeds the external field and that f_A, f_B, M_A and M_B have appropriate relative values. Stacey (1963) has considered such magnetostatic interactions between phases both in the mono-domain and multidomain state and concludes that self-reversal of TRM in nature by this mechanism must be very rare.

Uyeda (1958) investigated this mechanism experimentally by constructing a two component system (of magnetite and pyrrhotite) of suitable geometry and found that in very restricted circumstances self-reversal could sometimes occur. In natural samples or synthetics in which the geometry of the system of magnetic phases is allowed to develop in a 'natural' way (for example by oxidation of a titanomagnetite to a multi-phase intergrowth) a TRM, reversed by magnetostatic interactions, has not been found. However, there is sometimes a tendency to self-reversal, which may be exhibited as a reversed partial TRM (PTRM) which is acquired when the inducing field, applied on cooling through T_C or T_B is removed before the sample cools completely down to room temperature.

Creer *et al.* (1970) have produced reversed PTRM in basalts initially containing homogeneous titanomagnetites heat treated at 400 °C in air (Figure 11.17). The two magnetic phases are the parent Ti-rich titanomagnetite and finely exsolved Fe-rich titanomagnetite. The interaction field is weak, 0·14 Oe. The other exsolved phase, an ilmenite, plays no part in the reversal process. It remains to

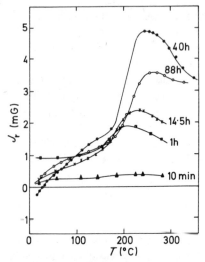

Figure 11.17 Partial self-reversal in Rauher Kulm (Germany) basalt oxidized in air at 400 °C in a field of 0·42 oersteds for the stated times. On cooling the samples from 400 °C to 20 °C in zero field a TRM is produced in the remaining parent phase which has a Curie point between 200 °C and 250 °C. The hump-shaped curves show that CRM and TRM are coupled anti-parallel. Ordinate = remanent magnetization measured at temperature T plotted along the abscissa (after Creer *et al.*, 1970).

be proved whether this oxidation process produced in laboratory times at 400 °C can occur in nature at ambient temperatures over geological times, but Harvard and Lewis's (1965) observation of the same type of PTRM in some Indian lavas which had not previously been subjected to laboratory treatment, suggests that it can.

(2) *Exchange coupling.* When crystallographic coherency exists between two phases (e.g. if there is a layer of oxygen ions common to both phases), atoms immediately on either side of the boundary layer may be coupled by an exchange interaction. Thus if the two phases have different magnetic ordering temperatures through which they are cooled in the presence of an external field, the phase with lower ordering temperature acquires a TRM which is independent of the external field but depends on the exchange field due to the atoms of the other phase. If the phase which orders first is an antiferromagnet and the second phase is a ferromagnet, the resultant TRM may be reversed or normal depending on the orientation of the spins of the antiferromagnet adjacent to the boundary layer. A further unusual property is that the aniso-tropy observed for the exchange coupled ferrimagnet is dictated by the anti-

ferromagnet which, if strongly anisotropic, results in a unidirectional aniso-tropy, so-called exchange anisotropy (Jacobs and Bean, 1963) shown by $\sin \theta$ torque curves and hysteresis loops shifted either parallel or antiparallel to the field-cooling direction. This effect leads to the interesting and well known natural occurrence of self-reversal found in a hypersthene hornblende dacite rock from Haruna in Japan containing haemo-ilmenite of composition $y \sim 0.5$. The cations of these crystals are partially ordered (see Section 11.2.8) and in this metastable state, small regions which are Fe-rich and other small regions which are Ti-rich exist in the lattice between ordered and disordered regions. The Curie points of the Fe rich regions are higher than the ordered part of the bulk of the grain (about 400 °C) which has the average proportion of cations: the Curie points of the Ti-rich regions are lower and these regions play no part in the self-reversing process. T_N for the disordered part of the (antiferromagnetic) structure is also slightly lower than T_c for the Fe-rich regions.

When the material is cooled down from above T_c in the geomagnetic field, the Fe-rich regions are the first to become magnetically ordered and the remanence on these small regions is aligned parallel to the field. These regions are coupled negatively by exchange interaction to that part of the bulk of the lattice which has the ordered (ferrimagnetic) cation distribution because ordering of Ti will take place mainly by migration within (111) layers (Figure 11.18). Thus at room

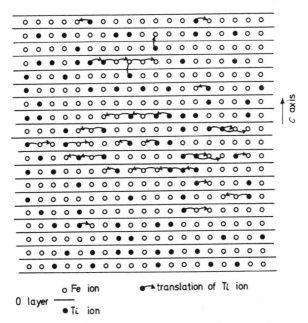

Figure 11.18a Two-dimensional representation of $Fe_{1.5}Ti_{0.5}O_3$ with the disordered cations positioned by means of random number tables. Translating the ions as shown in the diagram, results in the formation of regions of order and of regions enriched in iron ions within the disordered matrix.

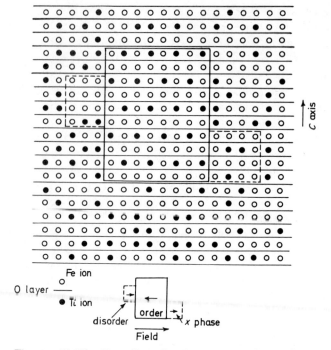

Figure 11.18b Two-dimensional representation of partially ordered $Fe_{1.5}Ti_{0.5}O_4$ showing development of iron-enriched and ordered zones (after Ishikawa and Syono, 1963).

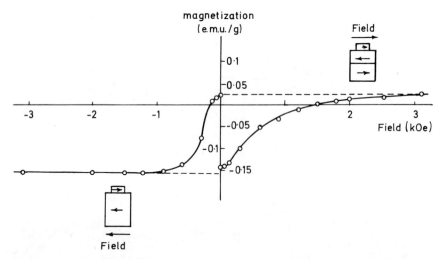

Figure 11.19 Remanent hysteresis loop in haemo-ilmenite ($y = 0.46$) with reverse TRM (after Ishikawa and Syono, 1963).

temperature the material has a reversed TRM. Ishikawa and Syono (1963) demonstrated the existence of superexchange interactions between the two phases by measuring shifted hysteresis loops (Figure 11.19) the ferrimagnetism of the ordered phase being coupled to the antiferromagnetism of the disordered phase present. After several heatings and coolings to 800 °C, the material loses its self-reversing properties due to the attainment of a completely ordered cation distribution.

CONCLUSION

During the present century research into magnetic materials has been directed increasingly towards technological applications particularly in the field of recording and memory devices described in succeeding chapters. The study of rock magnetism shows that nature has preceded man in the field of magnetic recording and that over millions of years the Earth has been storing up an infinite number of 'bits' of information about the geomagnetic field and the relative location of the continental masses. The retrieval of this information promises exciting and interesting research over many years to come.

REFERENCES

Aharoni, A., E. H. Frei and M. Schieber, 1962, *Phys. Rev.*, **127**, 439.
Akimoto, S., 1962, *J. Phys. Soc. Japan*, **17**, 706.
Anderson, P. W., 1959, *Phys. Rev.*, **114**, 1002.
Anderson, P. W., F. R. Merritt, J. P. Remeika and W. A. Yager, 1954, *Phys. Rev.*, **93**, 717.
Andratskii, V. P. and A. S. Borovic-Romanov, 1967, *Soviet Physics, J.E.T.P.*, 7-8181.
Armstrong, R. J., A. H. Morrish and G. A. Sawatsky, 1966, *Phys. Lett.*, **23**, 414.
Artman, J. O., J. C. Murphy and S. Foner, 1965, *Phys. Rev.*, **138**, A912.
Banerjee, S. K., 1963, *Phil. Mag.*, **8**, 2119.
Banerjee, S. K., 1965, *J. Geom. Geoelect.*, **17**, 357.
Banerjee, S. K., 1966, *Geophys. J. Roy. Astr. Soc.*, **10**, 4409.
Banerjee, S. K. and W. O'Reilly, 1967, *I.E.E.E. Trans.*, **MAG-2**, 463.
Banerjee, S. K., W. O'Reilly, T. C. Gibb and N. N. Greenwood, 1967, *J. Phys. Chem. Solids*, **28**, 1323.
Banerjee, S. K., W. O'Reilly and C. E. Johnson, 1967, *J. Appl. Phys.*, **38**, 1289.
Behar, I. and R. Gollonques, 1957, *Comptes Rendus Acad. Sci., Paris*, **244**, 617.
Berkowitz, A. E., W. J. Schuele and P. J. Flanders, 1968, *J. Applied Phys.*, **39**, 1261.
Besser, P. J., A. H. Morrish and C. W. Searle, 1967, *Phys. Rev.*, **153**, 632.
Bickford, L. R., J. M. Brownlow and R. F. Penoyer, 1957, *Proc. Inst. Electr. Eng.*, **B104**, 238.
Birss, R. R., 1966, *Symmetry and Magnetism* (North Holland: Amsterdam).
Blackman, M. and B. Gustard, 1962, *Nature*, **193**, 360.
Blake, R. L., R. E. Hesswick, T. Zoltai and L. W. Finger, 1966, *Amer. Min.*, **51**, 123.
Blasse, G., 1964, *Philips Res. Repts. suppl. No. 3*.
Bleil, U., 1971, *Z. Geophys.*, in press.
Blum, N., A. J. Freeman and L. Grodzins, 1965, *Bull. Amer. Phys. Soc.*, **138**, 465.
Borovik-Romanov, A. S., 1960, *Zh. eksper. teor. Fiz.*, **38**, 1088.
Braun, P. B., 1952, *Nature*, **179**, 1123.
Brown, W. F., 1959, *J. Appl. Phys.*, **30** Suppl., 130S.

686

Buddington, A. F. and D. H. Lindsley, 1964, *J. Petrology*, **5**, 310.
Chamalaun, F. H., 1964, *J. Geophys. Res.*, **69**, 4327.
Charlesworth, G. and F. A. Long, 1939, *Proc. Leeds Phil. Soc.*, **3**, 515.
Chevallier, R., 1951, *J. Phys. Rad.*, **12**, 172.
Chevallier, R. and S. Mathieu, 1943, *Ann. Phys. Paris*, **18**, 258.
Cohen, J., K. M. Creer, R. Pauthenet and K. Srivastava, 1962, *J. Phys. Soc. Japan*, **17**, B-1, 685.
Collinson, D. W., 1966, *Geophys. J. Roy. astr. Soc.*, **11**, 337.
Colombo, G., G. Fagherrazi, F. Gazzarini, G. Lanzavecchia and G. Sironi, 1964, *Nature*, **202**, 175.
Creer, K. M., 1961, *Geophys. J. Roy. astr. Soc.*, **5**, 16.
Creer, K. M., N. Petersen and J. Petherbridge, 1970, *Geophys. J. Roy. astr. Soc.*, **21**, 471.
Cullen, J. R. and E. Callen, 1971, *Phys. Rev. Letters*, **26**, 236.
David, I. and A. J. E. Welch, 1956, *Trans. Faraday Soc.*, **52**, 1642.
Domenicali, C. A., 1950, *Phys. Rev.*, **78**, 458.
Drabble, J. R., T. D. Whyte and R. M. Hooper, 1971, *Solid State Comm.*, **9**, 275.
Dunlop, D. J., M. Ozima and H. Kinoshita, 1969, *J. Geomag. Geoelectr*, **21**, 513.
Dzyaloshinsky, I., 1958, *J. Phys. Chem. Solids*, **22**, 617.
Eaton, J. A. and A. H. Morrish, 1969, *J. Appl. Phys.*, **40**, 3180.
Elder, T., 1965, *J. Appl. Phys.*, **36**, 1012.
Epstein, J. H., 1953, *M.I.T. Prog. Rept.*, XIV, 46–48.
Evans, B J. and S. S. Hafner, 1969, *J. Appl. Phys.*, **40**, 1411.
Ferguson, G. A. and M. Haas, 1958, *Phys. Rev.*, **112**, 1130.
Flanders, P. J., 1966, *Phil. Mag.*, **14**, 1.
Flanders, P. J., 1969, *J. Appl. Phys.*, **40**, 1247.
Flanders, P. J. and J. P. Remeika, 1965, *Phil. Mag.*, **11**, 1271.
Flanders, P. and W. Schuele, 1964, *Proc. Int. Conf. Magnetism (Nottingham)*, 594.
Fletcher, E. J., 1971, Ph.D. thesis, University of Newcastle upon Tyne.
Frankel, J. J., 1966, *Austral. J. Sci.*, **29**, 115.
Fujisawa, H., 1965, in Syono, 1965.
Gallagher, K. J., W. Feitnecht and U. Mannweiller, 1968, *Nature*, **217**, 1118.
Gallon, T. E., 1968, *Proc. Roy. Soc.*, **A303**, 511.
Geller, S., H. J. Williams, G. P. Espinoza and R. C. Sherwood, 1964, *Bell. Syst. Tech. J.*, **43**, 565.
Gustard, B., 1967, *Proc. Roy. Soc.*, **297**, 269.
Gyorgy, E. M. and H. M. O'Bryan, 1966, *Phys. Letters*, **23**, 513.
Haigh, G., 1957, *Phil. Mag.*, **2**, 505–520.
Hargrove, R. S. and W. Kundig, 1970, *Solid State Comm.*, **8**, 303.
Hauptman, Z. and A. Stephenson, 1968, *J. Phys. E.*, **1**, 1236.
Havard, A. D. and M. Lewis, 1965, *Geophys. J. Roy. astr. Soc.*, **10**, 59.
Hedley, I. G., 1968, *Phys. Earth planet. Int.*, **1**, 103–121.
Henry, W. E. and M. J. Boehm, 1956, *Phys. Rev.*, **101**, 1253.
Honda, K. and T. Sone, 1914, *Science Rept. Tohoku University*, **3**, 223–224.
Iida, S. and A. Tasaki, 1965, *Proc. Int. Conf. Magnetism, Nottingham*, Institute of Physics, 583–588.
Imbert, P. and A. Gerard, 1963, *Compte Rendus Acad. Sci. Paris*, **257**, 1054.
Ishikawa, Y., 1962, *J. Phys. Soc. Japan*, **17**, Suppl. B-1, 239.
Ishikawa, Y., 1967, *Phys. Letters*, **24A**, 725.
Ishikawa, Y. and Y. Syono, 1963, *J. Phys. Chem. Solids*, **24**, 517.
Ishikawa, Y. and Y. Syono, 1971, *Phys. Rev. Letters*, **26**, 1335.
Jacobs, I. S. and C. P. Bean, 1963, *G.E. Research Lab. rept.*, No. 63-RL-3224M.
Kaczer, J. and T. Shalnikova, 1964, *Proc. Int. Conf. Magnetism, Nottingham*, Institute of Physics, 589–593.
Kaye, G., 1961, *Proc. Phys. Soc. London*, **78**, 869.

Kawai, N. and F. Ono, 1966, *Phys. Lett.*, **21**, 297.

Krens, E., P. Szabo and G. Konczos, 1965, *Phys. Lett.*, **19**, 103–104.

Kume, S., 1962, *Ann. Geophys.*, **18**, 18.

Kushiro, I., 1960, *J. Geomag. Geoelectr.*, **11**, 148.

Lepp, H., 1957, *Amer. Mineral.*, **42**, 679.

Li, Y. Y., 1956, *Phys. Rev.*, **101**, 1450–1454.

Lin, S. T., 1961, *J. Phys. Soc. Japan*, **17**, Suppl. B-1, 226–230.

McNab, T. K., R. A. Fox and A. J. F. Boyle, 1968, *J. Appl. Phys.*, **39**, 5703.

Michel, A. and G. Chaudron, 1935, *C.R. Acad. Sci. Paris*, **201**, 1191.

Michel, A., G. Chaudron and J. Benard, 1951, *J. Phys. Rad.*, **12**, 189.

Mizushima, K. and S. Iida, 1966, *J. Phys. Soc. Japan*, **21**, 8.

Morin, F. J., 1950, *Phys. Rev.*, **78**, 819–820.

Morrish, A. H., 1965, *The Physical Principles of Magnetism* (Wiley: New York).

Morrish, A. H. and C. W. Searle, 1964, *Int. Conf. on Magnetism, Nottingham*, 574–577.

Moriya, T., 1960, *Phys. Rev.*, **120**, 91–98.

Nagata, T. and B. J. Carleton, 1969, *J. Geomag. Geoelectr.*, **21**, 427.

Nagata, T. and K. Kinoshita, 1967, *Phys. Earth Planet. Interiors*, **1**, 46.

Nagata, T. and K. Kobayashi, 1961, 'Rock Magnetism', Chapter 6, *Chemical Remanent Magnetization* (Marusen: Tokyo).

Nagata, T., M. Yama-Ai and S. Akimoto, 1961, *Nature*, **190**, 620–621.

Néel, L., 1949, *Ann. Geophys.*, **5**, 99–136.

Néel, L., 1951, *Ann. de Geophysique*, **7**, 8.

Néel, L., 1953, *Rev. Mod. Phys.*, **25**, 58–63.

Néel, L., 1955, *Adv. Phys.*, **4**, 191.

Néel, L., 1961, *Comptes Rendus*, I, **252**, 4075–4080, II, **253**, 9–12, III, **253**, 203–208.

Néel, L., 1962, *Comptes Rendus*, **254**, 598–602.

Néel, L. and R. Pauthenet, 1952, *C.R. Acad. Sci. Paris*, **234**, 2172.

Ogilvie, R. E., 1963, *M.I.T. Tech. Rept.*, No. 031–686, 5p.

Ono, K., L. Chandler and A. Ito, 1968, *J. Phys. Soc. Japan*, **25**, 174.

O'Reilly, W., 1968, *J. Geomag. Geoelectr.*, **20**, 381.

Palmer, P., 1963, *Phys. Rev.*, **131**, 1060.

Pauling, L. and S. B Hendricks, 1925, *J. Amer. Chem. Soc.*, **47**, 781–790.

Phillips, R., 1964, in 'Problems in Palaeoclimatology', ed. A. E. M. Nairn (Interscience: New York), p. 659.

Pickert, S. J., R. Nathans and H. A. Alperin, 1964, *Phys. Rev.*, **162**, 382.

Porath, H., 1968a, *J. Geophys. Res.*, **73**, 5959.

Porath, H., 1968b, *Phil. Mag.*, **17**, 603.

Porath, H. and C. B. Raleigh, 1967, *J. Appl. Phys.*, **38**, 2401.

Readman, P. W. and W. O'Reilly, 1970, *Phys. Earth Planet. Int.*, **4**, 121.

Readman, P. W. and W. O'Reilly, 1972, *J. Geomag. Geoelectr.* (Japan), **24**, 69.

Readman, P. W., W. O'Reilly and S. K. Banerjee, 1967, *Phys. Lett.*, **25A**, 446.

Riste, T. and L. Tenzer, 1961, *J. Phys. Chem. Solids*, **19**, 117.

Sakamoto, N. and M. Ozima, 1969, *Trans. Amer. Geophys. Union*, **50**, 133.

Sakamoto, N., P. I. Ince and W. O'Reilly, 1968, *Geophys. J. R. astr. Soc.*, **15**, 509.

Sawaoka, A. and N. Kawai, 1968, *J. Phys. Soc. Japan*, **25**, 133.

Sawatzky, G. A., J. M. D. Coey and A. H. Morrish, 1969, *J. Appl. Phys.*, **40**, 1402.

Schrader, R. and G. Buttner, 1963, *Zeit. anorg. allg. chem.*, **320**, 220–233.

Schroer, D. and R. C. Nininger, Jr., 1967, *Phys. Lett.*, **19**, 632–634.

Shull, C. G., W. A. Strausser and E. O. Wollan, 1951, *Phys. Rev.*, **83**, 333–345.

Schult, A., 1970, *Earth planet. Sci. Lett.*, **10**, 81.

Searle, C. W., 1967, *Phys. Lett.*, **25A**, 256–257.

Slonczewski, J. C., 1961, *J. Appl. Phys.*, **32**, 253S.

Stacey, F. D., 1963, *Phil. Mag. Suppl.*, **12**, 45–133.

Stephenson, A., 1969, *Geophys. J. R. astr. Soc.*, **18**, 199.

688

Strangway, D. W., R. M. Honea, B. E. McMahon and E. E. Larson, 1968, *Geophys. J. R. astr. Soc.*, **15**, 345–359.

Strickler, D. W. and R. Roy, 1961, *J. Amer. Ceram. Soc.*, **44**, 225.

Sunagawa, I. and P. J. Flanders, 1965, *Phil. Mag.*, **11**, 747–761.

Syono, Y., 1965, *Jap. J. Geophys.*, **4**, 71.

Syono, Y. and Y. Ishikawa, 1963a, *J. Phys. Soc. Japan*, **18**, 1230.

Syono, Y. and Y. Ishikawa, 1963b, *J. Phys. Soc. Japan*, **18**, 1231.

Syono, Y. and Y. Ishikawa, 1964, *J. Phys. Soc. Japan*, **19**, 1752.

Takada, T., N. Yamamoto, T. Shinjo, M. Kiyama and Y. Bando, 1965, *Bull. Inst. Chem. Res. Kyoto Univ.*, **43**, 406–415.

→ Takei, H. and S. Chiba, 1966, *J. Phys. Soc. Japan*, **21**, 1255–1263.

Tasaki, A. and S. Iida, 1961, *J. Phys. Soc. Japan*, **2**, 167.

Tasaki, A. and S. Iida, 1963, *J. Phys. Soc. Japan*, **18**, 1148.

Urquhart, H. M. A. and J. E. Goldman, 1956, *Phys. Rev.*, **101**, 1443.

Uyeda, S., 1958, *Jap. J. Geophys.*, **2**, 1.

Uyeda, R. and K. Hasegawa, 1962, *J. Phys. Soc. Japan*, **17**, Suppl. B-1, 391.

van der Woude, F., G. A. Sawatzky and A. H. Morrish, 1968, *Phys. Rev.*, **167**, 533.

van Houten, F. B., 1968, *Geol. Soc. Amer. Bull.*, **79**, 399.

van Oosterhout, G. W. and C. J. M. Rooijmans, 1958, *Nature*, **181**, 44.

Verhoogen, J., 1956, *J. Geophys. Res.*, **61**, 201.

Verwey, E. J. W. and P. W. Haayman, 1941, *Physica*, **8**, 979.

Vlasov, Ya. and N. V. Fedoseyeva, 1968, *Izv. Earth Phys.*, **5**, 108.

Vlasov, A. Ya., G. V. Kovalenko and N. V. Fedoseeva, 1967, *Izv. Phys. Solid Earth*, **2**, 129.

Weiss, P. and R. Forrer, 1929, *Ann. Phys.*, **12**, 279.

Whorlton, T. G., R. B. Bennion and R. M. Brugger, 1967, *Phys. Letters*, **24A**, 653.

Williams, H. J., R. C. Sherwood and J. P. Remeika, 1958, *J. Appl. Physics*, **29**, 1772.

Wilson, R. L., 1961, *Geophys. J. R. astr. Soc.*, **5**, 45.

Wohlfarth, E. P., 1963, in *Magnetism*, Vol. III, ed. G. T. Rado and H. Suhl (Academic Press: New York), pp. 351–390.

Yamamoto, M. and T. Iwata, 1964, *Proc. Int. Cong. Magn. Nottingham*, Institute of Physics, 581.

Zener, C., 1954, *Phys. Rev.*, **96**, 1335.

12 *Oxides for magnetic recording*

G. BATE

12.1 INTRODUCTION

Magnetic recording was invented by Poulsen (1900), who demonstrated that electrical signals could be written onto an iron wire which was wrapped around a drum. Stainless steel wire was used in later recorders, but in the 1940s recording surfaces composed of fine particles of iron oxide, at first on paper substrates and later on plastic films, gained ascendancy. The tapes did not twist in the machines and could be more easily and successfully spliced. Magnetic recording finds continually increasing application in audio, video, instrumentation and computers, and now accounts for the largest amount of money spent on magnetic products (in the U.S., about 280×10^6; cf. 180×10^6 for electrical steels (Jacobs, 1969)). There are several reasons for its popularity as a means of information storage, in addition to its economy, fidelity and reliability. First, it is simple to use: no development step is needed between the writing and reading processes and hence one can go back at any time and add to an incomplete record; that is, it is postable. Second, information is stored as a passive condition of the medium; it does not need to be regenerated continuously; it is non-volatile. Third, the stored information is usually quite unaffected by the environment, such as changes in temperature or pressure, or the presence of ionizing radiations or of electric or magnetic fields (smaller than the coercivity of the medium). Finally, the information can be erased and the same surface used to make another recording; the process is apparently infinitely reversible.

Recording surfaces take the form of tapes, discs, drums, strips and stripes, and belong to one of two classes. These are, first, particulate coatings in which small single-domain particles are separated from each other and attached to the substrate by a polymeric binder and second, thin, continuous metallic films. Only the first class will be considered here.

12.2 PRINCIPLES

The stored information is arranged on the recording surface in narrow bands or 'tracks' along which the writing and reading heads move. Almost invariably, the magnetization direction is longitudinal, that is, in the plane of the recording surface and along the track. Magnetization normal to the plane would be strongly opposed by demagnetization; magnetization across the tracks would lead to a smaller amount of emergent flux and would be difficult to read at any but the lowest densities. The writing head (Figure 12.1) consists of a magnetically

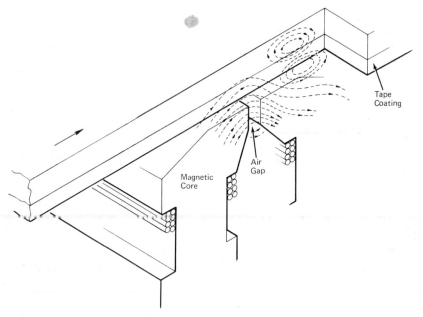

Figure 12.1 Diagram of the magnetic recording process.

soft core of nickel–iron laminations or ferrite blocks and has a narrow air gap (less than 1 mil or 25 μm). The input signal is applied to a coil of perhaps 100 turns wound around the core and results in flux which fringes around the air gap and penetrates the recording surface. If the fringing flux density is greater than the coercivity of the recording surface, the medium becomes magnetized in the direction of the flux and most of the magnetization is retained after the recording surface has moved away from the field of the head. Clearly, magnetic recording materials belong to the class of 'hard' magnetic materials ($H_c \geqslant$ 100 Oe) and must have properties similar to those of permanent magnet materials except that the coercivity must not be so high that the head's fringing field cannot overcome it.

The reading head is similar in structure to the writing head. Indeed, in disc files a single head is commonly used for both reading and writing. The reading process makes use of the fact that the flux from a magnetic dipole in the recording surface finds a path of lower reluctance around the core of the head than across the air gap (Figure 12.2). Some of the flux then follows this preferred path around the core and through the pick-up coil. Then, when the flux changes because of the relative motion of the head and the recording surface, a voltage,

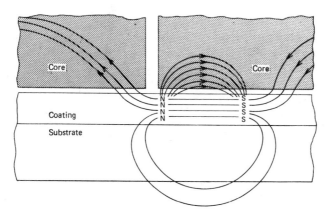

Figure 12.2 The reading process; showing some of the flux from a magnetized region in the recording surface following the low reluctance path around the reading-head core. This flux then passes through the reading coil.

which is proportional to the time rate of change of the flux around the core, is induced in the coil. As the separation between the recording surface and head increases, or as the wavelength of the recorded signal decreases, the inequality of the two reluctances (around the core and across the gap) becomes less pronounced. Less flux follows the core path and, in consequence, the output signal decreases. Most commonly, the relative motion between the recording surface and the heads is achieved by moving the medium past a stationary head. Occasionally the arrangement is reversed or, as in some video recorders, both the head and the medium are in motion. Recording applications fall into two broad categories, analogue and digital, which differ enough to require separate descriptions.

12.2.1 Analogue recording

When the signal to be stored (and reproduced) possesses continua both of frequency and of intensity, as in audio or instrumentation applications, analogue recording is used. It is desirable to produce an output signal that not only has a large signal-to-noise ratio, but also is linearly related to the input signal. This suggests that the magnetization intensity of the medium should be proportional to the applied field—a result which would clearly be difficult to achieve by using only the highly non-linear recording head field pattern and the magnetization curve of the medium. However, it was soon discovered that a close approximation to the desired linearity could be achieved by superimposing the signal to be recorded onto a sinusoidal carrier whose constant frequency was 5 to 10 times that of the highest frequency occurring in the signal. That this approach worked in practice there could be no doubt; but the question of how in detail it worked has provided both theorists and experimentalists with years of intriguing investigation and still remains to be answered.

The most promising model appears to be that using anhysteretic magnetization, which considers the simultaneous application to the recording surface of a constant field and an alternating field whose amplitude is initially larger than the constant field and ultimately allowed to decay to zero. The anhysteretic remanent magnetization is found to be proportional to the magnitude of the constant field regardless of the previous magnetic history of the medium. The proportionality, however, extends only to a value of the acquired remanence $\leqslant 1/2M_r$, where M_r is the remanent intensity after a saturating field H_s has been applied and removed. It is clear that this model can represent only an approach to reality when one remembers that during the actual recording process, the signal field is constant in neither magnitude nor direction as the medium moves past the recording head.

Mee (1964) has described a helpful way of picturing the recording process. As the medium moves along, a square-wave bias signal alone would generate in the recording surface a series of partially overlapping magnetized cylinders whose magnetization direction alternates in sign. The axes of the cylinders lie across the track. For a sinusoidal bias signal, an obvious modification of this picture is required. Then the addition of a signal field of lower frequency than the bias field will disturb the uniformity of the overlapping cylinders, causing some to expand and others to contract, so that different regions of the coating are magnetized to different depths depending on the amplitude of the input signal. If the bias field is increased from zero, the output signal increases and its harmonic distortion decreases until, at the optimum bias level, the magnetization varies in depth as the modulus of the input signal. If the bias field is increased beyond this point, the harmonic distortion increases and eventually the output signal begins to decrease as the magnetization induced by the two fields (signal + bias) extends completely through the magnetic surface.

The anhysteretic magnetization curve, unlike the normal magnetization curve, shows no point of inflection and is always concave to the field axis. Furthermore, the slope of the anhysteretic curve is initially very large; that is, the anhysteretic initial susceptibility is much greater than the normal initial susceptibility. Thus the use of a bias field leads not only to linearity between input and output signals, but also to an enhancement of the output signal. In addition to the shortcomings mentioned above, the simple anhysteretic magnetization model of the recording process ignores the interaction fields between the magnetic particles in the medium. These fields act somewhat as an additional bias field; however, they vary in magnitude and direction from point to point and also their whole pattern changes as the macroscopic magnetization of the recording surface changes. The effect of the interactions on the analogue recording process was discussed by Mee (1964) and by Daniel and Levine (1960). The model should also be further improved by taking account of the shape of the trailing edge field of the recording head and also the effect of recording wavelengths that are of the same order as the gap length of the reading head. When this is done, good agreement has been obtained between experimental recording results and those predicted by the anhysteretic model. At short wavelengths, however, it it difficult to satisfy the anhysteretic condition that the signal field be constant during the decay of the bias field. Further complications arise with short wavelengths; increased separation losses and unfavourable demagnetization geometry cause additional reduction in output signal and departures from the model.

Occasionally d.c. bias is used in applications where such high frequencies are to be recorded that the (still higher) a.c. bias frequency would exceed the capability of the recording head or the writing and reading circuits. The use of d.c. bias gives larger output signals and better resolution but also, unfortunately, considerably higher noise levels. Recording with no bias is also used with frequency-modulated signals since the depth of recording remains fixed by a writing current whose magnitude is constant; here amplitude linearity is irrelevant.

Noise may be present in the output signals in the form of random, unrelated pulses (background noise), unwanted signals (print-through), or random variations in the desired signal (d.c. noise and modulation noise). Background noise arises from the imperfect demagnetization of the magnetic medium on a scale equal to or larger than the magnetic particles. The alternating field used for erasure not only must be large enough to switch the magnetization of all but the hardest 1 % of the particles, but also must have a frequency high enough that during the passage of any particle through the erasing field the particle is subjected to several cycles of the field. If these conditions are met, background noise can be effectively reduced by using smaller magnetic particles.

Print-through occurs in tape recordings during the storage of the tape on reels. If the substrate is thin (0·5–1 mil), the leakage field from a portion of the tape recorded at a high dynamic level may penetrate the substrate and magnetize the adjacent layers of the tape coating. The thicker the tape substrate, of course, the less likely is print-through; and the effect can be further reduced by selecting a magnetic material which remains insensitive to fields smaller than those normally used to record on the material. Particles of synthetic magnetite (q.v.) have acquired the reputation of being more susceptible to print-through than are particles of γ-Fe_2O_3 of similar coercivity.

Modulation noise can be significantly larger than background noise and is most noticeable, in audio recording, when the recorded signal consists of a single frequency. This contribution to the noise (with the related d.c. noise) was studied by Eldridge (1964), who showed that it occurred in a repeatable way at definite locations along a tape and that it had two primary components: asperities on the surface, and magnetic inhomogeneities within the coating. The contribution from the surface irregularities can be many times greater than the other and has the same form with or without a.c. bias. With long-wavelength signals, this noise component predominates when the recording fields are less than those necessary to produce maximum output. The component due to magnetic non-uniformity is most important when a large uniform field is used to record, as in digital recording. The method of reducing modulation noise is easier to prescribe than to perform; it is to prepare recording media that are highly homogeneous and have smooth surfaces.

12.2.2 Digital recording

The information to be stored is first converted into binary form, so only two states of magnetization are needed. These may correspond to an a.c.-erased state and a single direction of magnetization or, more commonly, two opposite states of magnetization. Since the distinction between these two states will be greatest when they correspond to the maximum remanence condition, this form of recording is frequently, but imprecisely, called 'saturation recording'. The value of writing current is chosen that leads to the largest output signal, and at this point the medium is usually not magnetically saturated. Thus a recording surface storing digital information is composed of discrete longitudinal tracks along which the magnetization changes abruptly from a remanent state in one longitudinal direction to a similar state in the opposite direction. Erasure in digital recording usually involves the uniform magnetization of the recording surface by means of a constant field from a (wide-gap) head which is d.c. energized. A signal is read from the medium only when the magnetization changes direction; hence no signal will be obtained from a uniformly erased medium. The most obvious encoding scheme, in which magnetization in one direction corresponds to a 1 while magnetization in the other direction represents a 0, is never used. Long strings of 1s and 0s would give no output signal; or, conversely, no signal from the reading head would have to be interpreted as a string of 1s or 0s, depending on the sign of the last observed pulse. An alternative explanation, of course, could be that a speck of dirt had temporarily caused the separation between the reading head and the recording surface to increase enough to reduce the signal below the detectable limits. While recording density, access times and data rates are important in computing systems, above all the information must be recorded and retrieved reliably. Thus those encoding schemes are chosen which emphasize reliability without too much sacrifice in storage density. For example, in the NRZI scheme a flux change occurs only when a 1 is written. Thus, although the number of flux reversals per inch (fri) is equal to the number of bits per inch, the uncertainty mentioned above still exists when a long series of 0s is recorded. The phase-encoding scheme (PE) provides an output signal for every 1 and every 0, but it now takes two flux reversals to store one bit. To reduce confusion, when referring to storage density capabilities we shall give the number of flux reversals per inch of track rather than the number of bits per inch. The maximum density found in tape drives and disc files is at present about 3200 fri. This number is smaller by an order of magnitude than the maximum density used in instrumentation recorders, chiefly because much greater reliability is required in digital recording systems. Short surveys of encoding schemes have been given by Hoagland (1963) and by Pear (1967).

Linearity of the magnetization as a function of the applied field is quite unimportant in digital recording, but it is important that the remanent magnetization should change from $+M_r$ to $-M_r$ in a small distance compared with the bit spacing for good pulse resolution at high densities. The recording process is shown in the diagram in Figure 12.3. As a particular region in the coating moves past the head gap, it experiences a progressively weaker field. Eventually a field contour (H_1 in Figure 12.3) is reached beyond which the head field is no longer able to change permanently the state of magnetization of the region. Thus the trailing-edge field is particularly important in the recording process. The direction of the field along this critical contour changes with depth into the coating and, in general, the field has a perpendicular component as well as a longitudinal one. Demagnetization tends to oppose the establishment of a perpendicular component of magnetization, but Templeton and Bate (1964) showed by direct measurement that this can be as large as 10% of the longitudinal component. This agreed with measurements on a large-scale model of the recording process (Tjaden and Leyten, 1963, 1964).

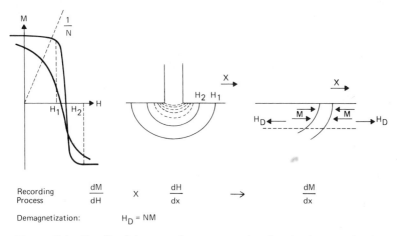

Figure 12.3   Details of the recording process; showing the demagnetization curves for two different recording surfaces (giving dM/dH), the constant longitudinal-field contours of the recording head (giving dH/dx) and a longitudinal cross-section of a recorded transition (giving dM/dx). The establishment of a partially demagnetized region is called 'recording demagnetization'. The expression $H_D = NM$ describes the subsequent self-demagnetization of the magnetized regions in terms of the geometrical factor N and the magnetization M (after Bate, 1965).

The transition zone between oppositely magnetized regions has a finite width, since neither the recording field as a function of position nor the magnetization of the medium as a function of the field is a step function. The creation of this partially demagnetized transition region is often called 'recording demagnetization', since at the trailing edge of the recording head field, the present field partially switches the magnetized region created by the previous field. When the surface has moved away from the recording head another demagnetizing field exists, which consists of two parts: self-demagnetization and adjacent-bit demagnetization. This field, which tends to broaden the transition zone still further, increases with the intensity of magnetization M

and with a geometrical factor N, which increases with increasing bit density and increasing thickness. N is difficult to calculate precisely and the magnetization is not constant throughout the thickness of the recording surface. The effect of H_D in reducing the remanent magnetization is shown in Figure 12.3. We see also that the coercivity H_c of a material is a rough measure of its ability to withstand demagnetizing fields (of internal or external origin). The inner of the two curves of M versus H represents the behaviour of a typical iron oxide recording surface, compared with the outer curve, which represents the more rectangular loop with higher coercivity of thin metallic recording surfaces. A given demagnetizing field (represented by a line of slope $1/N$) results in a more serious decrease in magnetization intensity in the former case.

One of the most important factors limiting the linear recording density is the phenomenon of peak-shift, which is illustrated in Figure 12.4 for two 1s

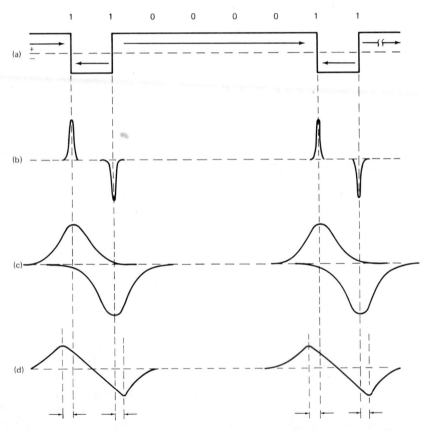

Figure 12.4 Peak-shift. (a) Idealized pattern of magnetization corresponding to the sequence 1 1 0 0 0 0 1 1 in the NRZI code. (b) The sharp output pulses from an ideal (narrow-gap) reading head. (c) Broadened pulses from a non-ideal head. This leads to (d) The output pulse doublets showing the peaks displaced outwards from the bit location.

(NRZI) preceded and succeeded by a series of 0s. At high packing densities (1000 fri or higher, with iron oxide tape 500 μin thick), the reading head begins to respond to the second transition before it has finished responding to the first. This pulse-crowding displaces the peaks of the pulse doublet outwards from their intended position. Thus a time error can be introduced into the reproduced pattern, and information would be lost if the shift equalled or exceeded half the natural bit period (the intended spacing of the two 1s). In practice even less peak-shift can be tolerated. Bate *et al.* (1964) showed that, with thin metallic recording surfaces, the output waveform of the two 1s could be predicted accurately to a density of at least 25,000 fri from the graphical superposition of the pulses obtained from isolated transitions.

Thus to obtain good resolution of high recording densities, narrow output pulses are required. This in turn implies a need for narrow transition zones, narrow reading-head gaps and small separation between the recording surface and the reading head (since separation tends to broaden the pulse). Narrow transition zones call for thin coatings (to minimize the demagnetizing field and the pulse broadening due to separation) and coercivities high enough to resist demagnetization but not so high that writing and erasure become difficult. One might suppose that a rectangular hysteresis loop would be ideal for a recording material. While it is true that thin metal coatings with excellent recording properties usually have rectangular loops, the situation with regard to the thicker particulate coatings is considerably less clear; it will be discussed later, when the effects of particle orientation are considered.

12.3 IRON OXIDES

From the beginning of the sudden growth in popularity of magnetic recording (c. 1945) to the present, one magnetic material has dominated the technology; that material is γ-Fe_2O_3 in the form of small single-domain particles. As more than 99% of commercially available tapes and discs make use of γ-Fe_2O_3 particles, it is very appropriate to consider this material first. The intrinsic properties have been described in Section 11.3.3: the mineral is known as maghemite.

12.3.1 Preparation

The earliest proposal to use iron oxide particles for recording purposes was apparently that of Ruben (1932). He suggested coating a flexible wire or strip with a dispersion of magnetite (Fe_3O_4) particles in a thermoplastic binder so that when the strip was heated the magnetite particles could reorient themselves in response to the field from the recording head. The earliest conventional recording tapes (in which the role of the head field is to switch the magnetization with respect to a stationary particle, rather than to rotate magnetization and particle like a compass needle) were probably those developed in Germany for the Magnetophon recorder before 1939. The particles were roughly spherical

698

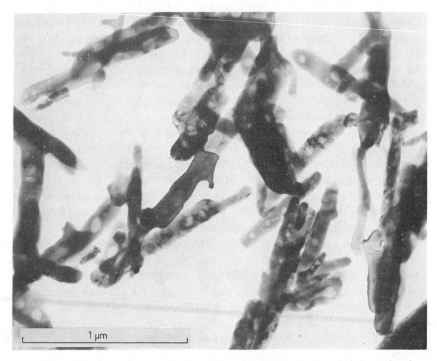

Figure 12.5a Electron micrograph of particles of γ-Fe$_2$O$_3$ showing their irregular shapes with dendrites and cavities.

and about 1 μm in diameter. A FIAT report (1947) gave the method of preparation, which consisted of precipitating roughly spherical particles of magnetite from an aqueous solution of ferrous sulphate and potassium nitrate with an excess of ammonium hydroxide. The magnetite particles were dried and oxidized without change of shape or size to give reddish-brown particles of γ-Fe$_2$O$_3$. When compared with the coercivity of today's acicular (needle-shaped) iron oxide particles, 260–320 Oe, the coercivity of the early spherical particles was low; it was usually 90 Oe and never greater than 150 Oe. Not unexpectedly, the output signal obtained from tapes made of these particles was low, presumably because of the low coercivity. The iron oxide particles now used typically have a length of 0·5–1·0 μm and an axial ratio of 6–7:1. Figures 12.5(a) and 12.5(b) show the presence of dendritic growth and the apparent porosity of these particles. The form of ferric oxide usually produced on the oxidation of the lower oxides is α-Fe$_2$O$_3$; special chemical processes must be used to prepare γ-Fe$_2$O$_3$.

12.3.1.1   Acicular particles of Fe_3O_4 and γ-Fe$_2$O$_3$

Many of the processes by which γ-Fe$_2$O$_3$ particles are prepared involve, as an earlier step, the preparation of particles of magnetite (Fe$_3$O$_4$), which are then converted to γ-Fe$_2$O$_3$ by controlled oxidation. It is thus convenient to discuss

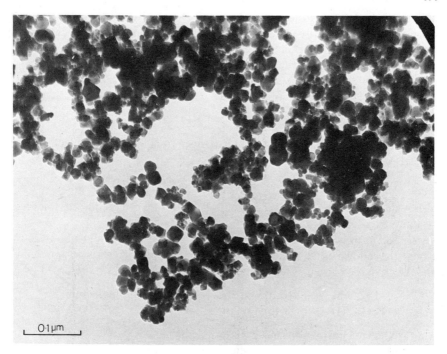

Figure 12.5b Electron micrograph of equiaxed particles of cobalt-substituted γ-Fe_2O_3.

both oxides together. Naturally occurring Fe_3O_4 is generally not pure enough, nor does it possess the correct crystallite size and shape, to be used in recording surfaces.

The starting material from which acicular particles of the iron oxides are prepared is alpha ferric oxyhydroxide (α-(FeO)OH = synthetic goethite), which can be obtained in acicular crystalline form. This is available commercially as a yellow pigment, or it can be made by a process described by Penniman and Zoph (1921) and later, in more detail, by Camras (1954). Sodium hydroxide solution and ferrous sulphate solution are agitated so that a fresh surface is continually exposed to the atmosphere and α-(FeO)OH is precipitated in colloidal form:

$$4FeSO_4 \cdot 7H_2O + O_2 + 8NaOH \rightarrow 4\alpha\text{-}(FeO)OH + 4H_2SO_4 + 30H_2O$$

The colloidal particles are next used as nuclei in the growth of larger crystals of α-(FeO)OH.

More ferrous sulphate is mixed with water and scrap iron and heated to 60 °C (140 °F) and then the seed material produced in the step described above is added and air bubbled through the mixture for about four hours. During this time α-(FeO)OH grows on the colloidal nuclei to produce light-yellow acicular crystals. The sulphuric acid formed in the process produces more ferrous sulphate by reacting with the iron. The crystals of α-(FeO)OH are then filtered, washed and dried and are seen in the electron microscope to be $0 \cdot 25$–$1 \cdot 5$ μm long and $0 \cdot 1$–$0 \cdot 3$ μm wide. Their composition and orthorhombic structure are essentially those of the mineral goethite (Section 11.1.4).

The yellow acicular crystals of α-(FeO)OH are dehydrated at 230–270 °C to give red acicular crystals of haematite:

$$2\alpha\text{-(FeO)OH} \rightarrow \alpha\text{-Fe}_2\text{O}_3 + \text{H}_2\text{O}$$

which are then reduced by heating in hydrogen at 300–400 °C to black acicular crystals of magnetite:

$$3\alpha\text{-Fe}_2\text{O}_3 + \text{H}_2 \rightarrow 2\text{Fe}_3\text{O}_4 + \text{H}_2\text{O}$$

although van Oosterhout (1965) believes that the iron oxide at this stage contains excess Fe_2O_3.

The last step is the reoxidation of the magnetite particles at about 250 °C to give the reddish-brown acicular particles of γ-Fe_2O_3. There is no appreciable change of shape or size of the particles in the series of steps represented by

Pigment → dehydrated oxide → magnetite → maghemite
$[\alpha\text{-(FeO)OH}]$ $[\alpha\text{-Fe}_2\text{O}_3]$ $[\text{Fe}_3\text{O}_4]$ $[\gamma\text{-Fe}_2\text{O}_3]$

That is, the transformations are pseudomorphic.

The process could be stopped after the preparation of magnetite particles. In principle these should show slightly better recording properties (high output, better resolution) than those of the more frequently used γ-Fe_2O_3. However, as we shall see later, magnetite particles are unpopular with tape manufacturers. By varying the processing steps it is possible to produce a (narrow) range of magnetic properties so that, to some extent, the particles can be tailored to fit the needs of a particular recording application. For example, Camras (1954) reported that if particles of γ-Fe_2O_3 were re-reduced to Fe_3O_4 and then reoxidized at about 200 °C, a powder having a coercivity of about 400 Oe (cf. the original value of 260 Oe) was produced. Cyclical oxidation-reduction procedures of this kind are believed to be used by manufacturers of oxides for recording. The exact processing steps are, of course, carefully protected industrial secrets.

12.3.1.2 γ-Fe_2O_3 particles by direct precipitation

The process described above, although the most common, does seem rather circuitous. A more direct method of preparing γ-Fe_2O_3 particles by precipitation was given by Baudisch (1933). In his first patent he describes how *gamma* ferric oxide hydrate (γ-(FeO)OH = lepidocrocite) can be made by dissolving chemically pure iron (from iron carbonyl) in hydrochloric acid (5–35%) and then adding an excess of pyridine or aniline (fairly strong bases) to the solution. Air is then bubbled through the solution, causing the precipitation of γ-(FeO)OH. This hydrated oxide is not ferrimagnetic but can be made so by gently heating to 150–280 °C, a fact known to von Kobell in 1838. Just as α-(FeO)OH dehydrates to give α-Fe_2O_3, so γ-$(FeO)OH$ gives γ-Fe_2O_3. There are no details available of the shape and size of the particles, but it is believed that this process leads to the production of non-acicular particles. The magnetic properties of Baudisch's particles unfortunately were not reported. More recently Bratescu and Vitan (1969) used an aqueous solution of ferrous sulphate, a mixture of hydrochloric and nitric acid as an oxidant, and ammonia as the precipitant. A gelatinous precipitate formed, darkened and became black and micro-crystalline. The reaction ended with the appearance of copious amounts of black foam. The black precipitate was settled by the addition of water, filtered, dried at room temperature, and then heated to 60 °C. A yellow-brown powder of γ-Fe_2O_3 was obtained which, it is claimed, was stable and of high magnetic susceptibility.

However, Sykora (1967) reported unsuccessful attempts to use direct processes to make particles of γ-Fe_2O_3 that were useful for magnetic recording. For example, he tried to dehydrate synthetic lepidocrocite (γ-$(FeO)OH \to (\gamma$-$Fe_2O_3)$) but always found the product to be significantly inhomogeneous. Furthermore, the shape anisotropy was not very pronounced and the particles showed a wide distribution of lengths. Sykora concluded that the only practical way to make oxides for recording purposes was the conventional one:

$$\begin{array}{cccc} \text{dehydration} & \text{reduction} & \text{careful oxidation} \\ \alpha\text{-}(FeO)OH \to & \alpha\text{-}Fe_2O_3 \to & Fe_3O_4 \to & \gamma\text{-}Fe_2O_3 \end{array}$$

12.3.1.3 Modified acicular particles

In a modification of the common method for preparing acicular particles of magnetite or of γ-Fe_2O_3 (Fukuda et al., 1962) a sheath of magnetite was grown onto the colloidal goethite, prepared as described by Penniman and Zoph (1921) and by Camras (1954). To the colloid were added solutions of ferric chloride ($FeCl_2 \cdot 5H_2O$) and of ferrous sulphate ($FeSO_4 \cdot 7H_2O$) in a mixture whose proportions were chosen to give $Fe^{2+}:Fe^{3+} = 1:2$ appropriate to magnetite. The mixture was heated in a sealed vessel containing nitrogen at 50 °C, caustic soda added during a period of five hours while the mixture was being vigorously agitated; and the temperature raised to 80 °C and maintained for one hour.

The black precipitate consisted mainly of particles whose cores were the colloidal goethite nuclei. Around the cores had grown shells of magnetite in such a way that the particles were spindle-shaped, with average length 0·7 µm, and average width 0·15 µm. The precipitate also contained an admixture of smaller magnetite particles which disappeared or were converted to the spindle-shaped particles if the temperature was increased to 80 °C, and held there for five hours. At this stage the magnetite particles could be used to make recording surfaces or they could be transformed into γ-Fe_2O_3 by oxidizing in air at 200–250 °C.

The seed particles need not be colloidal iron oxide. Fukuda mentions that acicular particles of zinc oxide having colloidal magnetite adsorbed on the surface (or indeed any needle-shaped particles having a thin coating of iron oxide) can be used. However, these alternative materials lead to more irregular spindles than do the iron oxide particles. The magnetic properties of the spindle-shaped particles are listed in Table 12.1; an increase in coercivity and remanence and a considerable increase in the slope of the initial magnetization curve are claimed. These increases are expected to result principally in improved signal output for the recording surface.

12.3.1.4   Plate-like particles of iron oxides

A method of making plate-like particles (length:width about 10:1; width:thickness about 5:1) was described in two patents by Ayers and Stephens (1962). A seeding colloid was made from a solution of $FeCl_2$ heated to 27 °C; $FeSO_4$ may also be used. The solution was agitated and a dilute solution of sodium hydroxide added slowly. Air was bubbled continuously through the liquid for about one hour, during which time the colour changed from dark blue to green to brownish-yellow to yellowish-tan. At this point the air flow was reduced but continued for a further hour before the preparation of the seeding material was complete. Ammonium hydroxide, aniline or pyridine may also be used in the process instead of sodium hydroxide.

As in the method described above for acicular goethite particles, the next step is the growth of the particle around the colloidal nuclei. The processing difference that Ayers and Stephens introduced at this point was the addition of zinc chloride to the mixture of ferrous chloride or sulphate, scrap iron, and seed material. As before, the mixture was oxidized in a stream of air at 60 °C for 24–48 hours and the particles were filtered, washed and dried. Rather surprisingly, the method produced flat, plate-like crystals of *gamma* ferric oxide monohydrate (lepidocrocite) rather than the acicular *alpha* ferric oxide monohydrate (goethite) which results from the Penniman–Zoph/Camras (P–Z/C) approach. The crystalline particles were quite uniform in size (about 1 μm long) and transparent under an optical microscope, while the acicular goethite particles produced by the P–Z/C method were opaque. The important difference between the two methods was the use by Ayers and Stephens of the zinc chloride, which they claimed acted as an inhibiting agent preventing the growth of the acicular goethite particles, zinc sulphate, sodium chloride and ammonium chloride having the same desirable effect. The plate-like particles were then reduced at 300–500 °C in hydrogen to form magnetite particles of the same size and shape. These particles could be made into recording surfaces claimed to show higher output with increasing audio frequency than those surfaces made from acicular magnetite. Alternatively the magnetite particles could be reoxidized to γ-Fe_2O_3 by being passed through a kiln into which air was pumped. The entrance end of the kiln was kept at 175–230 °C, and the discharge end at 230–285 °C. The transformations γ-$(FeO)OH \rightarrow Fe_3O_4 \rightarrow \gamma$-$Fe_2O_3$ were isomorphic as they were with the acicular particles.

Superior recording at high audio frequency was also claimed for the plate-like particles of γ-Fe_2O_3. The magnetic properties of these and the related magnetite particles are given in Table 12.1. The most likely reason for this superior

Table 12.1 Magnetic properties of materials (both H_c and M_r/M_s depend on the degree of dispersion and the alignment of the sample)

Oxide	σ_s (e.m.u./g)	θ (°C)	K_1 (ergs/cm³)	ρ(g/cm³) Bulk	ρ(g/cm³) Particles	H_c (Oe)	M_r/M_s	Shape of particles	Size of particles (µm)	Structure of particles
γ-Fe₂O₃	74	590	-4.64×10^4	5.07	4.60	75-150 250-365 325-375	0.46 0.80 0.75	Equiaxed Acicular Platelets	0.05-0.3 $l/w = 7$ $l = 0.2$-0.7 $l/w \simeq 10$ $w/t \simeq 4$	Inverse spinel with vacancies on a tetragonal superlattice; $a = 8.33$ Å, $c/a = 3$
Fe₃O₄	92 bulk 84 particles	575	-1.1×10^5	5.197	4.9-5.1	$\simeq 300$ 305-335 350-450	0.52 0.70 0.70	Equiaxed (natural) Acicular Platelets	$\simeq 1$ $l/w = 7$ $l = 0.2$-0.7 $l/w \simeq 10$ $w/t \simeq 4$ $l \simeq 1$	Inverse spinel; $a = 8.3963$ Å
δ-(FeO)OH	25 22			3.95		500 55		Hexagonal Platelets	$\simeq 0.6$ $\simeq 0.007$	Hexagonal; $a = 2.941 \pm 0.005$ Å; $c = 4.49 \pm 0.05$ Å
γ-Co$_x$Fe$_{2-x}$O₃			$\simeq +1 \times 10^6$		4.67	344-885	0.47-0.61	Acicular	$l/w = 5$ $l = 0.2$	Inverse spinel
$x = 0.04$ $x = 0.06$	50 44					400 515-600	0.70 0.70	Cubic Cubic	0.05-0.08 0.05-0.08	

Co$_x$Fe$_{3-x}$O₄
  $x = 0.04$
  $x = 1.0$
	80
	65-70
		570
		520
			$+3.0 \times 10^5$
			$+6.0 \times 10^5$
				5.185
				5.32
						600-800
						750-980
						$\simeq 1200$
							0.1
							0.65
							0.5
								Cubic
								Cubic
								Cubic
									$\simeq 0.2$
									0.004-0.016
									0.04-0.05
										Inverse spinel;
										$a = 8.395 \pm 0.005$ Å

CuFe₂O₄	25	455	-6.0×10^4	5.35		390		Cubic		Tetragonal; $a = 8.22$ Å; $c = 8.70$ Å
NiFe₂O₄ BaO·6Fe₂O₃	50-55 72 bulk 68 particles	585 450	-6.2×10^4 $+3.3 \times 10^6$	5.38 5.28		100-150 500-2000 5350	0.15-0.35 0.50	Cubic Platelets	0.013-0.050 $\leqq 1$ $d = 0.08$-0.150 $d/t \simeq 15$	Inverse spinel; $a = 8.34$ Å; Hexagonal; $a = 5.88$ Å; $c = 23.2$ Å
CrO₂ CrO₂ + Sb	100 bulk 89-92 particles 77-85	113.5 117 117	$+2.5 \times 10^5$	4.83	4.88-4.95 4.88-4.95	50-150 150-600	0.5-0.9	Equiaxed Acicular	$l \simeq 1$ $l = 0.2$-1.5 $w = 0.03$-0.1	Tetragonal; $a = 4.4218$ Å; $c = 2.9182$ Å

performance is that the values of coercivity are rather higher than those of the acicular particles.

12.3.1.4.1. Square plate-like particles of iron oxides. A method for making square, plate-like particles of γ-Fe$_2$O$_3$ was described by Goto and Akashi (1962). To an aqueous solution of ferrous sulphate was added sodium hydroxide in the presence of nitrogen. The precipitated ferrous hydroxide thus produced was heated in the mother liquor at 80 °C for five hours. This produced flattened, substantially square particles whose size was, on average, 1 μm × 1 μm × 0·1 μm. The shape was ascribed to the layered structure of Fe^{2+} and OH$^-$ ions in which the bond between the OH$^-$ layers is very weak.

Goto and Akashi described two methods by which the square platelets could be converted to γ-Fe$_2$O$_3$. In the first method they were heated in a closed vessel at 300 °C. The second method involved the usual series of processing steps: Fe(OH)$_2$ → α-Fe$_2$O$_3$ → Fe$_3$O$_4$ → γ-Fe$_2$O$_3$. Once again the transformations are isomorphic.

12.3.1.5 Cubic particles of iron oxide

Historically these are important as the first iron oxide particles to be used for magnetic recording, but their coercivity is considered far too low (75–150 Oe) for today's recording applications. They appear in the electron microscope as cubes with rounded corners or as rather deformed spheres. The method of preparation of the cubic particles should be described since, with slight modifications, it can be used to produce cubic particles in which cobalt ions replace some 2 to 10% of the iron ions and, in doing so, increase the coercivity of the particles by as much as an order of magnitude. These cobalt-substituted iron oxides will be considered later.

According to Krones (1955) an aqueous solution of ferrous sulphate was mixed with sodium or ammonium hydroxide, and the fine whitish-grey precipitate of ferrous hydroxide formed:

$$FeSO_4 + 2NaOH \rightarrow Fe(OH)_2 + Na_2SO_4$$

was then oxidized to magnetite at 70–90 °C with sodium nitrate, chosen because its oxidizing potential was insufficient to form α-Fe$_2$O$_3$:

$$3Fe(OH)_2 + \tfrac{1}{2}O_2 \xrightarrow{\ NaNO_3\ } Fe_3O_4 + 3H_2O$$

The magnetite particles were washed with deionized water, filtered, dried and then oxidized in air at 250–300 °C to give cubic particles of γ-Fe$_2$O$_3$.

It is possible to exercise some control over the size of the particles (from 0·05 to 0·3 μm) by varying the conditions of precipitation, such as concentration, pH and temperature.

Although these particles appear equiaxed in the electron microscope, the length : width ratio is probably greater than the value of 1·1 : 1 at which Osmond (1954) deduced that magnetocrystalline anisotropy and shape anisotropy were equal in γ-Fe$_2$O$_3$.

12.3.1.6 Reactions

12.3.1.6.1 Precipitation. The reactions described above allow a wide variety of shapes and sizes of iron oxide particles to be prepared, yet the methods show an almost bewildering similarity. For example, ferrous sulphate and sodium hydroxide reacted to give goethite in the P–Z/C process but gave ferrous hydroxide in the Krones method for cubic particles. Using apparently the same colloidal seeds and adding scrap iron, ferrous sulphate and air, the P–Z/C method grew particles of α-(FeO)OH, yet the Ayers and Stephens method yielded γ-(FeO)OH. Theses examples illustrate the importance of the detailed conditions of the reactions. In the first example the mixture (in the P–Z/C and the Ayers and Stephens methods) was strongly agitated to bring the reagents in frequent contact with the air. Thus the seed colloid formed was in a higher state of oxidation than the ferrous hydroxide precipitated in the Krones method. Moreover, the ferrous iron Fe^{2+} is very unstable in the presence of air and is usually transformed into a mixture of Fe^{2+} and Fe^{3+}. If ferrous ions are deliberately exposed to oxygen, the result is almost entirely ferric ions; this tendency is utilized in one form of radiation dosimeter.

In the second example, Ayers and Stephens added zinc chloride (alternatively zinc sulphate, sodium chloride or ammonium chloride) to the precipitation bath to inhibit the growth of acicular goethite particles and to encourage the growth of flat platelets. The mechanism of the action of the inhibitor was not specified. Similarly with other patented processes for preparing magnetic recording materials, we find recommendations that the baths be treated with this or that additive in order to enhance the properties of the particles; usually there is little or no attempt to explain the *modus operandi* of the additive. For example, Baronius *et al.* (1966) suggested treating the α-(FeO)OH with sodium silicate, zirconium hydroxonium chloride or, in general, Al^{3+}, Ti^{4+}, Zr^{4+} salts with the pH adjusted to be near incipient ionic hydrolytic polymerization. Apart from the addition of these ions, which presumably become a part of the final material, the oxidation and reduction procedures were conventional.

12.3.1.6.2 Oxidation-reduction. The reoxidation of the magnetite particles to γ-Fe_2O_3 must be performed carefully. Too high a temperature results in the formation of haematite (α-Fe_2O_3), which is (with $\sigma_s = 0.6$ e.m.u./g) useless for magnetic recording. Camras (1954) advised that the step be carried out at the lowest temperatures consistent with an economical processing time, usually from 200 °C to 250 °C. Higher temperatures progressively reduce the magnetic qualities; for example, a sample oxidized at 380 °C has only half the remanence of one oxidized at 285 °C, and at 665 °C the remanence has disappeared. More recently Klimaszewski and Pietrzak (1969) studied the transformation of Fe_3O_4 produced by heating in air at temperatures in the range 100–1300 °C. They used ferromagnetic resonance and plotted the linewidth and the resonance field against the treatment temperature. They found that the transformation of $Fe_3O_4 \rightarrow \gamma$-$Fe_2O_3$ begins at 100 °C and is complete at 250 °C, while the transformation of $Fe_3O_4 \rightarrow \alpha$-$Fe_2O_3$ starts at 250 °C and is finished at 500 °C.

Imaoka (1968) used both acicular and non-acicular particles in his study of the transformations:

$$Fe_3O_4 \rightarrow \gamma\text{-}Fe_2O_3 \rightarrow \alpha\text{-}Fe_2O_3$$

Although the first step took place between 100 °C and 250 °C (in agreement with Klimaszewski and Pietrzak) for both types of particles, the second step occurred at 250–400 °C for the non-acicular particles but at 560–650 °C for the acicular ones. Bando *et al.* (1965) determined that the $\gamma \rightarrow \alpha$ transformation took place in the range 400–500 °C for non-acicular particles and 450–500 °C for acicular ones. The magnetic consequences of this transformation are still puzzling; Gustard and Schuele (1966) converted acicular $\gamma\text{-}Fe_2O_3$ particles partially to $\alpha\text{-}Fe_2O_3$ at temperatures of 525–650 °C and found that the remanence increased by a factor of three (compared with its initial value) when 90 % conversion had been achieved.

David and Welch (1956) showed that 'specimens of magnetite which gave gamma ferric oxide on oxidation invariably contained appreciable percentages of water, while specimens prepared under dry conditions oxidized with great difficulty, never yielding the gamma oxide'. Furthermore, they confirmed the finding of Verwey (1935) that the gamma oxide itself contained a small amount of water that could not be removed without changing the characteristic structure of the material and transforming it to $\alpha\text{-}Fe_2O_3$. Healey *et al.* (1956) investigated the surface of a ferric oxide powder by water adsorption and calorimetry and found that part of the surface chemisorbed water, which could be released only by heating at temperatures of about 450 °C.

In magnetite powders prepared by the methods described in the preceding sections, it is more than likely that the particles contain water either from the solutions in which they were formed or as adsorbed water. However, natural magnetite in the form of bulk samples does not contain water. Elder (1965) showed that, while natural magnetite crystals ground in water to a particle size of less than 1 μm were converted to $\gamma\text{-}Fe_2O_3$ at 250 °C (more than 75 % conversion), larger particles, 25 μm or more, were converted only to $\alpha\text{-}Fe_2O_3$ under the same conditions. The effect of the small particle size Elder attributed to the stabilizing action of water adsorbed or hydrated on the fine-particle surface. As an additional demonstration of the effect, he showed that the smaller particles ground in dry acetone and heated in dry oxygen produced only $\alpha\text{-}Fe_2O_3$, whereas in an atmosphere of water vapour and oxygen they produced pure $\gamma\text{-}Fe_2O_3$. Clearly both small particle size and the presence of water are necessary for the formation of $\gamma\text{-}Fe_2O_3$. Figure 12.6 shows the result of Elder's measurements of M_s for oxidized Fe_3O_4 powders. The abscissa gives the divalent ion content $[Fe^{2+}]$, which was determined as FeO. On this scale magnetite ($FeO \cdot Fe_2O_3$) is represented by an FeO content of 31·3 weight per cent. The simple conversions to the metastable $\gamma\text{-}Fe_2O_3$ or the stable $\alpha\text{-}Fe_2O_3$ are shown by broken lines. The experimental points marked with circles refer to small particles (less than 1 μm) ground in water; the numbers beside the points

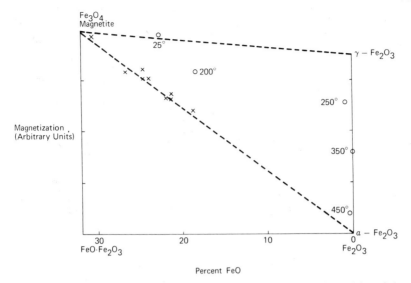

Figure 12.6 Oxidation of Fe_3O_4 to Fe_2O_3. The magnetization intensities of the oxidized powders are plotted as a function of the divalent ion $[Fe^{2+}]$ content. The temperatures marked are those at which the oxidation was carried out (after Elder, 1965).

indicate the temperatures at which oxidation (in flowing air for one hour) was carried out. In agreement with the later results of Klimaszewski and Pietrzak, oxidation was incomplete at temperatures below 250 °C, whereas at temperatures above 450 °C it was complete but α-Fe_2O_3 was formed. At about 250 °C oxidation was substantially complete, but magnetic measurements show that γ-Fe_2O_3 was formed. The experimental points marked 'x' correspond to the larger particles (25 µm or greater), which, on oxidation, formed only α-Fe_2O_3.

Colombo *et al.* (1964b) investigated the mechanism of oxidation of magnetites (by thermogravimetric, X-ray diffraction and differential thermal analyses) and concluded that it is a two-stage process. In the first stage a solid solution of γ-Fe_2O_3 in Fe_3O_4 always forms. In more detail, oxygen is adsorbed and ionized with the electrons supplied by the oxidation of Fe^{2+} to Fe^{3+}. This is followed by a diffusion of ferrous ions from the inside of the crystals of magnetite toward the surface, leading to the formation of a solid solution of γ-Fe_2O_3 in Fe_3O_4. Since the precipitated particles have a defect structure compared with natural magnetite, the diffusion rate of iron ions is high and oxidation proceeds rapidly until the conversion to γ-Fe_2O_3 is complete. These authors do not mention the important part played by water in the oxidation reactions of magnetite.

The more important iron oxide reactions are summarized in simplified form in Figure 12.7, and the physical characteristics of the products are given in Table 12.1.

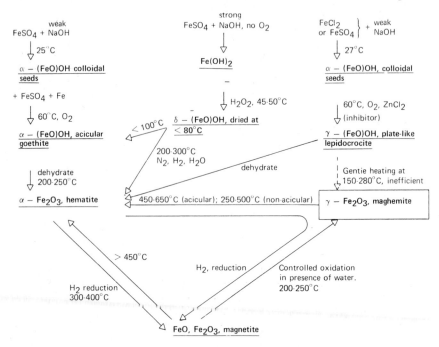

Figure 12.7 Iron oxide reactions; only those reactions are shown which are relevant to the preparation of particles for recording surfaces.

12.3.2 Structure and intrinsic properties of iron oxides

The crystal structures and intrinsic properties of iron oxides have been described in the preceding chapter and in Chapter 2.

12.3.3 Extrinsic magnetic properties of Fe_3O_4 and $\gamma\text{-}Fe_2O_3$

Some of the curious properties of naturally magnetized magnetite appear to have been known to the Chinese, perhaps as long ago as 1000 BC. Then in the Han Dynasty (206 BC to AD 220) the smooth stones which rotated freely on a polished board were used to discover the most suitable site for a building or a tomb. The earliest use of the lodestone as a compass for navigation is not known, but the idea seems to have originated in the Mediterranean in the eleventh century. Particularly strong lodestones were greatly treasured and some fine specimens were mounted in silver in Elizabethan times. With the advent of electromagnetism in the nineteenth century, the magnetic properties of iron oxides became principally of scientific rather than technological interest. It was not until thirty or forty years after the invention of magnetic recording that these materials, now used in synthetic form, were once more of great practical interest.

12.3.3.1 Extrinsic properties

The theory of the magnetic properties of fine particles in general has been the subject of comprehensive reviews by Wohlfarth (1959, 1963) and was outlined in Section 1.5. We shall consider here how these theories apply to the single-domain particles used in recording surfaces. Since the extrinsic properties of the particles depend most markedly on the shape and size of the particle, it is appropriate to consider this question first.

The particles prepared for use in recording surfaces are always intended to be single domains. The size at which the behavioural change between single-domain and multi-domain occurs is not at all well defined. Morrish and Yu (1955) calculated it for particles of Fe_3O_4 and γ-Fe_2O_3 by comparing the energy of the single-domain particle with the energy of multi-domain configurations which reduce the external flux and hence, that contribution to the magnetostatic energy. To simplify the model, only the terms for exchange energy and magneto-static energy were considered for the single-domain particles; magnetocrystalline energy was neglected. They found that particle shape was important also; the critical size for single-domain behaviour was larger for acicular particles than for spherical particles. Stoner and Wohlfarth (1948) had earlier reached the same conclusion from their calculations on particles of iron and nickel. When the particles were spherical, the critical dimension was 0·011 μm and 0·024 μm; but when they were elongated, with length:diameter = 10:1, the respective values for the lengths were 0·047 μm and 0·098 μm. Morrish and Yu (1955) also predicted that the critical size for particles of γ-Fe_2O_3 should be larger than that for Fe_3O_4 particles. Watt and Morrish (1960) checked this by reducing some large but still single-domain particles of γ-Fe_2O_3 to Fe_3O_4. The resultant particles tended to show multi-domain behaviour. An additional complication was introduced by Kondorskii (1952), who showed theoretically that multi-domain particles near the critical size might show reversal by mag-netization rotation, like single-domains, when compacted. Some experimental evidence that this happens was produced by Morrish and Watt (1957). Craik and McIntyre (1967) showed that even though a particle may be smaller than the critical size for single-domain behaviour, in the presence of applied fields a multi-domain state may have lower energy than the single-domain one.

At the other end of the single-domain range particles smaller than a certain critical size show superparamagnetic behaviour (Néel, 1949; Bean and Living-stone, 1959; see Sections 1.5 and 11.4.1). The colloidal particles of ferric oxide which are used to develop domain patterns (or recorded patterns on tape) and so make them visible are examples of superparamagnetic particles. Their diameter ranges from 20–200 Å—that is, below the limiting single-domain size of approximately 300 Å for ferric oxide (Berkowitz *et al.*, 1968). Since the magnetization direction in superparamagnetic particles is constantly changing, the particles are useless for recording. McNab *et al.* (1968) showed by the Mössbauer effect that particles of magnetite ranging in size from 100 to 160 Å were all superparamagnetic.

The range over which γ-Fe_2O_3 and Fe_3O_4 acicular particles show single-domain behaviour—that is, the useful range for recording—is roughly from 0·3 μm to 2 μm (particle length). In practice, of course, the particles show a distribution in both size and shape. The aim in making particles for recording surfaces should be to make the distribution as narrow as possible. Clearly any admixture of superparamagnetic or multi-domain particles must be avoided; but this is still not good enough. Even if all the particles in the sample were single-domains, the dependence of the switching field on particle size is suffi-ciently pronounced in the iron oxides (as in other hard magnetic particles) that the coercivity could range from less than 100 Oe to over 2000 Oe. Particles at the low end of the range would be very susceptible to disturbing fields and, therefore, to partial erasure or print-through; those at the upper end may not be switched by the field from the recording head and would therefore exercise an unwanted bias effect on the recording properties. In general, a wide range of magnetic properties within a sample produces a sheared hysteresis loop—particularly undesirable for digital storage.

There has been a considerable difference of opinion as to the morphology of γ-Fe_2O_3 particles. Osmond (1952) noted some evidence that the particles are composed of smaller crystallites intimately held together. Osmond (1953) then compared the transformation α-$(FeO)OH \rightarrow \alpha$-$Fe_2O_3$ with Ervin's (1952) study of the dehydration of diaspore [α-$(AlO)OH$] to give corundum [α-Al_2O_3], and concluded that the c axis of the hexagonal haematite becomes the [111] of magnetite and maghaemite (γ-Fe_2O_3). Even though a particle may be composed of several crystallites, Osmond extends Ervin's results to conclude that in all the crystallites the [111] axes would still be parallel to one another and 'most probably parallel to the long axis of the particle'. Campbell (1957), using the technique of selected-area electron diffraction, examined a large number of γ-Fe_2O_3 particles and concluded that although each particle had its long axis one of the low-order directions $\langle 111 \rangle$, $\langle 211 \rangle$, $\langle 221 \rangle$, etc., no one set of directions was preferred in any sample. Selected-area electron diffraction was also used by van Oosterhout (1960) to study the orientation of the long axis of acicular particles of α-$(FeO)OH$, γ-$(FeO)OH$, and γ-Fe_2O_3. He found that for the two oxyhydroxides, the orientation was [001], while for γ-Fe_2O_3 it was [110]. Van Oosterhout derives support for his conclusions from the work of Ervin (1952), which he interpreted rather differently than did Osmond (1953). He further asserts that Campbell's (1957) results do not rule out a $\langle 110 \rangle$ orientation.

Hurt et al. (1966) examined a number of samples of commercial gamma ferric oxide and found that the majority of particles were polycrystalline with a [110] long axis. Using dark-field electron microscopy, Gustard and Vriend (1969) compared the number of particles which were seen under bright field conditions to the number which remained visible in dark field and concluded that 40% (by volume) of the γ-Fe_2O_3 particles had [110] as their long axis.

12.3.3.1.1 Coercivity. The most commonly used particles in magnetic recording surfaces have, as we have seen, a pronounced acicularity. For predominating shape anisotropy we should have (Section 1.5.1) $H_c = (N_b - N)M_s$ with a maximum of $2\pi M_s$ for long cylinders (H parallel to long axis). Assuming ellipsoidal geometry (a reasonable first approximation) and substituting $M_s = 400$ e.m.u./cm^3 for γ-Fe$_2$O$_3$, to give 2500 Oe, greater by almost an order of magnitude than the experimental values, which usually range from 260 Oe to 320 Oe. Changing the orientation of the major axis away from the direction of the applied field reduces the coercivity monotonically, but the reduction ($\times 0.479$) for random orientation is still inadequate. More commonly encountered is a more or less random assembly of rather dissimilar particles; this led to a number of early attempts to reconcile the measured values of H_c with the distribution of particle shapes observed under the electron microscope.

The axial ratio of γ-Fe$_2$O$_3$ particles was found by Johnson and Brown (1958) to be at least 5; the value estimated from magnetic measurements followed by an 'unaveraging' process was at most 1.6. That is, the particles behaved magnetically as though they were less acicular than they appeared to be in the electron microscope. Osmond (1954) assumed a gaussian distribution of particles around a shape factor of 1.3 and derived $H_c = 205$, which he compared with his measured value of 210. However, Osmond claimed that the shape factor obtained by electron microscopy was also 1.3, which would make his particles not at all typical of the γ-Fe$_2$O$_3$ particles used in tapes or discs. Johnson and Brown decided that interactions between particles (the usual first assumption in this field, when theory and experiment do not agree) could not account for the lack of agreement between their experimental results and the Stoner–Wohlfarth theory. An alternative to this theory had been proposed by Jacobs and Bean (1955) and others, to account for the fact that the coherent process used in the Stoner–Wohlfarth theory almost invariably overestimated H_c. In the coherent process it is assumed that the atomic magnetic moments throughout the single-domain particle remain parallel during the switching—that is, they all rotate together. Jacobs and Bean showed that magnetization could be reversed with less expenditure of energy if the moments did not remain parallel during the reversal. They postulated a chain-of-spheres model in which the magnetization on adjacent spheres 'fanned' in opposite directions along the chain, much as a row of compass needles will move when the first one is turned. Wohlfarth (1959), using this model, calculated that the maximum value of the axial ratio, M, was 1.4. This was less than Johnson and Brown's value of 1.6, obtained from magnetic measurements, but much closer to it than to the value of 5 given by the electron microscope.

In another mode of incoherent reversal, the 'curling' mode, the rotation of the spins resembles the movement in a bundle of straws held at each end and twisted. In contrast with the simple coherent process, the incoherent modes show little tendency for magnetic poles to appear on the surfaces, and consequently the processes are 'softer'; that is, they occur at lower values of applied field.

If, in an aligned array of particles which rotate incoherently (say, by fanning), the angle between the applied field and the alignment direction is increased from zero, the switching field will at first increase with angle. This can be seen qualitatively by remembering the analogy to a chain of compass needles and considering the increased angles through which alternate needles must now rotate to reverse the magnetization. By contrast, on the coherent model the coercivity decreases monotonically with increasing angle. Bate (1961) measured the angular dependence of coercivity on tapes composed of partially oriented acicular particles of γ-Fe$_2$O$_3$, and found the dependence shown in Figure 12.8.

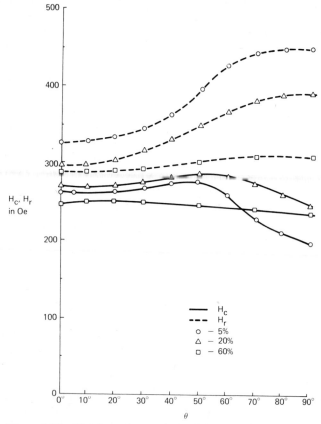

Figure 12.8 The dependence of coercivity, H_c and remanence coercivity, H_r of partially-oriented assemblies of acicular γ-Fe$_2$O$_3$ particles on the angle θ between the direction of orientation and the direction of measurement. The remanence coercivity is the field required in the reverse direction, to reduce the remanence of the sample to zero (after Bate, 1961).

This provided further evidence that these particles undergo magnetization reversal by an incoherent rotation.

In the comparison between theory and experiment of the coercivity shown by assemblies of γ-Fe$_2$O$_3$ particles, we have thus far considered only the contribution of the shape of the particles to the magnetic anisotropy. The possible role of magnetocrystalline anisotropy must now be considered. In spherical single-domain particles only magnetocrystalline anisotropy is effective, i.e. we ignore the possibility of magnetoelastic anisotropy in comparison. The expected coercivity is then given by $H_c = a|K_1|/M_s$. Here a lies between 0·64 and 2·00, the smaller value applying to a random assembly of particles and the larger to an assembly in which the magnetically easy axes are fully aligned. On substituting $a = 0·64$, $|K_1| = 4·7 \times 10^4$ (for γ-Fe_2O_3) and $M_s = 400$ e.m.u./cm^3, we find that $H_c = 75$ Oe. This is considerably smaller than the values of 260–320 Oe usually found for assemblies of γ-Fe_2O_3 particles, but it falls within the range of 50 to 150 Oe reported by Westmijze (1953) for the early equiaxed particles. In any case, the value of 75 Oe is not so low as to enable us to rule out immediately the possibility that magnetocrystalline anisotropy might play some part in determining the coercivity of these particles.

We should not expect shape-controlled coercivity to be particularly temperature-dependent. M_s increases only slowly with decreasing temperature (probably by 10% between room temperature and absolute zero) and $(N_b - N_a)$ is constant. On the other hand, K is usually a highly temperature-sensitive quantity and hence particles deriving their magnetic hardness from this source should show a pronounced temperature dependence of H_c. This is illustrated in Figure 12.9, in which are plotted the experimental curves of H_c versus T for a sample of conventional acicular particles of γ-Fe_2O_3 and for two samples of equiaxed particles of cobalt-substituted γ-Fe_2O_3 (in which predominantly magnetocrystalline anisotropy would be expected). Johnson and Brown (1958) found that the particle shape distribution they calculated was (not surprisingly) independent of temperature in the range 77 K to 293 K. From this they concluded

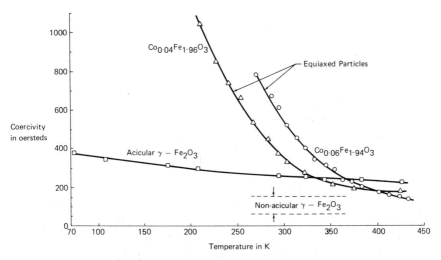

Figure 12.9 Coercivity as a function of temperature for acicular particles of γ-Fe_2O_3 and equiaxed particles of cobalt-substituted γ-Fe_2O_3. Electron micrographs of these particles are shown in Figure 12.5 (after Speliotis et al., 1965).

that the measured coercivity of their samples of γ-Fe_2O_3 particles was controlled by shape and was not affected to any great extent by magnetocrystalline anisotropy or by superparamagnetic effects.

In magnetite the easy axes are $\langle 111 \rangle$. If this were also true for γ-Fe_2O_3, for which the long particle axis is apparently [110], we should have a situation in which the two anisotropies, shape and magnetocrystalline, compete to determine the direction that the magnetization would take with respect to the long axis of the particle. If Takei and Chiba (1966) are correct, however, the easy axis of γ-Fe_2O_3 is [110] and so there would be no competition.

The question of the relative contribution of the two sources of anisotropy in γ-Fe_2O_3 particles seems to have been answered unambiguously by Eagle and Mallinson (1967). They took advantage of the fact that at 130 K $K_1 = 0$ (Bickford et al., 1957) and thus at this temperature H_c is determined by shape alone. The coercivity and saturation moment of two samples of commercial acicular particles of γ-Fe_2O_3 were measured over the temperature range from $-180\,°C$ to $+300\,°C$. The average particle length was 0·6 μm in one sample and 0·2 μm in the other; both had the same value of 5:1 for the average axial ratio. The 0·6 μm sample was then reduced to Fe_3O_4 and the coercivity measured at 130 K. From this measurement the factor of proportionality ('shape factor') between coercivity and intensity of magnetization could be determined. The shape factor will hold equally at all other temperatures and furthermore will be the same for particles of γ-Fe_2O_3 as for Fe_3O_4 if the mechanism of magnetization reversal is one of coherent rotation. For incoherent modes the shape factors will be slightly different, reflecting the differences in M_s between Fe_3O_4 and γ-Fe_2O_3. Eagle and Mallinson found that approximately 67 % of the coercivity at room temperature could be accounted for by shape anisotropy, and the remaining 33 % by magnetocrystalline anisotropy. Furthermore, the shape term was almost independent of particle diameter and agreed well with the chain-of-spheres model.

Imaoka (1968) used magnetic and X-ray diffraction measurements to study the transformations γ-$Fe_2O_3 \rightarrow \alpha$-$Fe_2O_3$ and α-$Fe_2O_3 \rightarrow Fe_3O_4$. In the first case he found that the coercivity of the sample increased steadily (from 260 Oe to 310 Oe), while M_s decreased from about 100 e.m.u./g to near zero. The 'squareness' ratio M_r/M_s remained throughout at 0·4 to 0·5. In the reduction reaction H_c increased from 370 Oe to 390 Oe and then decreased to about 240 Oe as the magnetization increased. Imaoka interpreted his results as due to the heterogeneous reactions which caused each particle to contain many crystallites (Bando et al., 1965, gives some electron microscope evidence for this) and supposed that the nucleation and growth of the crystallites within each particle caused the magnetic behaviour to be progressively that of

superparamagnetism $\rightarrow$ single-domain $\rightarrow$ multi-domain

Effect of particle volume on coercivity. In comparing the theoretical coercivity for an assembly of single-domain particles of γ-Fe_2O_3 with the experimental results, we assumed that each particle had an M_s equal to the bulk value of 400 e.m.u./cm^3.

None of the formulae for H_c, for coherent rotation (Section 1.5.1) include particle size as a variable, and the conclusion is that, as long as the particle size lies within the bounds of single-domain behaviour, the coercivity does not depend on it. However, experimental results on a wide range of powders of both metals (Kneller and Luborsky, 1963) and oxides (Okamoto et al., 1966; Eagle and Mallinson, 1967; Berkowitz et al., 1968) invariably show a pronounced variation of coercivity on the average particle size of the sample.

The more sophisticated models of incoherent reversal (Brown, 1959; Shtrikman and Treves, 1959, 1963), consider other reversal processes, such as 'buckling' and 'curling'. Curling has already been described; buckling is a modification of the chain-of-spheres 'fanning' mode of Jacobs and Bean (1955), in which the magnetization angle varies more gradually with respect to the applied field direction. The incoherent modes predict that the coercivity of single-domain particles should increase as the particle radius decreases. This is in qualitative agreement with the experimental results. For example, the measurements of Berkowitz et al. (1968) show a rather indistinct upward trend of 250 Oe at 700 Å (crystallite size) to 330 Oe at 400 Å, below which the coercivity decreases since the particles are now superparamagnetic. The upward trend of coercivity is shown more clearly in the results of Eagle and Mallinson (1967) on equiaxed particles of cobalt-doped γ-Fe$_2$O$_3$ (Figure 12.10). The theoretical prediction of the size dependence of coercivity for the coherent rotation mode, the curling mode, and the buckling mode is shown in Figure 12.11.

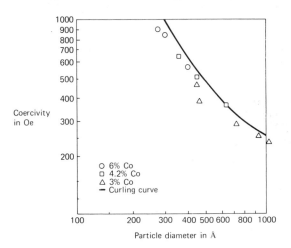

Figure 12.10 Coercivity as a function of particle diameter for equiaxed particles of cobalt-substituted γ-Fe$_2$O$_3$. The line was calculated on the assumption that magnetization reversal occurred by the curling process (after Eagle and Mallinson, 1967).

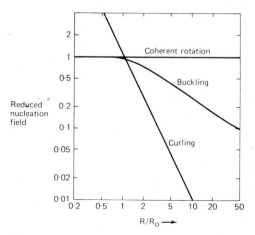

Figure 12.11 Calculated results of the field needed to begin magnetization reversal in an infinite cylinder as a function of its radius R, where R_0 is a characteristic radius (after Shtrikman and Treves, 1963).

Effects of particle interactions. In the theories of single-domain behaviour discussed above, the particles are treated, principally for analytic simplicity, as being so far apart that they do not interact with one another. In practice this is far from being the case. Figure 12.12 is a photomicrograph of a partially oriented assembly of γ-Fe_2O_3 particles coated at a very low loading (5% by volume, compared with 30% to 40% for commercial tape coatings). It is clear that considerable chaining of the particles has occurred during the orienting process and that we cannot expect such an assembly to behave like a collection of non-interacting particles. Even when great care is taken to prepare samples of dry powders for electron microscopy, it is relatively rare to see a single isolated particle. Figure 12.13 is an electron micrograph of a particle treated with colloidal iron oxide that had been prepared according to the recipe of Craik and Griffiths (1958). Either the particle is lying across a smaller particle, or it has dendritic growths. The colloidal particles are attracted to the regions of maximum field strength—i.e. to the poles. It was found that the colloidal distribution was the same whether the sample had been a.c. demagnetized or not, a result which further confirms the single-domain nature of the γ-Fe_2O_3 particles used in recording surfaces.

The theoretical problem of taking quantitative account of particle interactions has been described by Wohlfarth (1963) as 'an n-particle problem but with an unknown Hamiltonian'. As a first step it is reasonable to consider a two-particle model (Néel, 1958, 1959). There are then three remanent states for the pair, $+M_r, 0, -M_r$, and two critical fields, one to switch the particles into a state of antiparallel magnetization and a larger one to switch them parallel again so that both are now switched with respect to the original direction of magnetiza-

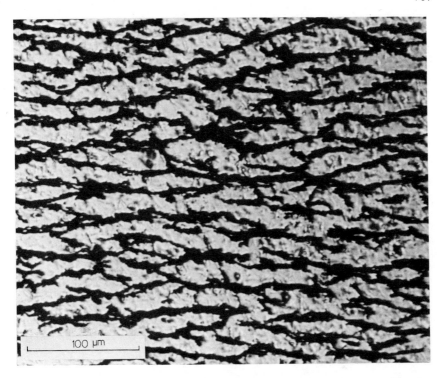

Figure 12.12 Photomicrograph of a partially-oriented assembly of acicular γ-Fe$_2$O$_3$ particles at a loading of approximately 5% by volume.

tion. Using this simple model, Shtrikman and Treves (1960) obtained improved agreement between experimental and theoretical remanence curves.

Brown (1962) criticized the 'local field' method of nucleation field calculations on the ground that it tests the stability of one particle while holding all the others constant. Bertram and Mallinson (1969) avoided this objection in their rigorous calculation of the nucleating field for a pair of interacting anisotropic dipoles whose axes were parallel though the particles were not necessarily side by side. In contrast to the independence of nucleating field and packing fraction found in the previous 'local field' models, they found a monotonically decreasing coercivity with increasing packing density.

Most attempts to consider larger numbers of interacting particles use the approach of Preisach (1935). The basic assumption here was that a particle, which was assumed to have a symmetrical hysteresis loop when isolated, exhibited a loop of the same size but translated parallel to the field axis when subjected to the field of neighbouring particles. The particle then behaves as though it had two different coercivities (H_+ and H_-), one for each of the two switching directions. The Preisach diagram is a three-dimensional histogram whose axes correspond to the switching fields in the positive and the negative directions and to the fraction of particles in the sample with a particular

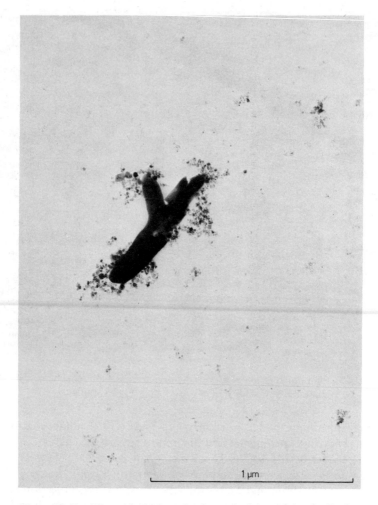

Figure 12.13 Electron-micrograph of an acicular particle of γ-Fe_2O_3 to which colloidal iron particles were applied.

combination of H_+ and H_-. Complications arise since, during the journey of the sample around the hysteresis loop, the interaction field experienced by any one particle is constant in neither magnitude nor direction. Thus it appeared that a different Preisach diagram might be needed for every point on the loop and the usefulness of the diagram would disappear. However, Bate (1962) was able to prove experimentally that the diagram is statistically stable over most of the remanent conditions covered by the hysteresis loop. This means that only one diagram is needed.

Della Torre (1965) measured the remanent magnetization of γ-Fe_2O_3 samples shaped as a sphere and as prolate and oblate ellipsoids of different

eccentricities. By choosing the proper shape it was possible to obtain a statistically stable distribution of interaction fields whose magnitude could be deduced
(approximately 150–2500 Oe). Once more the chain-of-spheres model seemed
to fit the results better than the Stoner–Wohlfarth model. Waring (1967)
observed that experiments on statistical stability do not allow one to distinguish
among the microscopic models. He proposed a model in which bundles of
particles were hypothesized. Within each bundle, interaction took the form of
a demagnetizing field, but between bundles the interaction was weak or absent.
Further support for this view was obtained by Waring and Bierstedt (1969)
from experiments on mixtures of CrO_2 and γ-Fe_2O_3 particles. The model led
to a stable Preisach diagram and also to a better agreement with the shape of
the anhysteretic magnetization curve determined experimentally. This curve
was analysed by Jaep (1969) by a method that took account of the time dependence of a particle switching; each particle was assigned a switching probability
which depended upon the applied field and its period. The analysis predicted
an initial susceptibility that, though large, was not infinite as in previous
calculations.

The possibility that single-domain particles in permanent magnet materials
may behave cooperatively in 'interaction domains' was discussed by Craik
and Lane (1969) on the basis of Bitter pattern observations. A similar behaviour
was found theoretically by Moskowitz and Della Torre (1967), who treated an
$n \times n \times n$ cube of aligned particles each of which was assigned a dipole moment
and a switching field. They found that coherent switching of a string of particles
could occur as a result of the sudden large change in the interaction field at a
particle when its neighbour reverses.

In the last decade a great deal of work, both theoretical and experimental,
has been devoted to unravelling the complex problems of particle interactions.
Papers in the first half of this period were reviewed by Wohlfarth (1964).
Because the task is so complex, we must expect little progress until new analytical
tools or experimental approaches are developed.

12.3.3.1.2 Remanence and particle alignment.

12.3.3.1.2 Remanence and particle alignment. A random assembly of uniaxial
single-domain particles is expected to have a reduced remanence (i.e., remanent
intensity/saturation intensity) of 0·5. For, if an applied field at an angle θ to the
preferred axis of a particle of moment μ is reduced to zero, the magnetization
will relax through an acute angle to the nearest of the two preferred directions.
The remanent moment (measured along the field direction) will be $\mu \cos \theta$, and
the average value of the reduced remanence will be 0·5. This value is also found
experimentally for iron oxide particles, in the single-domain region used in
magnetic recording surfaces. In this application the remanent intensity is
important, since the output of a tape and the signal-to-noise ratio, as least at
low recording densities, can be enhanced by increasing the remanence and thus
providing more flux to be sensed by the reading head. Consequently the acicular
γ-Fe_2O_3 particles in tapes are usually partially aligned during the manufactur-
ing process, by passing them through a strong magnetic field (1000 Oe or

greater) while the coating is still fluid. In this way the reduced remanence (measured along the alignment direction) can be increased from 0·5 to 0·7–0·8, the value attained with any coating ink and process depending on the size, shape, packing fraction and degree of dispersion of the particles. Even without deliberate orientation the reduced remanence of a tape sample is higher than the value of 0·5 found for a random sample. This is because some directionality is introduced during the application and the drying of the coating.

Perfectly aligned and isolated single-domain particles would have a reduced remanence of 1·0 along the alignment direction, and a value of zero at 90° to this direction. Actual tape samples containing 35–40% by volume of γ-Fe$_2$O$_3$ particles give values of reduced remanence of 0·7–0·8 and 0·4–0·5, respectively, in the two principal directions. Imperfect particle alignment is the principal reason for this discrepancy, but undoubtedly the presence of multi-domain particles and the interactions between particles are contributing factors. Since particle alignment leads to greater remanence and thus to higher output signals (as least, at densities below about 3000 fri) and probably also to a smoother signal envelope, it is important to be able to produce well-aligned coatings. However, as the reduced remanence measurements show, the degrees of alignment now being achieved leave much to be desired. Unfortunately it is not simply a matter of applying a large magnetic field to the particles during the drying phase of the process. Even a field of about 1000 Oe would be largely ineffective unless the particle agglomerates had been broken down. The agglomerates form and re-form with great readiness, principally because of the mutual attraction of the particles. Consequently the problem of obtaining a good particle dispersion must be addressed first. The most common method is to combine the oxide with some or all of the other ingredients of the coating ink and then to agitate mechanically in a ball-mill or similar device. The other ingredients include the binder resin(s), whose purpose is to cover the particles with a tough, plastic film which separates the particles from each other and binds them to the substrate. Solvents are added to aid in mixing and to reduce the ink viscosity to the level where it can be coated as a thin, uniform film. Chemical dispersing agents, lubricants and carbon black are also added, the purpose of the last being to lower the electrical resistivity of the coating and hence reduce the possibility of the building up of electrostatic charge on the tape.

An alternative way of increasing the remanence would be to increase the proportion of magnetic particles in the coating. The requirement that the particles be coated with binder, so that they adhere firmly to one another and to the substrate, places an upper limit on the packing fraction, which, for the size and shape of γ-Fe$_2$O$_3$ particles usually found in tapes, is usually 38–40% of the total coating volume (or roughly 70% by weight). Adams and Knees (1962) suggested that the particle loading could be increased by adding fatty diamine salts to the particle-binder mixture after the ball-milling, apparently to displace adsorbed bases from the surface of the particles and thus to allow them to be wetted more effectively by the binder. Smaller and Newman (1969) concluded that an increase in particle loading results not only in greater signal output but also in reduced harmonic distortion at a given signal level.

Probably the most critical step in the preparation of a coating ink is the mixing. Two principal methods are used, both of which owe their origin to the process of paint making. In the first the ingredients, together with some steel or ceramic balls, partially fill a cylindrical tank which rotates about a horizontal axis. The contents are carried part of the way up one side of the tank and then fall to the bottom. In the second method (e.g. that of Szegvari, 1955), the vessel remains stationary and upright while the coating material and the balls are stirred by means of a rotating axle to which hardened steel arms are radially attached. In either method the purpose is to develop very high shear forces to break up the agglomerates and to force the binder resins and solvents into contact with the individual particles. The choice of ball size is important, and since the larger balls are more effective in breaking up large agglomerates and the smaller ones in producing a homogeneous final mix, the ball size is often sequentially reduced during the milling process. Sometimes the final stage involves milling with sand particles, which are later removed by filtration. The time of milling required to produce a uniform ink can vary from minutes to days. The process must be controlled empirically since, although the magnetic and physical characteristics of the mixture improve at first as a function of time, overly prolonged milling can lead to a deterioration of the properties. This is caused partly by a breaking up of the particles themselves and partly by their re-agglomeration as the layer of dispersant on the particles is removed. A well dispersed oxide-binder-solvent mixture can have a shelf-life of weeks, but usually it is used as soon as possible. The various processes of coating the magnetic ink onto the substrate have been described by Daniel and Eldridge (1967).

The point in the drying phase at which the orienting field is applied is important. If it is applied too late, the viscosity of the drying coating does not allow good alignment; if it is applied too early, disalignment can occur after the coating has left the orienting field. The magnetic fields used are usually from 1000 Oe to 5000 Oe and can be produced by one or more permanent magnets or by an electromagnet (e.g. Speed, 1957) or a solenoid. In the latter case both d.c. and, surprisingly, a.c. fields have been found effective. Not infrequently the two are combined, since the a.c. field produces considerable particle agitation, which is apparently helpful in maintaining a good dispersion.

The problem of the movement of a magnetic particle in a viscous medium in the presence of a magnetic orienting field was treated by Newman and Yarbrough (1968). They concluded that high magnetic fields are not necessary for efficient orientation. The best orienting field was found to be $H = 2K_u/M_s$, where K_u is the relevant uniaxial anisotropy constant: $K_u = (N_b - N_a)M_s^2/2$ for shape anisotropy. Newman and Yarbrough (1969) then extended their calculations to cover the cases of non-interacting assemblies of particles, originally with either a planar random distribution or a spatially random distribution, moving in a Newtonian fluid. They concluded that the major barrier to the extension of their equations to cover the effects of interactions among particles was finding the proper functional form for the applied field. Interaction fields are important and perhaps beneficial, since samples with a 30 % (volume) loading were found experimentally to orient two to three times faster than those whose loading was only 5 %.

This theory was extended by Lissberger and Comstock (1970), who found that, in the high field mode ($H > H_c$), the orientation time decreased with increasing magnetic field.

Although these analyses are interesting, one cannot help concluding that, in the understandable interests of analytical simplification, resemblance to the real world has unfortunately been lost. An electron microscopic examination of γ-Fe_2O_3 particles, in dry powder form or in a coating, invariably shows matted clumps of particles, which are obviously prevented from rotating freely. Thus a more accurate theory of the orienting process must consider the effect of the applied field on these clumps. This is obviously a much more complex problem. However, an attempt at this type of analysis was made by Sládek (1970), who treated the field-induced deformation of agglomerates and the orientation of particles within them.

A convenient parameter for use in assessing the degree of particle orientation in a tape is the 'orientation ratio', the ratio of M_r, measured along the direction of orientation to that measured across the orientation direction but still in the plane of the coating. This ratio usually has the value of 1·7–1·8 : 1, and occasionally goes as high as 2 : 1. However, even in relatively well oriented tapes there is little evidence of crystallographic orientation if the coatings are examined by X-ray diffraction.

Daniel and Eldridge (1967) commented on an interesting but undesirable effect of improper particle alignment. If the orienting magnet is tilted so that the strong field in the centre of the gap is applied at an angle to the tape (rather than parallel to it), the signal output will be greater when the tape is run in one direction over the head than in the other direction. This happens because, in the direction of higher output, the particles are oriented along the field contours of the trailing edge of the writing head. In the opposite direction these contours lie across the average orientation direction and produce a smaller remanent magnetization of the tape and thus a smaller output.

If the level of the output signal from a tape is of prime importance, then the most appropriate direction for particle alignment is clearly along the path that the head takes over the tape; usually this is along the length of the tape. In high-density digital recording, however, resolution of the adjacent pulses is at least of equal importance. Bate et al. (1962) showed that the resolution of adjacent pulses could be improved appreciably by orienting the particles (or the particle chains) at a large angle to the direction of tape travel. The outward shift of two adjacent pulses was steadily reduced as the alignment angle θ was increased from 0° (i.e. along the direction of tape travel) to about 70° (as shown in Figure 12.19). The output signal also decreased with increasing angle at a somewhat lower rate and so a compromise between signal level and pulse resolution must be made. It was found that at $\theta \simeq 45°$ the peak shift had decreased by more than 50% in a normally oriented tape and the output was rather less than 20% below its value at $\theta = 0°$.

In some video recorders the head travels at an angle of $45°$ with respect to the length of the tape. Since high output is of concern, the particles in these tapes are also oriented at $45°$. Discs, however, are usually coated by a spinning technique and the drying time of the coating is short; for this reason, as well as the difficulty in applying a sufficiently strong field of the right shape, the orientation step is usually omitted.

12.3.4 γ-Fe$_2$O$_3$ versus Fe$_3$O$_4$ particles

The use of γ-Fe$_2$O$_3$ particles for magnetic recording is very much more common than that of Fe$_3$O$_4$ particles, and yet the reasons for this partiality are difficult to see if one considers simply the usual magnetic properties of the two materials. Actually M_s for Fe$_3$O$_4$ is about 20% higher than that for γ-Fe$_2$O$_3$, which should lead to a higher signal output, at least at low recording densities. Furthermore, the coercivity of Fe$_3$O$_4$ particles is usually higher by about 15% than the coercivity of γ-Fe$_2$O$_3$ particles of the same size and shape and this would tend to balance the adverse effect of the higher magnetization intensity at high recording densities. As we saw above, the most common method for making γ-Fe$_2$O$_3$ particles involves as the penultimate step the preparation of pseudomorphic particles of Fe$_3$O$_4$. The final oxidation step must be carefully controlled, and therefore its elimination should both improve the yield and reduce the cost of the magnetic particles. Particles of natural magnetite have been used (1945–50) in recording surfaces, and they are also used in the United States Senate as ink-blotting sand (which is, in a sense, a recording application!).

Nevertheless there is considerable reluctance in the industry to use Fe$_3$O$_4$ particles, for reasons that are by no means satisfactorily explained in the literature. Natural magnetite apparently cannot easily be converted to γ-Fe$_2$O$_3$, but rather becomes α-Fe$_2$O$_3$. That the oxidative transformation, Fe$_3$O$_4 \rightarrow \gamma$-Fe$_2$O$_3$, does occur with synthetic magnetite is evidence for the defect structure of this material. This could mean an additional predisposition to oxidation over and above that resulting from the lower oxidation state of Fe$_3$O$_4$. Thus the long-term stability of information stored on Fe$_3$O$_4$ surfaces may be suspect, but this is not at all well established. Second, magnetic accommodation effects appear to be more pronounced in Fe$_3$O$_4$ particles than in γ-Fe$_2$O$_3$. Accommodation is the phenomenon in which a stable hysteresis loop is not attained the first time around; the field must be cycled several times before a reproducible value of magnetization intensity is reached. This means that it would be difficult to demagnetize a recording surface showing the effect by simply applying an alternating field and letting it diminish to zero. The erasure problem would, of course, be more serious for analogue than for digital recording.

The third disadvantage is that tapes made of Fe_3O_4 particles are traditionally regarded as the more subject to print-through. When tape is wound on a reel the recordings of information on adjacent layers are separated only by the thickness of the substrate (usually $1-1.5$ mils, $25-38$ μm). Each recorded pattern then experiences a magnetic field from the neighbouring layers of tape and the field may be strong enough to change the state of magnetization of the pattern. This can be particularly noticeable and objectionable in the case, for example, of audio tapes, where a low-level signal is sometimes heard from what should be a silent, i.e. a demagnetized, part of the tape. The effect is enhanced by storing at high temperatures tapes whose M_s intensity is high and whose substrate is thin; long recorded wavelengths are worse than short ones.

Since the fields involved are small compared with those used in writing, a tape will be resistant to print-through if its remanence is not very susceptible to small fields. Thus, ideally, the curve of remanence as a function of applied field shows negligible remanence until a critical field of about $150-200$ Oe is reached, after which it should rise steeply (and linearly for anlaogue recording) to its maximum value.

Print-through does have a positive aspect. In the presence of an applied field the transfer of information from a recorded tape to an unrecorded one may be enhanced. Thus it is possible, at least in principle, to make several copy tapes at once in very simple equipment. The effect of the transfer field is apparently to bias the copy tape and thus enhance response of its magnetic material to the smaller field from the recorded master tape. The conditions for optimizing the effect have been discussed by Sugaya et al. (1969) and by Morrison and Speliotis (1968).

In conclusion, though it is undeniable that almost all present-day iron oxide recording surfaces are made of γ-Fe_2O_3 (the exception being that a few instrumentation tapes are made with magnetite particles), and though there is considerable distrust of magnetite as a reliable recording material, the published reasons for the preference of γ-Fe_2O_3 are at best speculative.

12.4 IRON–COBALT OXIDE PARTICLES

The search for materials that offer better recording performance than those of the iron oxides considered above has usually been focused on two properties: smaller particle size and higher coercivity. Fortunately these properties frequently occur together, but diminished size would be useful per se, since it results in reduced background noise—always an important consideration in high-quality audio and instrumentation recordings. In addition, smaller particles usually imply smoother recording surfaces, which allow the reading head to come into more intimate contact with the tape and thus help in the reproduction of high frequencies or high bit densities. On the other hand, the proportion of particle surface area to volume increases with decreasing particle size, which makes it more difficult to obtain coatings at higher loadings. Furthermore, to avoid the dangers of thermal instability, the average particle size must not be close to the lower end of the stable single-domain range, since there is always a finite width to the size distribution function. Higher coercivity of course means a greater ability to withstand demagnetization and a higher information packing density, but the coercivity must not be high enough to impair the ability to write and erase.

Both of the desired properties (and, unfortunately, some undesirable ones) are found in particles of iron oxide in which cobalt has been chemically substituted for iron during the deposition process.

12.4.1 Cobalt-substituted γ-Fe$_2$O$_3$

Jeschke (1954) and Krones (1960) described a method of preparing single-domain particles of cobalt-substituted γ-Fe$_2$O$_3$ which was similar to the method by which Krones (1955) prepared cubic particles of iron oxide, except that a mixture of cobalt sulphate and ferrous sulphate was used as starting material. Krones found that the coercivity of the equiaxed particles, whose average size was 0·1 μm, increased from 100 Oe for no added cobalt to 800 Oe for particles containing 10 at. % cobalt. Nobuoka et al. (1963) prepared similar particles by co-precipitating cobalt and iron hydroxides from an alkaline solution of ferric chloride (FeCl$_3$.6H$_2$O) and cobaltous nitrate (Co(NO$_3$)$_2$.6H$_2$O). The mixture was heated at 130 to 170 °C under pressure in an autoclave to produce a fine-grained precipitate, which was rinsed, dried and heated in a reducing atmosphere at 350–400 °C to give particles of cobalt-substituted γ-Fe$_2$O$_3$. These were then re-oxidized at 200–550 °C to yield cobalt-substituted γ-Fe$_2$O$_3$ in which the cobalt content was 0·2–0·3% ($H_c \leqslant 300$ Oe; $M_r/M_s = 0.5$).

Arshinkov et al. (1961–62) precipitated a mixed iron-cobalt oxalate from an aqueous solution of ferrous sulphate and cobaltous nitrate by the addition of oxalic acid. The precipitate was rinsed, dried at 105 °C, and finally decomposed to the oxide at 385 °C. Powders containing 2–30% cobalt made in this way had the following magnetic properties:

Cobalt content (at. %)	H_c (Oe)	Br (gauss)	(BH) max. ($\times 10^6$ gauss Oe)
2	275	1193	0·14
9	1100	1680	0·60
30	830	1070	0·22

Not unexpectedly, the result of substituting cobalt in γ-Fe$_2$O$_3$ is to reduce the saturation magnetization intensity roughly in proportion to the extent of the doping.

One undesirable property of cobalt-substituted iron oxides is the pronounced temperature dependence of coercivity, as shown in Figure 12.9. This is undesirable because, if a recorded medium encounters high temperatures either in use or during storage, the decrease in coercivity would allow increased demagnetization, with the consequence of a reduction or perhaps even a total loss of the recorded information.

Measurements by Eagle and Mallinson (1967) on 3–6% cobalt-doped γ-Fe_2O_3 (Figure 12.10) led them to conclude that the coercivity of the particles was independent of K/M in this composition range. In contrast, cobalt dopings below 3% yielded anisotropy constants proportional to the cobalt content. The measured coercivities agreed with the nucleating field (Figure 12.10) calculated from the curling equation

$$H_n = \left[2\pi(1\cdot39)\left(\frac{d_0}{d}\right)^2 - \frac{4}{3}\pi \right]M + 0\cdot64\frac{K}{M}$$

with the critical value $d_0 = 150$ Å assumed. From this agreement Mallinson and Eagle (1967) concluded that the mechanism of magnetization reversal in their equant particles was óne of incoherent rotation.

An interesting difference between particles that are magnetically hard because of magnetocrystalline anisotropy, such as the equiaxed particles of cobalt-substituted γ-Fe_2O_3, and those that derive their hardness from their shape, is their relative sensitivity to interaction fields. With the latter particles the presence of neighbours may completely change the flux distribution around a given particle and hence change its effective demagnetizing factors. In contrast, for a system of non-interacting particles Wohlfarth (1958) showed that the following relation between different remanence values should hold:

$$M_D(H) = M_R(\infty) - 2M_R(H)$$

where $M_R(\infty)$ is the maximum remanent intensity after a saturating field has been applied along the positive direction. If this field is followed by a field H in the negative direction, then $M_D(H)$ is the new remanent intensity. Once the sample has been demagnetized in a diminishing alternating field, it can be magnetized to an intensity $M_R(H)$ in a field $(+H)$.

Wohlfarth's relation was deduced on the assumption that the particles do not interact. Bate (1962) used the discrepancy between the experimental values of $M_R(H)$ and those deduced from the formula as a measure of the importance of particle interactions in the sample. The difference in behaviour between samples of acicular particles of γ-Fe_2O_3 (20% loading) and equiaxed particles of cobalt-substituted γ-Fe_2O_3 (40% loading) is shown in Figure 12.14. From this it can be clearly seen that, even though the loading is lower for the acicular particles, the discrepancy between $M_R(H)$ calculated and $M_R(H)$ experimental is greater, implying the greater importance of interactions.

There is no question that the small, equiaxed particles of cobalt-substituted γ-Fe_2O_3 do show recording properties rather superior to those of the more common acicular particles of γ-Fe_2O_3. In particular, the signal output level is higher and stays higher out to greater frequencies, and the peak-shift is lower. Both of these improvements can be attributed to the higher coercivity, usually 400–600 Oe. The smaller particle size may help the smoothness of the signal envelope and might be responsible for some of the improvement in performance. The equiaxed shape is not believed to confer any advantages *per se*, but this is

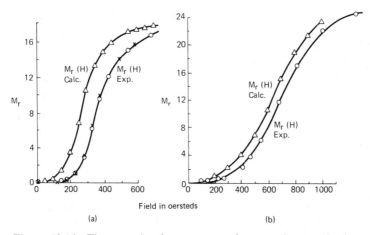

Figure 12.14 The growth of remanence after a.c. demagnetization; showing the experimental points and those calculated from the descending remanence curve for: (a) γ-Fe_2O_3 particles, 20% volume loading, (b) 3% cobalt-substituted γ-Fe_2O_3 particles, 40% volume loading. The difference between the curves is in each case a measure of the strength of particle interactions (after Bate, 1962, and Speliotis *et al.*, 1965).

not at all well established. It is interesting to speculate whether, in recording surfaces, a coercivity derived from magnetocrystalline anisotropy is in any way preferable to one derived from shape anisotropy, or vice versa. In single-domain particles having many easy directions of magnetization these directions will clearly be separated from each other by relatively small angles, and therefore, as Wohlfarth and Tonge (1957) showed, the ratio M_r/M_s will always be higher than for a random spatial array of uniaxial particles (see Section 1.5.1). However, this may not be entirely an unmixed blessing. The trailing-edge field of the recording head has a substantial perpendicular component. This fact, together with the particle's large number of easy directions (some of which will be near the perpendicular), will produce a perpendicular component of magnetization and, in consequence, asymmetry of the output pulse. Since M_r/M_s is normally higher with the equiaxed particles, orientation is normally omitted.

The recording performance of tapes made of equiaxed particles of cobalt-substituted γ-Fe_2O_3 particles was reported by Morrison and Speliotis (1966). The particles used had an average size of about 500 Å, a coercivity of 580 Oe, and $M_r/M_s = 0.70$, and were coated at a loading of 37% by volume to a thickness of 88 µin. The temperature dependence of coercivity was similar to that shown in Figure 12.9. The graph of output as a function of recorded density did extend noticeably further than that of the acicular γ-Fe_2O_3 tape used as control; however, the coating on the latter was 75% thicker and this was no doubt responsible for part of the difference. The cubic particles also gave a superior result for peak-shift, which was not found until the density reached 2000 fri (cf. 500 fri for the control tape); here again the difference in the thickness of the two tapes would play a part.

728

Tapes recorded at different densities were stored at elevated temperatures. For densities between 100 and 1000 fri the loss in signal after storage for 1 hour at 160 °C was 45 %. At higher densities the loss appeared to be proportionately less and this result may reflect the changing roles of coercivity and remanence with increasing density. For example, at low densities a high remanence means a high output, but at densities above 1000 fri, where pulse crowding and demagnetization become important, a high remanence means more demagnetization, broader pulses, and lower output. Thus, although the magnetic properties of the oxide change reversibly with temperature, recorded information can be erased. This happens because a recorded tape is in a state of high magnetostatic energy which is reduced by erasure, and the tape reverts to this preferred state when the temperature becomes high enough.

Higher temperatures (100 °C or higher) might be encountered during transportation or storage and it is also conceivable that they could be generated by friction as the tape passes over the head. Morrison and Speliotis ran a cobalt-substituted γ-Fe$_2$O$_3$ tape in the form of a loop for many thousands of revolutions. Figure 12.15 shows that the output eventually decreased to 25 % of its

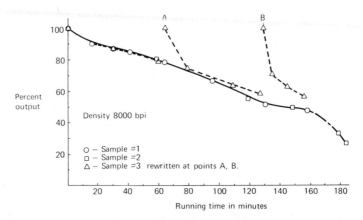

Figure 12.15 The output of tapes made of equiaxed particles of cobalt-substituted γ-Fe$_2$O$_3$ as a function of the running time over the head (after Morrison and Speliotis, 1966).

original value. Points A and B on the graph show that when the tape was re-recorded, the output regained its original level. One possible and simple explanation of this result is of course that part of the recorded surface was worn away, and it is difficult to measure thickness precisely enough to be able to rule this out. Morrison and Speliotis favoured the explanation that heat generated by friction raised the temperature enough to cause partial erasure. In contrast, Daniel and Eldridge (1967, p. 67) believed that the effect was due to a stress-induced instability in the oxide.

12.5 CHROMIUM DIOXIDE, CrO_2

Michel and Bénard (1935, 1943) showed that the decomposition of chromyl chloride, CrO_2Cl_2, yielded about 50 % of CrO_2 along with Cr_2O_3 and that the chromium dioxide was ferromagnetic and had the tetragonal rutile structure with $a = 4\cdot41$ Å and $c = 2\cdot86$ Å. They measured the Curie temperature as 121 °C and the magnetic moment per chromium atom as $2\cdot07 \pm 0\cdot03\beta$. Rode and Rode (1961) reported that chromium monochromate $Cr_2(CrO_4)_3 = CrO_{2\cdot40}$ was also ferromagnetic, but more weakly so than their CrO_2 ($\sigma_s = 12$ e.m.u./g; cf. 85 e.m.u./g). They found that the Curie temperature of $CrO_{2\cdot40}$ was about 80 °C.

12.5.1 Preparation of CrO_2 particles

The three principal methods are thermal decomposition of CrO_2Cl_2, thermal decomposition of pure CrO_3 under oxygen pressure, and thermal decomposition of aqueous CrO_3 under oxygen pressure. The first was probably the earliest method of making chromium dioxide, but the product was not pure CrO_2. Cox (1963) used CrO_2 particles as seeds and increased the yield of chromium dioxide in the decomposition of the chromyl to about 90%. CrO_2 particles can also be formed by vapour deposition from chromyl chloride (Darnell, 1961).

The methods now being used almost invariably involve high pressures. Very pure single-phase chromium dioxide can be obtained from the thermal decomposition of CrO_3 in the presence of water at high pressure (Swoboda et al., 1961). Equimolar amounts of CrO_3 and water were enclosed in a thin-walled platinum container, which in turn was enclosed in a thick-walled pressure vessel. The mixture was heated to 400–525 °C at a pressure of 500–3000 atmospheres; the reaction time was 5–10 minutes. Oxygen was needed to stabilize the CrO_2 and was produced by the decomposition reaction $CrO_3 \xrightarrow[\text{pressure}]{\text{heat}} CrO_2 + \frac{1}{2}O_2$. It was found that, in general, the higher the reaction temperature, the higher the pressure needed to produce CrO_2, and that high temperatures, high pressures, and long reaction times tended to result in large particles being formed. Platinum containers are most commonly used because of their chemical inertness, but Kubota et al. (1963) found that platinum, antimony, and tellurium tend to reduce the grain size of CrO_2.

Powder samples of chromium dioxide have also been made by the thermal decomposition of *anhydrous* CrO_3, with the oxygen again coming from the decomposition of the CrO_3 (Ariya et al., 1953). Kubota (1961) used temperatures from 400 °C to 560 °C, pressures of 497–627 kg/cm², and a vessel of stainless steel; the reaction time was four hours. In this way Kubota (1960) obtained CrO_2 particles with $\sigma_{so} = 133\cdot7 \pm 0\cdot2$ e.m.u./g and $T_c = 117$ °C.

Claude et al. (1968) heated CrO_3 with $CaCl_2.2H_2O$ in a thick glass tube to 360 °C for 24 hours, the pressure reaching a maximum of 110 atmospheres. The sealed tube was then cooled in liquid nitrogen and cautiously opened to release the free chlorine. The particles were finally washed with hydrochloric acid and found to consist of 90 % CrO_2 having the tetragonal structure with $a = 4.421$ Å and $c = 2.916$ Å. The reaction was probably

$$CrO_3 + Cl_2 \rightarrow CrO_2Cl_2 + \tfrac{1}{2}O_2$$

The chromyl chloride then decomposes to give CrO_2. In the electron microscope the particles were found to be acicular, with lengths of 5–10 μm and widths of 0.5–1.0 μm. Claude et al. (1968) measured the saturation magnetization of their CrO_2 particles and found $n_\beta = 1.96$; $T_c = 116\,°C$.

A process for preparing CrO_2 of very uniform particle size was described by Arthur and Ingraham (1964). They decomposed hydrothermally chromium oxides with compositions between Cr_3O_8 and Cr_2O_5. The improvement in the particle uniformity was ascribed to the relative insolubility of the oxides, which was 1 % or less in water at 25 °C. Cr_3O_8 was prepared by heating CrO_3 in oxygen at 250 °C for 2 to 5 days; if CrO_3 was heated at 360 °C for 8 to 24 hours, then Cr_2O_5 was formed. Alternatively the starting oxides can be obtained from the thermal decomposition of chromic nitrates $Cr(NO_3)_3.9H_2O$ at 150–380 °C and atmospheric pressure. The chromium oxides were then broken down to particles less than 2 μm in diameter in a ball mill and converted to CrO_2 by hydrothermal decomposition between 330 °C and 400 °C and pressures not exceeding 800 atmospheres for times from 1 to 10 hours. The decomposition was carried out in the presence of water or an aqueous solution of nitric acid in amounts up to 50 % by weight of the starting oxide. The use of additives, though not essential to obtain useful magnetic properties, enabled particles to be made whose properties covered a wider range of H_c, M_r/M_s and T_c. The particles prepared without additives were acicular with length of about 0.5–1.0 μm, widths of 0.02–1.0 μm, and axial ratio of 3 to 10; their coercivities ranged from 220 Oe to 440 Oe and M_r/M_s was about 0.05.

12.5.1.1 The use of catalysts and additives

The method described by Swoboda et al. (1961) yielded particles that were dark grey and acicular, with $l = 3$–10 μm and $w = 1$–3 μm. More highly acicular particles of smaller size were made by adding metal oxides as catalysts. The most effective were found to be RuO_2 (Oppegard, 1959) and Sb_2O_3 (Ingraham and Swoboda, 1960). Another advantage of the use of Sb_2O_3 was that the reaction temperature and pressure could be reduced as low as 300 °C and 50 atmospheres. X-ray diffraction patterns of the particles produced with the aid of these catalysts showed only the presence of a tetragonal, rutile crystal structure with a 4.41 Å and $c = 2.91$ Å.

Kubota et al. (1966) added tellurium compounds to the CrO_3, which was then decomposed under oxygen pressure generated by the reaction $CrO_3 \rightarrow CrO_2 + \tfrac{1}{2}O_2$. As long as the tellurium content did not exceed 10 % of the amount of chromium plus tellurium, the product was pure CrO_2. Balthis (1969) used a mixture of CrO_3, Cr_2O_3 and sodium dichromate $Na_2Cr_2O_7.2H_2O$ in the proportions 2:1:5. The starting materials were ground dry under nitrogen, hermetically sealed in a platinum tube at a pressure of 1000 atmospheres, heated to 265 °C for 4 hours and 400 °C for 6 hours, and cooled. The resulting CrO_2 particles were not more than 2·5 µm long and 0·5 µm wide; the coercivity was 246 Oe, σ_s 90.3 e.m.u./g, and M_r/M_s was 0·44. Variations in the proportions of the starting material and in the processing steps produced a wide range of sizes of CrO_2 particles having the properties

$$37 \text{ Oe} \leqslant H_c \leqslant 542 \text{ Oe}$$

$$79\cdot3 \text{ e.m.u./g} \leqslant \sigma_s \leqslant 82\cdot9 \text{ e.m.u./g}$$

$$0\cdot17 \leqslant M_r/M_s \leqslant 0\cdot49$$

12.5.2 Structure and properties of CrO_2

The structure of chromium dioxide, unlike that of $\gamma\text{-Fe}_2O_3$, is well established. Wilhemi and Jonsson (1958) summarized earlier measurements and reported their own results on powder samples. They found the tetragonal rutile structure with $a = 4\cdot423$ Å and $c = 2\cdot917$ Å. The observed density of $4\cdot83$ gm/cm^3 was in close agreement with the value of $4\cdot89$ gm/cm^3 calculated for a unit cell having two formula units of CrO_2. Cloud et al. (1962) measured the lattice constants of a single crystal of CrO_2 by X-rays and found $a = 4\cdot4218$ Å and $c = 2\cdot9182$ Å. Small single crystals (about $0\cdot1$–$0\cdot4$ mm) were also obtained by Swoboda et al. (1961), who used the hydrothermal method to decompose CrO_3. Platinum sheets that had been stretched and annealed to produce crystallites measuring about 2 mm were used as substrates and the decomposition was allowed to proceed for several days.

The space group of CrO_2 is $P4/mnm$ (D_{4h}^{14}); the chromium ions are situated at the corners and the body centre of the tetragonal unit cell. Thus the number of chromium ions per unit cell is $(8 \times \frac{1}{8}) + 1 = 2$. Oxygen atoms are found at (u, u, o), $(\bar{u}, \bar{u}, o)$, $(u - \frac{1}{2}, \frac{1}{2} - u, \frac{1}{2})$ and $(\frac{1}{2} - u, u - \frac{1}{2}, \frac{1}{2})$; Siratori and Iida (1960) determined, from X-ray measurements on the intensities of the (111) and (210) reflections, that the oxygen parameter $u = 0\cdot294$. Cloud et al. (1962) found u to be $= 0\cdot301 \pm 0\cdot04$. The temperature dependence of c and a was measured by Siratori and Iida (1960) on a powder sample; they observed an anomalous decrease in c with increasing temperature. The existence of this effect was confirmed by Darnell and Cloud (1965) on a single crystal of CrO_2; the temperature dependence of c was the same for increasing as for decreasing temperatures ruling out the possibility of chemical degradation on heating.

The density of a number of powder samples of CrO_2 (obtained by decomposition of anhydrous CrO_3 under high pressure) was measured by Kubota (1961), who obtained values ranging between $4\cdot875$ and $4\cdot955$ gm/cm^3.

Gustard and Vriend (1969) showed by dark field electron microscopy that 67 % of their CrO_2 particles had a [001] acicular axis. Furthermore they found that preferred crystallographic orientation was easily observed by X-rays in CrO_2 tapes but not in tapes made of $\gamma\text{-Fe}_2O_3$ particles.

Oriented layers of CrO_2 were grown epitaxially on {100}, {110}, {210}, and {001} surfaces of single-crystal TiO_2 of the rutile form and on the {0001} planes of single-crystal Al_2O_3 and $\alpha\text{-Fe}_2O_3$ by De Vries (1966). The method used was

732

the thermal decomposition of CrO_3 at temperatures up to 425 °C and 4000 psi of oxygen for 1 to 48 hours.

Figure 12.16 shows the appearance of CrO_2 particles in the electron microscope.

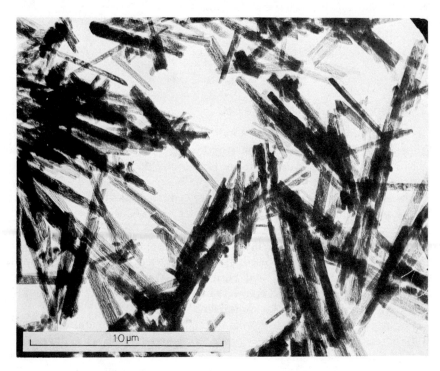

Figure 12.16 Electron micrograph of particles of CrO_2.

12.5.2.1 Intrinsic magnetic properties of CrO_2

The formula CrO_2 suggests that the chromium ions are in the Cr^{4+} state, for which a magnetic moment per ion of 2β is expected. Therefore, since the material is ferromagnetic rather than ferrimagnetic, the saturation magnetization per unit mass can be calculated from $\sigma_{so} = (n_\beta \cdot 5587)/M$, where M is the molecular weight, 84·01 for CrO_2. Using $n_\beta = 2$, we find $\sigma_{so} = 133$ e.m.u./g. Guillaud et al. (1944) decomposed chromyl chloride, $CrO_2.Cl_2$, and obtained a product containing about 50% CrO_2. When they extrapolated the magnetization measurements to 100% CrO_2, the value of σ_{so} obtained was 138·3 e.m.u./g or $2·07 \pm 0·03\beta$. Claude et al. (1968) found $n_\beta = 1·96$ for particles prepared from CrO_3 with $CaCl_2.2H_2O$. Kubota (1960) prepared CrO_2 by the thermal decomposition of anhydrous CrO_3 and found $\sigma_{so} = 133·7 \pm 0·2$ e.m.u./g; hence $n_\beta = 2·01 \pm 0·03$. Measurements on the single crystals of Swoboda et al. (1961) gave $\sigma_s = 100 \pm 1$ e.m.u./g at room temperatures; fields of 10,000 were found adequate to achieve saturation. These authors also reported that for microcrystalline powders measured at room temperature in fields of 4000 Oe, values of σ of 89 to 92 e.m.u./g were obtained. Extrapolation to infinite field yielded $\sigma_s = 98$–100 e.m.u./g (20 °C).

Chromium dioxide has been found difficult to saturate; Darnell and Cloud (1965) found that the magnetization reaches 94% to 98% of σ_s in 10,000 Oe. This is much larger than the normal anisotropy fields, about 1000 Oe, used by Rodbell et al. (1967). Darnell and Cloud ascribed it to the lower symmetry experienced by the surface chromium atoms, which results in generally larger and randomly arranged anisotropies. They measured $n_\beta = 2.00 \pm 0.5$ per chromium ion.

The addition of the catalysts Sb_2O_3 and RuO_2 was found by Swoboda et al. (1961) to reduce both the specific magnetization and also the rate of approach to saturation. From measurements at 4000 Oe, they found that σ was 90 e.m.u./g when no additive was used but was reduced to 77–85 e.m.u./g when 0.2% to 2.0% catalyst σ was added. Darnell (1961) states that this reduction is greater than could be accounted for by simple non-magnetic dilution.

The Curie temperature was found by Guillaud et al. (1944) to be 121 °C. The measurements of Swoboda et al. (1961) on powder samples gave $T_c \simeq 126$ °C. Darnell and Cloud (1965) determined T_c for large particles of CrO_2 from graphs of σ versus H at different temperatures and obtained 119 °C; Kubota (1960) found $T_c = 117$ °C, and Claude et al. (1968) obtained $T_c = 116$ °C for their samples. Apparently the use of catalysts in small amounts has little effect on the Curie temperature; however, the modified CrO_2 particles were reported by Swoboda et al. (1961) as rather less thermally stable than those made without a catalyst. The magnetization of CrO_2 in the form of a compressed powder as a function of the applied field and temperature was measured in detail by Kouvel and Rodbell (1967) near the Curie temperature. From these measurements they concluded that $T_c = 113.5$ °C.

Cloud et al. (1962) found from measurements on a small single-crystal sphere of CrO_2 that the directions of easy magnetization lie in the (100) planes and at an angle of approximately 40° to the tetragonal axis; they found the anistropy constant K to be approximately 2.5×10^5 erg/cm^3. The direction of easiest magnetization was reported by Darnell and Cloud (1965) to lie in the a–c plane at 30° to the c axis. Neutron diffraction studies referred to by Cloud et al. (1962) supported the 40° inclination. However, later work by Rodbell (1966) and Rodbell et al. (1967) gave the c axis as the easy axis and $K = 2.5 \times 10^5$ erg/cm^3.

12.5.2.2 Extrinsic magnetic properties

The influence of the catalysts used in preparing CrO_2 particles was detectable at concentrations as low as 0.1% by weight of the CrO_3 starting material. Most noticeably they affected the particle size, and therefore the coercivity. Swoboda et al. (1961) found that as the amount of catalyst was increased, the particle size decreased until particles 0.2–1.5 μm long and 0.03–0.1 μm wide were formed at a catalyst concentration of 2%. The coercivity of the large particles ($l = 3$–10 μm, $w = 1$–3 μm) prepared without the use of catalysts was 57 Oe, whereas the addition of 2% Sb_2O_3 resulted in smaller, more acicular particles ($l = 0.1$–$2.0\,\mu m$, $w = 0.04$–$0.08\,\mu m$) with $H_c = 349$ Oe. Darnell (1961) calculated the critical, upper size for single-domain behaviour by the method used by Morrish and Yu (1955) for iron oxide particles. The method consists of finding the size at which the magnetostatic energy of a single-domain particle is equal to the domain wall energy of a two-domain particle. Using $M_s = 490$ e.m.u./cm^3 (at 20 K), $K = 3.0 \times 10^5$ erg/cm^3, and an axial ratio of 5:1, Darnell found that the largest minor axis dimension for single-domain behaviour was $0.46\,\mu m$. From his magnetic measurements he concluded that CrO_2 particles having diameters less than $0.2\,\mu m$ and axial ratios of 5:1 show single-domain behaviour; he explained the difference between the calculated and the observed values as due to the approximate nature of the calculations and to the distribution of particle shape and size in the experimental samples.

For a spatially random assembly of single-domain CrO_2 particles with an axial ratio of 5:1, the theoretical coercivities are, for coherent notation

$$\langle H_c \rangle = 0.96 \frac{K}{M_s} = 585 \text{ Oe}$$

(assuming only magnetocrystalline anisotropy)

$$\langle H_c \rangle = 0.48(N_b - N_a)M_s = 1230 \text{ Oe}$$

and for the chain-of-spheres model

$$\langle H_c \rangle = 0.18\pi M_s(6K_n - 4L_n) = 590 \text{ Oe}$$

where K_n, L_n are parameters that depend on the length and axial ratio of the particle. Darnell (1961) tested the experimental results of H_c versus T against the predictions of models based on crystalline anisotropy and shape anisotropy, and found that the closest fit was obtained with the shape anisotropy curve; i.e. the coercivity tended to follow M rather than K/M. He ascribed the difference between the measured and the theoretical values of H_c for single-domain particles to the presence of multi-domain particles. The dominance of shape anisotropy was further suggested by measurements of the anisotropy field from torque curves. This field (2150–2500 Oe) was much too large to be explained by magnetocrystalline anisotropy ($2K/M_s = 1200$ Oe). The values of the rotational hysteresis integral (also obtained from the torque curves),

$$\int_0^\infty \frac{W}{M_s} d(1/H)$$

were found to lie between 1.10 and 1.36, suggesting incoherent rotation.

12.5.3 Recording properties

By varying the additives and the processing conditions, it has been possible to make particles of CrO_2 having coercivities from less than 100 Oe to more than

650 Oe. The material has a σ_s of 90–100 e.m.u./g; compared with 74 e.m.u./g for γ-Fe_2O_3. In addition, chromium dioxide particles look comparatively clean and free of dendrites and the other assorted excrescences so typical of acicular γ-Fe_2O_3 particles, as we can see by comparing Figures 12.5 and 12.16. This makes the particles easier to disperse and thus easier to orient (see, for example, Arrington, 1963). A tape made of acicular γ-Fe_2O_3 particles is considered to be well oriented if M_r/M_s along the orientation direction is 0·8; with chromium dioxide particles, values in excess of 0·9 can be achieved. We should expect then that a superior recording surface could be made from CrO_2 particles; the higher moment density and orientation and greater coercivity imply larger output signals at low densities, and the coercivity should also improve high-density performance.

Recording measurements in support of these expectations were reported by Speliotis (1968), who compared the performance of two tapes. One was of acicular γ-Fe_2O_3 and was 500 µin thick; H_c = 299 Oe, M_s = 68·5 e.m.u./cm^3, and M_r/M_s = 0·76. The other was of CrO_2 particles and was 285 µin thick; H_c = 465 Oe, M_s = 114 e.m.u./cm^3 and M_r/M_s = 0·91. The density response curves for the two tapes are shown in Figure 12.17. In one case the head used

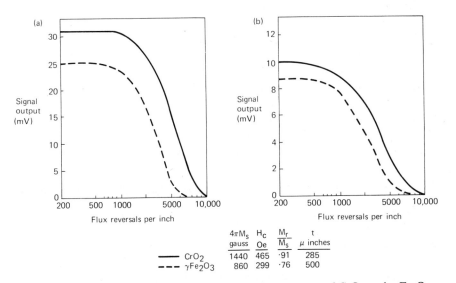

	$4\pi M_s$ gauss	H_c Oe	$\frac{M_r}{M_s}$	t µ inches
—— CrO_2	1440	465	·91	285
--- γFe_2O_3	860	299	·76	500

Figure 12.17 Signal output versus recording density for tapes of CrO_2 and γ-Fe_2O_3: (a) Writing-head gap = 500 micro-inches, reading-head gap = 250 micro-inches, writing current = 120 mA; (b) Writing-head gap = 90 micro-inches, reading-head gap = 90 micro-inches, writing current = 80 mA (after Speliotis, 1968).

had a writing gap of 500 µin and a reading gap of 250 µin, and in the other the gaps were both 90 µin. We notice that, even though the CrO_2 tape is thinner, its output at low densities is appreciably higher than that of the iron oxide tape with either head. At the higher densities, of course, the thinness of the CrO_2 tape helps the resolution, and thus the output. Unfortunately the numbers of turns on the two reading heads was not given, and so it is impossible to make quantitative comparisons of their outputs.

Peak shift was measured for the common pattern of two 1s, following and followed by four 0s, at a density of 3000 fri. When the chromium dioxide tape was read, the separation between the two pulses corresponding to the 1s was found to be increased by 23%, compared with an outward shift of 61% for the iron oxide tape. Thus in recording characteristics the CrO_2 tape is superior to the γ-Fe_2O_3 tape. However, this does not necessarily imply that the CrO_2 tape would perform better on existing devices, since those devices were designed around the characteristics of γ-Fe_2O_3 recording surfaces. This is the problem of compatibility; if we take a new recording surface and modify its properties so that it will work on existing machines, its performance may be no better than that of the older material it replaces. If, on the other hand, we design a new surface to have inherently superior recording characteristics, it probably will not work on the machines. The only solution apparently is to design and develop together the new recording surface and the device on which it is to be used.

The Curie point of chromium dioxide is much lower than that, for example, of γ-Fe_2O_3 (113.5 °C versus 590 °C). In fact, the temperature is so low as to occasion concern that accidental erasure might result from a relatively modest increase in the ambient temperature during use, transportation, or storage. However, as we saw above, the dominant anisotropy appears to be that of shape, and thus the coercivity of the particles should be proportional to the intensity of magnetization. Consequently an increase in temperature toward the Curie point will produce an equivalent decrease in magnetization and coercivity. That is, the balance between demagnetization field (αM) and the ability to resist demagnetization (H_c) will remain substantially unchanged, and no erasure will result (unless, of course, the temperature reaches the Curie point).

12.6 RECORDING PERFORMANCE AND MAGNETIC PROPERTIES

Ideally the selection of new particulate recording materials would be based on measurements of the magnetic properties of the powders, since these measurements are in general simple to make and the results are unambiguous. However, this approach presupposes that the relationships between the magnetic properties and the performance of the recording surfaces are well established; this is unfortunately not so. There is no standard way of measuring the recording properties of experimental recording surfaces. When a new particle is being developed, the initial recording results are obtained from measurements on a short length of hand-drawn tape (say), which is run on a 'loop-tester'. This apparatus is a greatly simplified tape drive on which a continuous loop of the sample tape is held at a supposedly constant tension by means of a vacuum column. Usually the wrap angle of the tape over the head is variable, as is the tape velocity. Workers in different laboratories can and do get conflicting results when working with tape samples which are ostensibly the same, on loop-testers operating at the same speed with heads having similar gaps. Among the variables responsible for these differences are the rise and fall characteristics of the flux in the writing head, the level of the writing current, the efficiency of the reading head (that is, the ratio of the flux passing around the core to that going across the gap) as a function of frequency, the path taken by the tape over the head (and particularly the separation between the tape and the head during writing and reading), the roughness of the tape and the head, the remanent magnetization of the writing head, and possibly several other factors whose importance is not yet recognized. In an attempt to minimize the contribution of the reading head and amplifier to the results, most experiments are made at low tape velocities (30 in/s or less) and with reading-head gaps and head-to-medium distances that are small relative to the bit dimensions.

It has been common for some years to characterize recording performance in terms of the width and amplitude of a single isolated pulse on the assumption that the response of more complex patterns can be derived by the linear superposition of such pulses. Unfortunately, there is no universally accepted definition of pulse width; two in common use are the width at half the pulse height and that at one fourth the pulse height. Furthermore, the detailed shape of the pulse, and not just the pulse width, must be considered when attempting to predict the output wave function of a complex pattern. There is by no means general agreement on the validity of the superposition approach (see for example Mallinson and Steele (1969) and Beaulieu (1969)) and the experimental measurements on isolated pulses are usually supplemented by a measurement of the output of an all 1s pattern as a function of density. From this curve the density at which the output has fallen to one half of its value at low densities is extracted and used as a figure of merit. Figure 12.18 shows an experimental

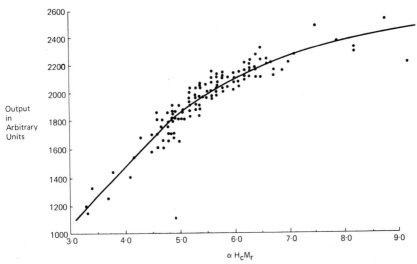

Figure 12.18 Signal output versus (coercivity H_c) × (remanent magnetization, M_r) for a number of samples of γ-Fe$_2$O$_3$ recording surfaces.

curve, obtained by the author, relating the output at 2000 fri to the product H_cM_r for a large number of samples of similar $\gamma\text{-Fe}_2\text{O}_3$ tapes. This is the 'energy product' commonly used to characterize the energy stored in a permanent magnet. The line drawn through the points was obtained by using a computer to make a statistical analysis of the experimental results. Its mathematical form output $= A - B/(H_cM_r)$ (where A and B are constants), has not yet occurred in theoretical papers on magnetic recording, but no doubt it will appear eventually!

Until recently the phenomenon of peak-shift, which is largely responsible for limiting linear recording densities, was treated as though it was simply as a result of pulse-crowding (Templeton and Bate, 1964). The peak-shift, which occurred in a two-bit pattern, appeared to correlate quite well with the ratio M_r/M_c in a series of measurements made by Bate (1965) on strips of tape cut at different angles to the direction of particle orientation. The results are shown in Figure 12.19. By this means it was possible to study the effects of the

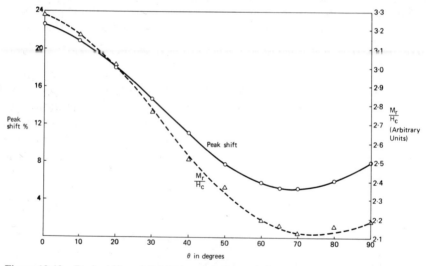

Figure 12.19   Peak-shift and (M_r/H_c) versus the angle between the direction of particle orientation and the direction of measurement for a tape made of $\gamma\text{-Fe}_2\text{O}_3$ particles (after Bate, 1965).

variation of M_r/H_c in a series of samples which were identical in thickness, particle loading and roughness. Then Iwasaki and Suzuki (1968) observed the asymmetry in the shifts of two adjacent pulses and explained it by considering the succession of fields that a particular region in the recording surface experiences during its passage over the recording head. Beaulieu (1969) studied a more complex pattern of 16 bits as a function of the thickness and the magnetic parameters of the tape and of the writing current. He also observed a pronounced asymmetry and concluded that linear superposition alone would not explain his results.

It now appears that there are at least two different types of peak-shift. One type is strongly dependent on the details of the writing process and on the particular pattern being written, and less dependent on the magnetic properties of the medium. The other type is more obviously a result of simple pulse-crowding and can be predicted by linear superposition. It can be minimized by choosing a medium having small thickness and remanent magnetization and high coercivity, although the precise functional dependence of pulse shape on these parameters is still unclear (see Bate and Alstad, 1969), and a search for it may be in vain (Mallinson, 1969). Finally, it should be recognized that obtaining particles having desirable magnetic properties is only one aspect, and probably a relatively minor aspect, of the total job. The difficult, time-consuming, complex and interrelated questions of surface chemistry, polymer chemistry, rheology and wear can really begin only at the point where magnetic development ends.

REFERENCES

Adams, P. and L. E. Knees, 1962, *U.S. Patent* 3,042,639.

Amar, H., 1959, *J. Appl. Phys.*, **30**, 139S.

Ariya, S. M., S. A. Shchukarev and V. B. Glushkova, 1953, *Zhurnal Obshchei Khimii*, **23**, 1241.

Arrington, C. H., Jr., 1963, *U.S. Patent* 3,080,319.

Arshinkov, I., P. Pesher and O. Tsurnorechki, 1961–1962, *Izr. Nauchnoizsled. Inst. Kinematogr. Radio (Sofia)*, **3**, 31.

Arthur, P., 1960, *U.S. Patent* 2,956,955.

Arthur, P. and J. N. Ingraham, 1964, *U.S. Patent* 3,117,093.

Ayers, J. W. and R. A. Stephens, 1962, *U.S. Patents* 3,015,627; 3,015,628.

Balthis, J. H., Jr., 1969, *U.S. Patent* 3,449,073.

Bando, Y., M. Kiyama, T. Takada and S. Kachi, 1965, *Japan J. Appl. Phys.*, **4**, 240.

Baronius, W., F. Henneberger and W. Geidel, 1966, *East German Patent* 48,590.

Bate, G., 1961, *J. Appl. Phys.*, **32**, 261S.

Bate, G., 1962, *J. Appl. Phys.*, **33**, 2263.

Bate, G., 1965, *I.E.E.E. Trans. Mag.*, **MAG-1**, 193.

Bate, G. and J. K. Alstad, 1969, *I.E.E.E. Trans. Mag.*, **MAG-5**, 821.

Bate, G., H. S. Templeton and J. W. Wenner, 1962, *IBM J. Res.*, **6**, 348.

Bate, G., J. R. Morrison and D. E. Speliotis, 1964, *Proc. Int. Conf. Mag. Recording (London)*, 7.

Baudisch, O., 1933, *U.S. Patents* 1,894,749; 1,894,750.

Bean, C. P. and J. D. Livingston, 1959, *J. Appl. Phys.*, **30**, 120S.

Beaulieu, T. J., 1969, *I.E.E.E. Trans. Mag.*, **MAG-5**, 259.

Berkowitz, A. E., W. J. Schuele and P. J. Flanders, 1968, *J. Appl. Phys.*, **39**, 1261.

Bertram, H. N. and J. C. Mallinson, 1969, *J. Appl. Phys.*, **40**, 1301.

Bickford, L. R., J. M. Brownlow and R. F. Penoyer, 1957, *Proc. Instn. Elec. Eng.*, **104B**, Suppl. No. 5, 238.

Bratescu, V. and P. Vitan, 1969, *Rumanian Patent* 51,610.

Brown, W. F., Jr., 1959, *J. Appl. Phys.*, **30**, 62S.

Brown, W. F., Jr., 1962, *J. Appl. Phys.*, **33**, 1308.

Campbell, R. B., 1957, *J. Appl. Phys.*, **28**, 381.

Camras, M., 1954, *U.S. Patent* 2,694,656.

Claude, R. (Mlle.), G. Lorthioir and C. Mazières, 1968, *Comptes rendus, Ser. C*, **266**, 462.

740

Cloud, W. H., D. S. Schreiber and K. R. Babcock, 1962, *J. Appl. Phys.*, **33**, 1193.

Colombo, U., G. Fagherazzi, F. Gazzarrini, G. Lanzavecchia and G. Sironi, 1964a, *Chim. Ind. (Milan)*, **46(4)**, 357.

Colombo, U., G. Fagherazzi, F. Gazzarrini, G. Lanzavecchia and G. Sironi, 1964b, *Nature*, **202**, 175.

Cox, N. L., 1963, *U.S. Patents* 3,074,778; 3,078,147.

Craik, D. J. and P. M. Griffiths, 1958, *Brit. J. Appl. Phys.*, **9**, 280.

Craik, D. J. and R. Lane, 1969, *Brit. J. Appl. Phys. Ser. 2*, **2**, 33.

Craik, D. J. and D. A. McIntyre, 1967, *Proc. Roy. Soc.*, **A302**, 99.

Daniel, E. D. and D. F. Eldridge, 1967, *Magnetic Recording in Science and Industry*, ed. C. B. Pear, Jr. (Reinhold: New York), p. 58.

Daniel, E. D. and I. Levine, 1960, *J. Acous. Soc. Amer.*, **32**, 258.

Darnell, F. J., 1961, *J. Appl. Phys.*, **32**, 1269.

Darnell, F. J. and W. H. Cloud, 1965, *Bull. Soc. Chim. France*, **4(4)**, 1164.

David, I. and A. J. E. Welch, 1956, *Trans. Farad. Soc.*, **52**, 1642.

Della Torre, E., 1965, *J. Appl. Phys.*, **36**, 518.

De Vries, R. C., 1966, *Mat. Res. Bull.*, **1**, 83.

Eagle, D. F. and J. C. Mallinson, 1967, *J. Appl. Phys.*, **38**, 995.

Elder, T., 1965, *J. Appl. Phys.*, **36**, 1012.

Eldridge, D. F., 1964, *I.E.E.E. Trans. Comm. and Electronics*, **83**, 585.

Ervin, G., 1952, *Acta Cryst.*, **5**, 103.

FIAT, 1947, *Final Report No. 923* (H.M. Stationery Office: London).

Fukuda, S., H. Goto, G. Akashi and T. Miyake, 1962, *U.S. Patent* 3,046,158.

Fukuda, S., T. Miyake, G. Akashi and M. Seto, 1962, *U.S. Patent* 3,026,215.

Goto, H. and G. Akashi, 1962, *U.S. Patent* 3,047,428.

Guillaud, C., A. Michel and J. Bénard, 1944, *Comptes rendus*, **219**, 58.

Gustard, B. and W. J. Schuele, 1966, *J. Appl. Phys.*, **37**, 1168.

Gustard, B. and H. Vriend, 1969, *I.E.E.E. Trans. Mag.*, **MAG-5**, 326.

Healey, F. H., J. J. Chessick and A. V. Fraioli, 1956, *J. Phys. Chem.*, **60**, 1001.

Hirsch, A. A. and Z. Eliezer, 1966, *Physica (Netherlands)*, **32**, 591.

Hoagland, A. S., 1963, *Digital Magnetic Recording* (Wiley: New York).

Hurt, J., A. Amendola and R. E. Smith, 1966, *J. Appl. Phys.*, **37**, 1170.

Imaoka, Y., 1968, *J. Electrochem. Soc. Japan* (Overseas Suppl. Edn.), **36(1)**, 15.

Ingraham, J. N. and T. J. Swoboda, 1960, *U.S. Patent* 2,923,683.

Iwasaki, S. and M. Matsumoto, 1967, *Sci. Rep. Res. Inst., Tohoku Univ.*, B-(Elect. Comm.), **19**, 89.

Iwasaki, S. and T. Suzuki, 1968, *I.E.E.E. Trans. Mag.*, **MAG-4**, 269.

Jacobs, I. S., 1969, *J. Appl. Phys.*, **40**, 917.

Jacobs, I. S. and C. P. Bean, 1955, *J. Appl. Phys.*, **29**, 537.

Jacobs, I. S. and C. P. Bean, 1955, *Phys. Rev.*, **100**, 1060.

Jaep, W. F., 1969, *J. Appl. Phys.*, **40**, 1297.

Jeschke, J. C., 1954, *E. German Patent* 8,684.

Johnson, C. E. and W. F. Brown, 1958, *J. Appl. Phys.*, **29**, 1699.

Klimaszewski, B. and J. Pietrzak, 1969, *Bull. Acad. Pol., Sci. Ser., Math., Astron., Phys.*, **17(1)**, 51.

Kneller, E. F. and F. E. Luborsky, 1963, *J. Appl. Phys.*, **34**, 656.

Kondorskii, E., 1952, *Izvest. Akad. Nauk S.S.S.R., Ser. Fiz.*, **16**, 398.

Kouvel, J. S. and D. S. Rodbell, 1967, *J. Appl. Phys.*, **38**, 979.

Krones, F., 1955, *Mitt. Forschungslab. Agfa.*, **1**, 289.

Krones, F., 1960, *Technik der Magnetspeicher* (Springer-Verlag: Berlin), p. 479.

Kubota, B., 1960, *J. Phys. Soc. Japan*, **15**, 1706.

Kubota, B., 1961, *J. Am. Cer. Soc.*, **44**, 239.

Kubota, B., T. Nishikawa, H. Chiba and M. Sugimura, 1966, *U.S. Patent* 3,243,260.

Kubota, B., T. Nishikawa, A. Yanase, E. Hirota, T. Mihara and Y. Iida, 1963, *J. Am. Cer. Soc.*, **46**, 550.

Lissberger, P. H. and R. L. Comstock, 1970, *I.E.E.E. Trans. Mag.*, **MAG-6**, 512.

MacDonald, R. E. and J. W. Beck, 1969, *J. Appl. Phys.*, **40**, 1429.

Mallinson, J. C., 1969, *I.E.E.E. Trans. Mag.*, **MAG-5**, 91.

Mallinson, J. C. and C. W. Steele, 1969, *I.E.E.E. Trans. Mag.*, **MAG-5**, 186.

McNab, T. K., R. A. Fox and A. J. F. Boyle, 1968, *J. Appl. Phys.*, **39**, 5703.

Mee, C. D., 1964, *The Physics of Magnetic Recording*, Ch. 2 (North-Holland: Amsterdam).

Michel, A. and J. Bénard, 1935, *Comptes rendus*, **200**, 316.

Michel, A. and J. Bénard, 1943, *Bull. Soc. Chim. Fr.*, **10**, 315.

Morrish, A. H. and L. A. K. Watt, 1957, *Phys. Rev.*, **105**, 1476.

Morrish, A. H. and S. P. Yu, 1955, *J. Appl. Phys.*, **26**, 1049.

Morrison, J. R. and D. E. Speliotis, 1966, *I.E.E.E. Trans. Elect. Comput.*, **EC-15**, 782.

Morrison, J. R. and D. E. Speliotis, 1968, *I.E.E.E. Trans. Mag.*, **MAG-4**, 290.

Moskowitz, R. and E. della Torre, 1967, *J. Appl. Phys.*, **38**, 1007.

Néel, L., 1947, *Comptes Rendus Acad. Sci., Paris*, **224**, 1550.

Néel, L., 1949, *Comptes Rendus Acad. Sci., Paris*, **228**, 664.

Néel, L., 1958, *Comptes Rendus Acad. Sci., Paris*, **246**, 2313.

Néel, L., 1959, *J. Phys. Radium*, **20**, 215.

Newman, J. J. and R. B. Yarbrough, 1968, *J. Appl. Phys.*, **39**, 5566.

Newman, J. J. and R. B. Yarbrough, 1969, *I.E.E.E. Trans. Mag.*, **MAG-5**, 320.

Nobuoka, S., T. Ando and F. Hayama, 1963, *U.S. Patent* 3,081,264.

Okamoto, S., S. Tochihara and Y. Imaoka, 1966, *Denki Kagaku*, **34**, 749.

Oppegard, A. L., 1959, *U.S. Patent* 2,885,365.

Oppegard, A. L., F. J. Darnell and H. C. Miller, 1961, *J. Appl. Phys.*, **32**, 184S.

Osmond, W. P., 1952, *Proc. Phys. Soc., London*, **65B**, 121.

Osmond, W. P., 1953, *Proc. Phys. Soc., London*, **66B**, 265.

Osmond, W. P., 1954, *Proc. Phys. Soc., London*, **67B**, 875.

Pear, C. B., Jr., 1967, *Recording in Science and Industry* (Reinhold: New York).

Penniman, R. S., Jr. and N. M. Zoph, 1921, *U.S. Patent* 1,368,748.

Poulsen, V., 1900, *U.S. Patent* 661,619.

Preisach, F., 1935, *Zeits, Physik*, **94**, 277.

Rodbell, D. S., 1966, *J. Phys. Soc. Japan*, **21**, 1224.

Rodbell, D. S., R. C. De Vries, W. D. Barber and R. W. DeBlois, 1967, *J. Appl. Phys.*, **38**, 4542.

Rode, T. V. and V. E. Rode, 1961, *Zhur. Physcheskoi Chem.*, **35**, 2475.

Ruben, S., 1932, *U.S. Patent* 1,889,380.

Shtrikman, S. and D. Treves, 1959, *J. Phys. Radium*, **20**, 286.

Shtrikman, S. and D. Treves, 1960, *J. Appl. Phys.*, **31**, 58S.

Shtrikman, S. and D. Treves, 1963, *Magnetism*, Vol. III (Academic Press: New York), Chapter 8.

Siratori, K. and S. Iida, 1960, *J. Phys. Soc. Japan*, **15**, 210.

Sládek, J., 1970, *I.E.E.E. Trans. Mag.*, **MAG-6**, 506.

Smaller, P. and J. Newman, 1969, *I.E.E.E. Trans. Mag.*, **MAG-5**, 327.

Smith, D. O., 1956, *Phys. Rev.*, **102**, 959.

Speed, W. C., 1955, *U.S. Patent* 2,796,359.

Speliotis, D. E., 1968, *I.E.E.E. Trans. Mag.*, **MAG-4**, 553.

Speliotis, D. E., J. R. Morrison and G. Bate, 1965, *Proc. Int. Conf. Magnetism, Nottingham*, p. 623.

Stoner, E. C. and E. P. Wohlfarth, 1958, *Phil. Trans. Roy. Soc.*, **A240**, 599.

Sugaya, H., F. Kobayashi and M. Ono, 1969, *I.E.E.E. Trans. Mag.*, **MAG-5**, 437.

Swoboda, T. J., P. Arthur, N. L. Cox, J. N. Ingraham, A. L. Oppegard and M. S. Sadler, 1961, *J. Appl. Phys.*, **32**, 374S.

Sykora, V., 1967, *Industrie Chemique Belge*, **32**, Spec. no. Pr. 2, 428.

Szegvari, A., 1955, *U.S. Patent* 2,719,009.

Takei, H. and S. Chiba, 1966, *J. Phys. Soc. Japan*, **21(7)**, 1255.

Templeton, H. S. and G. Bate, 1964, *I.E.E.E. Trans. Comm. and Electronics*, **83**, 429.

Tjaden, D. L. A. and J. Leyten, 1963/64, *Philips Tech. Rev.*, **25**, 319.

Van Oosterhout, G. W., 1960, *Acta Cryst.*, **13**, 932.

Verwey, E. J. W., 1935, *Zeits. Krist.*, **91**, 65.

Von Kobell, F., 1838, *Grundzuge der Mineralogie*, 304.

13 *Magnetic bubbles*

A. H. BOBECK

13.1 INTRODUCTION

Cylindrical magnetic domains, as described in Chapter 9, have recently been utilized in memory and logic 'bubble' devices (Bobeck, 1967; Bobeck *et al.*, 1969; Perneski, 1969). In the course of developing these devices it has been necessary to investigate many magnetic oxide materials and to develop techniques to evaluate these materials. In this chapter we will (1) consider the general material requirements for bubble materials, (2) introduce techniques developed specifically for evaluating these materials and (3) compare specific materials in terms of storage densities, data rates, etc. To this end it will be necessary to review the static and dynamic behaviour of bubble domains.

13.2 STATIC PROPERTIES OF CYLINDRICAL DOMAINS

In Figure 13.1 strip domains in a 55 μm thick platelet of terbium orthoferrite made visible by the Faraday effect form cylindrical bubble domains when a

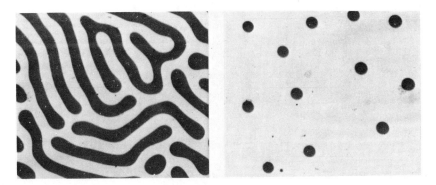

Figure 13.1   Strip domains (left) seen by means of Faraday rotation of polarized light represent regions magnetized alternately inward and outward in a 55 μm thick platelet of terbium orthoferrite. Domains are typically 90 μm in width. When a 50 Oe bias field is applied perpendicular to the orthoferrite sheet (right) a cylindrical domain or 'bubble' precisely 35 μm in diameter is formed for each properly oriented single-walled strip domain.

50 Oe field is applied perpendicular to the platelet surface. Static properties of these cylindrical domains have been discussed in Chapters 1 and 9 (and see, for example, Thiele, 1969). It is assumed that the magnetization orients along the uniaxial axis perpendicular to the platelet except, of course, through the domain wall itself, that is, $H_k \gg 4\pi M_s$ (see also Craik and Cooper, 1972). Furthermore the domain wall width is assumed to be much less than the domain diameter.

The geometry of a cylindrical domain is illustrated in Figure 13.2. Using the method of Bobeck, the total energy can be expressed as the sum of the applied field, domain wall and magnetostatic energies.

$$\xi_T = 2M_s H_A \pi r^2 h + 2\pi r h \sigma_w - \xi_D \qquad (13.1)$$

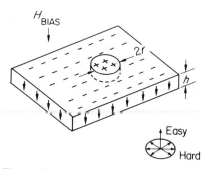

Figure 13.2 Geometry of a cylindrical domain in a uniaxial platelet. The domain wall width is assumed much less than $2r$.

Taking the partial derivative with respect to the radius r and normalizing to the bias field H_A (applied normal to the platelet surface) results in, for a stable domain,

$$0 = H_A + \frac{\sigma_w}{2rM_s} - \frac{\partial\xi_D/\partial r}{4\pi M_s hr} \qquad (13.2)$$

The term $\sigma_w/2rM_s$ designated H_w represents a field generated by the domain wall energy. H_A, the bias field, and H_w, the effective wall energy field, are opposed by a magnetostatic field $\partial\xi_D/\partial r/4\pi M_s hr$ which we call H_D.

$$\frac{H_D}{4\pi M_s} = \frac{2}{\pi}\left[-\frac{2r_0}{h} + \sqrt{1 + (4r_0^2/h^2)}E(k, \pi/2)\right], \qquad (13.3)$$

where $E(k, \pi/2)$ is the complete elliptic integral of the second kind and

$$k^2 = \frac{1}{1 + (h^2/4r_0^2)}$$

Equation (13.2) applies over the range in which circular domains are stable. Thiele has shown that for a given magnetic material and platelet thickness there exists an upper limit to the domain diameter. We designate this diameter $2r_{S-B}$ with the subscript reminding us of the onset of a reversible bubble to domain strip instability.

The terms of (13.2) are plotted in Figure 13.3. We have adjusted the bias field so that the bubble diameter is $2r_{S-B}$ and have designated that particular field H_{S-B}. If the bias field is now increased by an amount H_0 then the domain diameter will reduce to $2r_c$. Any further increase in the bias will result in bubble collapse. It is important to note the significance of the intercept 'a'. Under special conditions it is possible to observe bubbles at a diameter $2r_c^*$. However, at this diameter they are unstable and with the slightest perturbation either recover to the diameter $2r_{S-B}$ or collapse. In practice $r_c^* \sim r_c/2$.

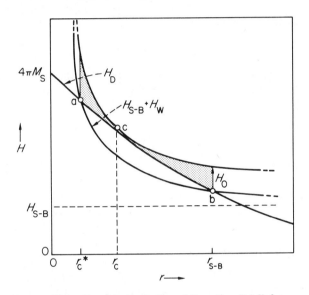

Figure 13.3 Graphical solution of Equation (13.2) for a bias field H_{S-B}. Intersection 'b' is the stable solution while 'a' is unstable. When the bias field is increased to $H_{S-B} + H_0$ the cylindrical domain decreases to the minimum stable radius r_c.

The interplay of the various fields effective on the domain wall of a bubble domain are readily apparent from a study of Figure 13.3. Using this presentation, however, it is rather cumbersome to extract parameters such as σ_w and M_s from bubble domain measurements. Rather the force equation of Thiele

$$\frac{l}{h} + \frac{d}{h} \frac{H_A}{4\pi M_s} - F\left(\frac{d}{h}\right) = 0 \tag{13.4}$$

plotted in Figure 13.4, is found to be more useful. In this expression l is a material length parameter defined as $l = \sigma_w/4\pi M_s^2$. Also included in Figure 13.4 are Thiele's 'stability functions' $S_0(d/h)$ for bubble domain collapse and

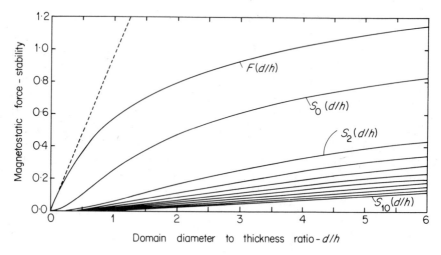

Figure 13.4 The magnetostatic radial force function $F(d/h)$ and stability functions, S_0–S_{10} ($S_1 = 0$), are functions of domain diameter to thickness ratio, d/h.

$S_2(d/h)$–$S_{10}(d/h)$ for cylinder to strip instability. $S_0(d/h)$ is derived from $F(d/h)$; $S_2(d/h)$ results from a perturbation analysis. To appreciate the significance of $S_2(d/h)$ it should be noted that the demagnetizing energy is reduced if a domain becomes elliptical whereas the applied field and domain wall energies are at a minimum with the domain circular.

To convert from 'fields' to 'forces' we multiply the former by the domain wall area $2\pi rh$. This has the effect of converting H_w, the wall energy field term with a $1/r$ dependence, to a force term independent of r.

A graphical solution to (13.4) may be obtained by constructing a line on Figure 13.4 whose intercept with the vertical axis is l/h and whose slope is $H_A/4\pi M_s$. The intersections of this straight line with the F curve are then the solutions to the force equation. Study of the force equation shows that stable isolated cylindrical domains exist only in the presence of an applied field having a magnitude between zero and $4\pi M_s$ and polarity tending to collapse a domain.

The theory of Thiele tells us the following:

(1) The diameter of the smallest stable domain realizable is approximately $4l$.
(2) The platelet thickness h which yields the smallest bubble domain is πl.
(3) A bias field $H_A = (0.3)4\pi M_s$ is needed to support the minimum diameter bubble domain.

The dependence on thickness of the domain diameter is illustrated in Figure 13.5. Note that even for thicknesses many times the so-called optimum the domain diameter is not significantly increased. There is, however, a real size penalty with thicknesses below optimum. Domains in thick platelets tend to remain round whereas those in thin platelets often assume irregular shapes and falsely convey the impression of a high material coercivity.

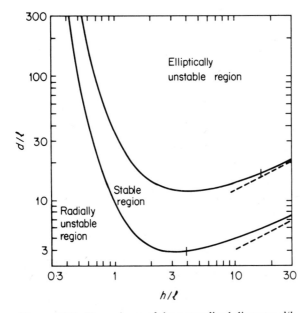

Figure. 13.5 Dependence of the normalized diameter d/h on the normalized thickness h/l. The upper curve defines the bubble to strip instability and the lower curve bubble collapse.

It is well to remember (Figure 13.6) that the field external to a uniformly magnetized platelet is zero and that the field within that platelet is $4\pi M_s$. This can immediately be derived from the equation $H_z = 2M_s\theta$ illustrated in Figure 13.7. A uniformly thick 'permanent magnet' film since it has no external field cannot be used to supply the bias field necessary to support bubble domains. Liu *et al.* (1971) have shown that a permanent magnet film uniformly magnetized

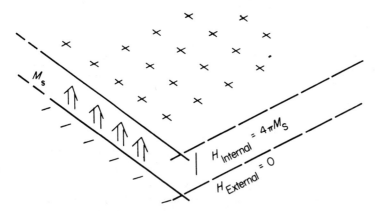

Figure 13.6 The internal magnetostatic field within a uniformly magnetized platelet is $4\pi M_s$, the external field is zero.

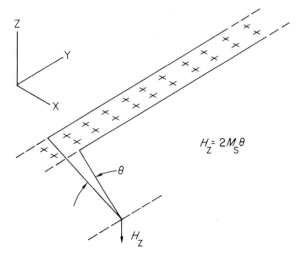

$$H_z = 2M_s\theta$$

Figure 13.7 Normal magnetostatic field component H_z generated by a strip of magnetic charge is proportional to the angle θ.

normal to the film surface, exchange coupled to a bubble domain platelet can supply an effective bias field H_A of amplitude $\sigma_w/2hM_s$ where σ_w, h and M_s are parameters of the domain wall material.

13.3 METHODS TO DETERMINE MATERIAL PARAMETERS BASED ON STATIC BUBBLE DOMAIN THEORY

Magnetic moment, uniaxial anisotropy and domain wall energy represent a complete set of material parameters from which static bubble behaviour can be predicted. The first two can be readily measured on bulk samples with, for example, a torque magnetometer and a B–H characteristic plotter. Such methods are often either inconvenient or impractical when applied to extremely thin, small area platelets such as are frequently encountered in the search for domain wall materials. On the other hand domain wall energy is generally determined by a fit of experimental data to predictions based on a theoretical study of the observed domain configuration.

Let us now consider, as an example, the use of Figure 13.4 to determine graphically the material parameters for $Gd_{2.34}Tb_{0.66}Fe_5O_{12}$ garnet provided as a (111) platelet of thickness $h = 15\,\mu m$ wafered from beneath a (211) facet of flux grown crystal. The nature of the graphical solutions are indicated in Figure 13.8 which has been modified to include a direct reading bias field axis. By direct observation using the Faraday effect, the experimental data on cylindrical domain collapse and runout presented in Table 13.1 was obtained. The remaining tabulated parameters were determined as follows. Consider first the column 'Bubble to strip'. Calculate $2r_{s-B}/h = 1.35$, Construction abc, using

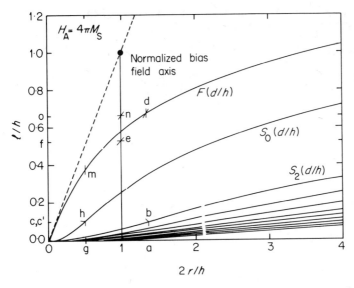

Figure 13.8 Constructions used to obtain the parameters of Table 13.1. Construction abcdef concerns bubble to strip instability; ghc'mno bubble collapse. The direct reading bias field axis simplifies the use of the curves.

stability curve $S_2(d/h)$, gives $l/h = 0.096$. Construction cdef gives $H_A/4\pi M_s + l/h = 0.52$. Parameters πl, $4\pi M_s$ and σ_w are then readily calculated since $h = 15$ μm and $\sigma_w = 4\pi l M_s^2$. For column 'Bubble collapse' calculate $2r_c/h = 0.5$. From construction ghc' (hopefully c = c') $l/h = 0.10$ and from c'mn $H_A/4\pi M_s + l/h = 0.66$. The two columns of data are in good agreement. To obtain H_k, the uniaxial anisotropy field, one must know the exchange constant A for this garnet. The published value for YIG (Rado and Suhl, 1963) is $A = 4.4 \times 10^{-7}$ ergs/cm. We use instead $A = 1.0 \times 10^{-7}$ since this gives rather good agreement among bubble, direct and torque measurements of H_k. Since $\sigma_w = 4\sqrt{AK_u}$ and $H_k = 2K_u/M_s$ it follows that $H_k \sim 5500$ Oe.

Table 13.1

		Bubble to strip	Bubble collapse
Experimental	H_A	57 Oe	75
	$2r$	20.0 μm	7.5
Calculated	πl	4.6 μm	4.8
	$4\pi M_s$	135 gauss	137
	σ_w	0.21 ergs/cm^2	0.23
	K_u	2.8×10^4 ergs/cm^3	3.3×10^4
	H_k	5000 Oe	6100

13.4 MANIPULATION OF CYLINDRICAL DOMAINS TO OBTAIN MATERIAL PARAMETERS

13.4.1 Domain wall coercivity

As already stated, domains are maintained in the cylindrical form by a uniform bias field applied normal to the platelet surface. An increase in the bias field decreases the domain diameter and vice versa. However, if a gradient field is applied the domain may respond by moving.

Consider, as shown in Figure 13.9, a cylindrical domain of diameter $2r$ in a uniform gradient field. The domain will experience a force attempting to move

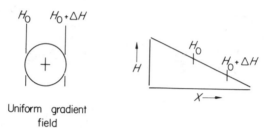

Uniform gradient
field

Figure 13.9 Cylindrical domain of diameter $2r$ positioned in a uniform gradient field. The force on the domain drives it towards a position of reduced average field.

it toward a position of reduced bias. To overcome wall coercivity, H_c, the following condition must be met: $\Delta H > 8H_c/\pi$. One method to produce a gradient field is to interact one domain with another. In the case of domains widely separated the far field of a domain can be approximated as that of a dipole and the following relationship derived (see Figure 13.10):

$$H_c/4\pi M_s = 3\pi r_0^3 h/8l_{12}^4 \qquad (13.5)$$

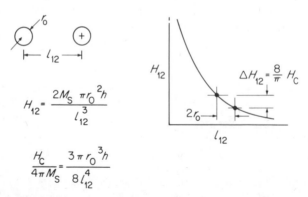

$$H_{12} = \frac{2M_s \, \pi r_0^2 h}{l_{12}^3}$$

$$\frac{H_c}{4\pi M_s} = \frac{3\pi r_0^3 h}{8 l_{12}^4}$$

Figure 13.10 Two domains, mutually repelled in material whose coercive force is H_c, reach stable separation l_{12}.

(13.4) provides a useful method for measuring H_c, the domain wall coercivity. For example, let us consider again the 15 μm thick garnet platelet described in Section 13.3. By means of a permalloy thin wire probe the separation between a pair of 15 μm diameter domains was first reduced below the stable minimum and the probe withdrawn (Figure 13.11). The bubble domains then stabilized

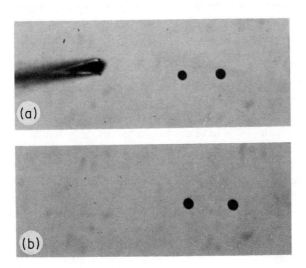

Figure 13.11 A permalloy wire probe is manoeuvred to reduce the domain separation below the unperturbed minimum stable separation and withdrawn (a). The domains then separate somewhat and recover to identical diameters (b). Bias field is 115 Oe.

at a separation of 52·5 μm. Entering these data into (13.5) gives $H_c = 0·17$ Oe. This value is typical of that obtained for garnets and orthoferrites. Thulium orthoferrite, $TmFeO_3$, with $H_c \sim 0·01$ Oe had the smallest coercivity of any material measured.

13.4.2 Domain wall mobility

The study of the domain wall transient behaviour in magnetic platelets is complicated by the two-dimensional nature of domain patterns. Nevertheless, experimenters have developed techniques for measuring domain wall mobilities (Section 9.5.2). In this section we shall restrict the scope of the discussion to cover those techniques which deal specifically with bubble domains.

Rossol (1969) has measured domain wall mobility by stroboscopic observation of a bubble domain subjected to a sinusoidal modulation of the stabilizing bias field. Rossol states that 'direct optical observation removes ambiguity as to the number and configuration of domain walls present and provides a means of assessing the quality of the material where walls are being moved'. Indeed, it is one of the important attributes of bubble domain measurement techniques that they can start with a specific and easily reproduced domain configuration.

In the case where the domain wall velocity $\dot{\chi}$ for a straight wall is linearly related to drive field H, one can define a domain wall mobility $\mu = d\dot{\chi}/dH$. For a domain caused to pulsate in diameter by a sinusoidally varying bias field the simple relaxation equation $\beta\dot{\chi} + \alpha\chi = 2MH \exp(i\omega t)$ yields $\mu = 2M/\beta = x_0\omega_c/H$. Here β is a viscous damping parameter, α is the restoring force, $x_0 = 2MH/\alpha$ is the low frequency wall displacement and ω_c is the frequency at which the displacement reduces to $s_0/\sqrt{2}$. With this cylindrical domain technique and a similar straight wall technique, Rossol (1970) has measured mobilities of the orthoferrites obtaining a high of 5000 cm/s/Oe for $YFeO_3$ to a low of 200 cm/s/Oe for $HoFeO_3$. He finds that they are well characterized by the simple relaxation model.

Rossol and Thiele (1970) have reported a technique by which a bubble domain in $Sm_{0.55}Tb_{0.45}FeO_3$ orthoferrite travels around the circumference of a circular permalloy film in response to a rotating in-plane field. An increase in lag angle is noted as the frequency of the in-plane rotating field is increased and the data analysed to provide a domain wall mobility. This method is much too cumbersome to find general use; however, it does give an experimental confirmation of the relation between domain wall mobility μ and bubble domain velocity. Thiele has shown that $\dot{\chi}$, the velocity of a bubble domain, is given by

$$\dot{\chi} = \mu\Delta H/2 \tag{13.6}$$

In this equation ΔH is the field differential across the diameter of the domain (see Figure 13.9).

We now describe a method used by Bobeck et al. (1971a) to determine domain wall dynamics. In this method a bubble domain is driven by a pulse field which, for the duration of the pulse, increases the bias field and drives the domain toward collapse. By noting both the duration and amplitude of the pulse field to just collapse the domain, a mobility or, more specifically, a wall velocity versus drive plot can be generated. In this technique velocity data are taken over a very wide dynamic range.

Refer once again to Figure 13.3. Assume that the bias field is adjusted to H_{S-B}. Application of a pulsed magnetic field H_p as per Figure 13.12 will drive the bubble domain from its starting diameter $2r_{S-B}$ toward zero diameter. Because of the losses associated with the domain wall motion we expect a gradual decrease in diameter. We can then plot as in Figure 13.13 the amplitude of the pulse field H_p versus the reciprocal of the pulse width $1/T$ to just exceed $2r_c^*$. From the slope of the $1/T$ versus H_p line and the distance a wall travels a domain wall mobility can be derived.

From a study of Figures 13.3 and 13.12 it is seen that the effective field driving the domain wall inward is not constant but rather proceeds smoothly from H_p to H_p-H_0 and back again. Using a quadratic approximation it can be shown that a deviation from a linear $1/T$ versus H_p characteristic can be expected at

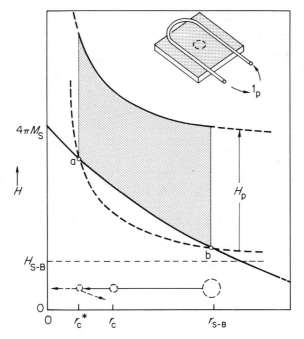

Figure 13.12 Responding to H_p, the instantaneous increase in the bias field, the bubble shrinks in diameter at a rate related to the domain wall mobility. If the pulse field is removed before the bubble diameter reaches $2r_c^*$ then the bubble will return to its original size, $2r_{S-B}$.

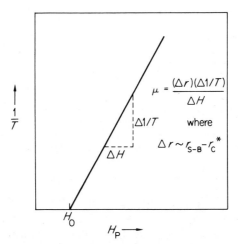

Figure 13.13 Details for obtaining the mobility from a plot of $1/T$ versus H_p.

the lower pulse drives. As an example, experimental data taken over the range $H_0 < H_p < 2H_0$ should have a slope $4/\pi$ times the actual mobility.

The experiment requires a current pulser that can be triggered manually, an oscilloscope to monitor the amplitude and duration of the current pulse, a single or multiple turn jig and a polarizing microscope. A bubble is manoeuvred within the conductor loop by means of a magnetic probe. The pulse amplitude (or duration) is then slowly increased until the bubble disappears. It is important that the bubble should actually collapse and not merely escape outside the conductor loop. This can be ensured by choosing a conductor geometry which produces a saddle field point. A pair of 0·001 inch diameter wires separated 0·015 inch has proved to be entirely satisfactory.

Data taken on $PrFeO_3$ orthoferrite and $Gd_{2.34}Tb_{0.66}Fe_5O_{12}$ garnet are given in Figures 13.14 and 13.15, respectively. In each case the domain wall velocity is a linear function of the applied field and a mobility can be calculated.

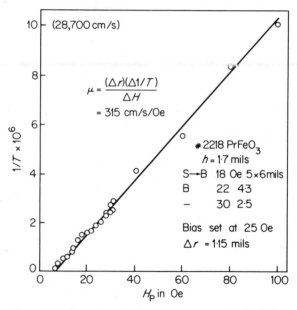

Figure 13.14 A domain response curve typical of that obtained for the orthoferrites. Notice that in this example the bias field had been set well above the strip-bubble value. The corresponding Δr is then computed using Thiele's curves.

Not all materials can be characterized by a mobility (Asti et al., 1965). The hexagonal ferrites, when measured by this technique, display rather non-linear $1/T$ versus H characteristics. Such non-linear behaviour has been noted by other observers (see Section 9.5.1.6). Furthermore, there is an upper limit to the domain wall velocity. Data on two hexagonal ferrites are given in Figures 13.16

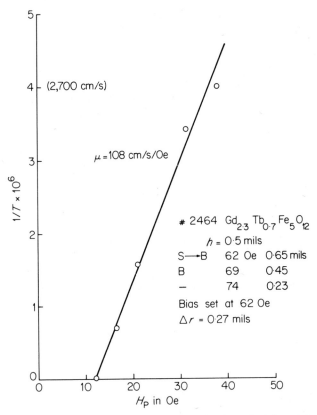

Figure 13.15 Domain response curve for the uniaxial garnet $Gd_{2.3}Tb_{0.7}Fe_5O_{12}$.

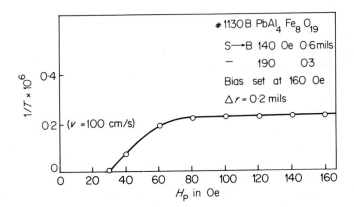

Figure 13.16 A domain response curve for the hexagonal ferrite $PbAl_4Fe_8O_{19}$. This non-linear curve can best be characterized by an initial mobility $\mu_{in} = 3$ cm/s/Oe and a limiting wall velocity $v_{max} = 110$ cm/s.

and 13.17. The highest limiting velocity, 600 cm/s, was achieved in a platelet of barium ferrite.

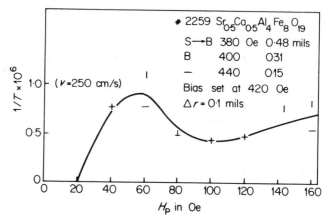

Figure 13.17 Occasionally, as with this flux grown hexagonal ferrite, the velocity versus drive relationship exhibits a curious overshoot.

13.5 GENERAL MATERIAL REQUIREMENTS

Cylindrical domains can be introduced into platelets of such diverse magnetic materials as MnBi, FeF_3, $\alpha\text{-}Fe_2O_3$, $FeBO_3$, $Li_{0.5}Fe_{2.5}O_4$ and the general material classes of the orthoferrites, hexagonal ferrites and the garnets (Kurtzig and Guggenheim, 1970; Wolfe et al., 1970; Charap and Nemchik, 1969; Kozlowski and Zietek, 1966).

What all these materials have in common, at least when they support bubble domains, is a uniaxial anisotropy either intrinsic or induced.

In a hexagonal ferrite such as $PbFe_{12}O_{19}$ bubble domains less than a micron in diameter readily form. Hexagonal ferrites have a very high magnetic moment (≈ 4000 gauss) and spontaneously nucleate reversal domains. On the other hand, in the orthoferrites, it is usually necessary to 'cut' strip domains to introduce bubble domains. Orthoferrites, which have a rather low magnetic moment (≈ 100 gauss) and a very high anisotropy field, do not spontaneously nucleate new bubble domains. Both the hexagonal ferrites and the orthoferrites have large *intrinsic* uniaxial anisotropies.

On the other hand the trigonal magnetic oxide, $\alpha\text{-}Fe_2O_3$, has an easy plane of magnetization and only if a properly oriented platelet of this material is stressed will it support bubble domains (F. B. Hagedorn, private communication). The domains in this case are hundreds of microns in diameter, too large for $\alpha\text{-}Fe_2O_3$ to be considered as a practical bubble material.

Just what then are the requirements of a bubble material. The most general requirement can be stated as follows: *The magnetization should orient normal to the surface of the platelet and be supported by a uniaxial anisotropy.* Mathematically we can restate this as $H_k > 4\pi M_s$ keeping in mind that this expression is to be used more as a rule of thumb rather than a well-defined limit.

Of the materials with an intrinsic uniaxial anisotropy, the orthoferrites meet this requirement easily and the hexagonal ferrites and garnets marginally. These materials, especially the garnets, can be modified to meet specific requirements since they lend themselves to extensive material engineering.

A list of other more general requirements of a useful material follows:

(1) The material must be available as large, essentially defect free single crystals, platelets or epitaxial films. Thickness, magnetization and wall energy should be constant to within 1% over at least a square centimetre.

(2) Thin platelets should support bubble domains 5–10 μm in diameter. In terms of the material length this is restated as $l(= \sigma_w/4\pi M_s^2) \approx 2$ μm. Practical considerations such as bubble domain generation indicate that the saturation magnetization falls within a 100–200 gauss range.

(3) Circular, not elliptical, bubble domains are desired. Magnetization reversal should be by 180° domain wall with the domain wall energy substantially constant around the perimeter of a bubble domain.

(4) Bubble domains must be mobile. If a high data rate is to be achieved with low drive power, a low domain wall coercivity and a high domain wall mobility are desired. A megabit data rate with 8 μm bubbles will require a mobility of 800 cm/s Oe. An approximate upper limit to the domain wall coercivity is 1 Oe.

(5) The material should be optically transparent at the thickness required ($\approx \pi l$) and possess a useful Faraday rotation. Although not a necessity, the ability to observe domain behaviour enables one to quickly judge crystal quality and expedites the design and testing of devices.

(6) Bubble domain properties should be insensitive to temperature change. A most important parameter is the magnetic moment. Bubble diameter (d) is related to material constants by the expression $d \approx (AK_u)^{\frac{1}{2}}M_s^{-2}$. Operating magnetic field bias is strongly dependent on M_s.

Many of the above requirements are really just guidelines to aid in the search for improved bubble materials. Gianola *et al.* (1969) compared the static properties of bubble domains in a variety of materials using a format similar to that of Figure 13.18. An ideal situation in which the platelet thickness $h = \pi l$ and the operating bubble diameter $d = 8l = \sqrt{32AH_k/\pi^2 M_s^3}$ is assumed. Furthermore $A = 1.0 \times 10^{-7}$ ergs/cm (see Section 13.3) is used as the exchange constant for all materials. This simplifies the presentation yet gives precision in the garnets where it is needed. For Fe and Ni the cubic anisotropy constants K_1 are treated as though they are uniaxial anisotropy constants K_u.

Clearly one constraint is that the domain wall width l_w cannot exceed the bubble diameter. The indicated preferred material sector is much less exact as to its origin but is based rather on device considerations such as storage density, ease of domain subdivision and the effect of coercivity. A magnetization

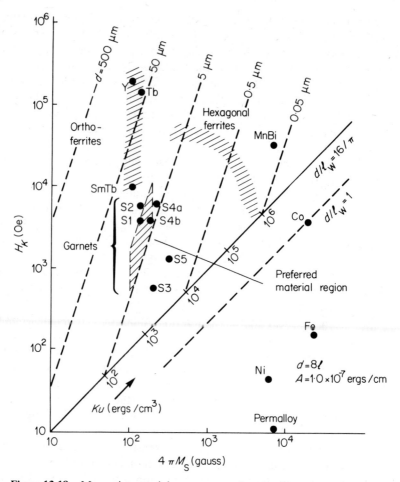

Figure 13.18 Magnetic materials are compared on an H_k versus $4\pi M_s$ plot. The preferred material region as it pertains to bubble devices is determined by considerations of storage density and device performance. Note the position of permalloy, a uniaxial metallic alloy.

$4\pi M_s = 150$ gauss and an operating diameter $d = 8\,\mu m$ are reasonable objectives.

Of the materials indicated only the garnets and possibly the composite orthoferrites would seem to be serious candidates. The garnets designated S1, S2, etc., are the uniaxial garnets discussed in Section 13.6.2.

13.6 BUBBLE DOMAIN MATERIALS

Although the list of room temperature magnetic materials is extensive, those with a chance of fulfilling the requirements outlined in the previous section are few. Only the orthoferrites, hexagonal ferrites and garnets have been singled out for extensive evaluation. The static stability of cylindrical domains in hexagonal ferrites were first studied by Kooy and Enz (1960). Recent effort has shown that these hexagonal ferrites do meet the requirements of a bubble material except for their very low and non-linear mobility. As a result the present materials activity centres on orthoferrites and garnets.

13.6.1 Orthoferrites

Orthoferrites are usually grown from molten salt solutions (fluxes) following the method of Remeika (1956). Okada *et al.* (1971) reported the growth of yttrium orthoferrite rods 8 mm in diameter using a floating zone technique. Blank *et al.* (1971) prepared yttrium orthoferrite from the melt by the Bridgman technique.

Virtually all of the information reported in this section was obtained on orthoferrites grown by the flux method. Although the flux method occasionally produces platelets several mils thick with the uniaxial magnetization axis normal to the surface, most often it is necessary to slice larger crystals into plates and then polish the plates to the desired thickness. Chemical etch, high temperature anneal, sputter etch and Syton polish were some of the procedures used to reduce or eliminate the strains introduced by this polishing.

An excellent paper by Treves (1962) summarizes many of the properties of the orthoferrites. A pair of antiparallel, or nearly so, spin systems cant approximately 0·5 degree giving rise to $4\pi M_s$ typically 100 gauss. The uniaxial anisotropy axis at room temperature is the [001] axis in all orthoferrites except $SmFeO_3$ where it is the [100] axis.

The combination of a low magnetic moment and high anisotropy field results in simple domain structures being observed in these materials. Orthoferrites are sufficiently transparent in the red to enable direct visual observation of domain behaviour by means of the Faraday effect.

We are, of course, concerned with the suitability of these materials for use in domain wall devices. To this end it was necessary to examine bubble domain size, coercivity, mobility and uniformity. A summary of data on orthoferrites is presented in Table 13.2. As expected materials with high moments possess small diameter bubble domains and vice versa. This follows from the material parameter $l = (\sigma_w/4\pi M_s^2)$ which has an inverse square dependence on M_s. The domain wall energy density σ_w, obtained as per Section 13.3, is about 2 ergs/cm^2 and varies only slightly in single rare earth ion (and yttrium) orthoferrites.

Orthoferrites of composition $R_x Sm_{1-x} FeO_3$ were developed to provide materials with greater storage densities. In these materials the bubble diameter d is reduced by operating on the uniaxial anisotropy constant K_u. Earlier it was stated that $SmFeO_3$ is the lone exception to the observation that the easy axis of magnetization aligns along the [001] axis. It was reasoned by Sherwood *et al.* (1967) that a partial substitution of Sm in other rare earth orthoferrites should result in a reduced anisotropy constant K_u and this was indeed the case. It is

Table 13.2 Experimental and calculated material parameters for single element and mixed orthoferrites

Rare earth	$4\pi M_s$	M_s	Experimental			Calculated	
			$2r$ (mils)	Field (Oe)	Thickness (mils)	πl (mils)	σ_w (erg/cm²)
Y	105	8·4	3·0	33	3·0	2·5	1·8
La	83	6·6			not available		
Pr	71	5·7			not available		
Nd	62	4·9	7·5	3·2	2·0	4·4	1·1
Sm	84	6·7	6·0	3·0	1·1	2·9	1·3
Eu	83	6·6	5·5	10·5	2·0	3·7	1·6
Gd	94	7·5	3·7	16	2·4	2·9	1·7
Tb	137	10·9	1·7	51	2·2	1·4	1·7
Dy	128	10·2	2·0	32	1·6	1·7	1·8
Ho	91	7·3	4·5	12	2·1	3·3	1·7
Er	81	6·5	6·0	8	2·0	3·9	1·6
Tm	140	11·2	2·3	37	2·3	1·9	2·4
Yb	143	11·4	3·8	41	3·0	3·0	3·9
Lu	119	9·5	7·5	10·5	2·0	4·3	3·9
$Sm_{0.6}Er_{0.4}$	83	6·6	1·0	33	1·8	0·80	0·35
$Sm_{0.55}Tb_{0.45}$	108	8·6	0·75	61	2·0	0·40	0·30

seen that at room temperature the composition $Sm_{0.55}Tb_{0.45}FeO_3$ has a wall energy of 0·3 ergs/cm² (reduced from 1·7 ergs/cm² for $TbFeO_3$) and $l = 2·5$ μm. However, the magnetic parameters of this material are rather sensitive to temperature changes.

Umebayashi and Ishikawa (1965) and Rossol have reported domain wall mobilities of 200–5000 cm/s Oe for the orthoferrites. In spite of these rather high mobility figures, higher than those of any other reported bubble material, the orthoferrites have fallen from favour. As bubble materials they have been severely handicapped by their rather large bubble diameters. Furthermore, present day single crystal techniques have been unable to produce platelets or epitaxial films which hold more than 10^3 bubble domains.

13.6.2 Uniaxial flux grown garnets

The formula for rare earth iron garnets can be written as $A_3Fe_5O_{12}$, where A can be Y and at least in part La, Bi or rare earth metal ions. The A ions occupy the dodecahedral sites, the iron ions the tetrahedral and octahedral sites. It is well known that the dodecahedral sites can also be filled with combinations of rare earth ions and that the tetrahedral and octahedral Fe^{3+} sites can be partially substituted with Al^{+3} and Ga^{+3}. These substitutions allow adjustment of the magnetization, the temperature coefficient of the magnetization, the magnetostriction and, as explained later, the uniaxial behaviour in flux grown crystals.

This section shall be most concerned with those rare earth ion combinations in the dodecahedral sites which minimize either one or both of the λ_{111} or λ_{100} magnetostrictions. In addition to minimizing the effect of strains such as those introduced during crystal growth or processing, the possibility exists that the rare earth ions will either short-range or long-range order during the growing process. Néel, Taniguchi and Yamamoto, and Chikazumi and Oomura independently explained induced uniaxial anisotropy in permalloy on the basis of a pseudo-dipolar interaction between non-random pairs (Chikazumi, 1964). Rosencwaig and Tabor (1971) have applied a similar analysis to explain the induced uniaxial anisotropy in the flux grown garnets. Their model predicts that as little as a 1 % ordering of rare earth sites can account for the observed anisotropy.

Room temperature garnet data of magnetostriction from Iida (1967) and magnetization, lattice constant, compensation temperature and crystalline anisotropy constant from von Aulock (1965) are presented in Table 13.3. The

Table 13.3

	λ_{111} $(\times 10^{-6})$	λ_{110} $(\times 10^{-6})$	λ_{100} $(\times 10^{-6})$	$4\pi M_s$ (gauss)	Lattice constant (Å)	Compensation temperature (K)
Sm	−8.5	−1.1	+21.0	1675	12.53	—
Eu	+1.8	+6.6	+21.0	1172	12.52	—
Gd	−3.1	−2.3	0.0	56	12.47	286
Tb	+12.0	+8.2	−3.3	198	12.43	248
Dy	−5.9	−7.6	−12.5	378	12.39	225
Ho	−4.0	−3.9	−3.4	882	12.37	140
Y	−2.4	−2.2	−1.4	1767	12.36	—
Er	−4.9	−4.2	+2.0	1241	12.35	30
Tm	−5.2	−4.7	+1.4	1397	12.33	—
Yb	−4.5	−4.0	+1.4	1555	12.30	—
Lu	−2.4	−2.2	−1.4	1815	12.27	—

magnetostriction constant λ_{110}, which is related to λ_{100} and λ_{111} by the equation $\lambda_{110} = \frac{3}{4}\lambda_{111} + \frac{1}{4}\lambda_{100}$, is also tabulated for convenience. Under the assumption that the 'λ_{111}'s and 'λ_{100}'s of a garnet such as $Tb_1Tm_2Fe_5O_{12}$ are determined by linear weightings of the respective magnetostrictive constants, it is easily seen that the bulk magnetostriction of this particular garnet is substantially zero. The above example of a binary garnet is a special case and in general three rare earth ions are needed. Eu or Tb must be one of the rare earths selected to make λ_{111} zero.

Essentially all of the information on flux grown uniaxial garnets has originated at Bell Telephone Laboratories (Bobeck et al., 1970). Garnets grown by van Uitert et al. (1970) were evaluated for bubble domain applicability. These included single rare earth ion garnets as well as low magnetostriction combinations. Platelets of {111} plane orientation were prepared, this orientation being chosen since K_1, the first-order cubic crystalline anisotropy constant is negative ($\sim -10^4$ ergs/cm^3) making the $\langle 111 \rangle$ axes supposedly equivalent easy axis of magnetization. Often regions within these {111} platelets exhibited sufficient uniaxial anisotropy to support bubble domains. Platelets cut from zero magnetostriction garnets exhibited the largest and most uniform uniaxial regions. Barns (1971) has reported that the cubic cell of one of these garnets, $Gd_{2.34}Tb_{0.66}Fe_5O_{12}$, is slightly distorted. Furthermore the uniaxial anisotropy persists until annealing temperatures of 1200 °C are reached (F. B. Hagedorn, private communication).

LeCraw *et al.* (1971) discovered that the growth induced uniaxial anisotropy in sectors beneath (110) facets was sufficiently large in $Eu_2Er_1Ga_{0.7}Fe_{4.3}O_{12}$ garnet to result in a $\langle 100 \rangle$ easy axis. Examples of materials with $\langle 110 \rangle$ easy axes have also been found.

The growth induced uniaxial anisotropy is intimately related to the morphology of garnet crystals. Growth habits of flux grown crystals produce {211} and {110} facets as illustrated in Figure 13.19 with the $\langle 111 \rangle$ axes the fast

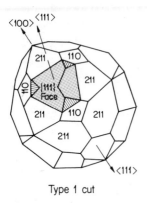

Type 1 cut

Figure 13.19 Flux grown garnet crystals display {211} and {110} facets. A slice has been taken perpendicular to the $\langle 111 \rangle$ axis most nearly normal to a set of three natural {211} faces. This is defined as a Type 1 cut.

growth directions. It was observed in specific garnet formulations that the sectors directly beneath {211} and {110} facets extending well toward the centre of the crystals, when sliced for example as shown in Figure 13.19 produce magnetically uniaxial platelets. However, each {211} and {110} sector had to be wafered in a specific manner.

Six types of cuts have been defined. Type 1, illustrated in Figure 13.20(a), requires slicing perpendicular to the $\langle 111 \rangle$ axis which is nearest to normal to the {211} facet under consideration. Slicing perpendicular to the unique $\langle 111 \rangle$ axis contained in a {211} facet is defined as a Type 2 cut and is illustrated in Figure 13.20(b). Cuts 3, 4 and 5, by symmetry considerations, are the only three required for a {110} facet. The Type 6 cut, $\langle 110 \rangle$ easy axis contained in the {211} facet, completes the definitions. In the case of Types 1 and 2 it should be

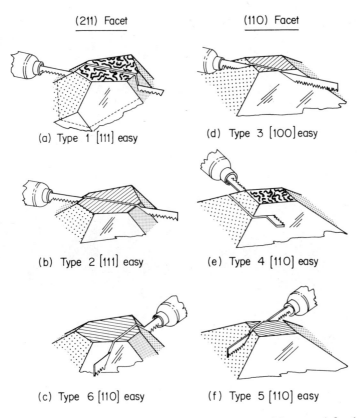

(211) Facet

(110) Facet

(a) Type 1 [111] easy

(d) Type 3 [100] easy

(b) Type 2 [111] easy

(e) Type 4 [110] easy

(c) Type 6 [110] easy

(f) Type 5 [110] easy

Figure 13.20 Six types of cuts, three for each natural facet are defined. Most device work has been done with platelets produced by Type 1 and Type 2 cuts.

noted that the induced uniaxial anisotropy axis is often as much as 20° away from the indicated $\langle 111 \rangle$ axis (D. H. Smith, private communication).

Photographs of domain patterns characteristic of Type 1 garnet $Gd_{0.94}$-$Tb_{0.75}Er_{1.31}Al_{0.5}Fe_{4.5}O_{12}$ are given in Figure 13.21. The uniaxial domain behaviour evident was photographed in a platelet sliced from material directly beneath a {211} facet of a flux grown crystal. Garnet compositions from which other uniaxial platelets were processed are listed in Table 13.4 where a representative set of six garnets and their properties have been tabulated. From measurements of bubble domain collapse (typical data presented in table) and bubble runout the material length l, magnetization M_s and wall energy σ_w were determined. K_u and H_k were calculated on the assumption that the exchange constant $A = 1.0 \times 10^{-7}$ ergs/cm. The mobility was determined using the bubble collapse method of Section 13.4.2.

Note that the domain mobilities, though lower than those of the orthoferrites, are high enough to be useful because of the $1.5–10$ μm diameter bubble domains

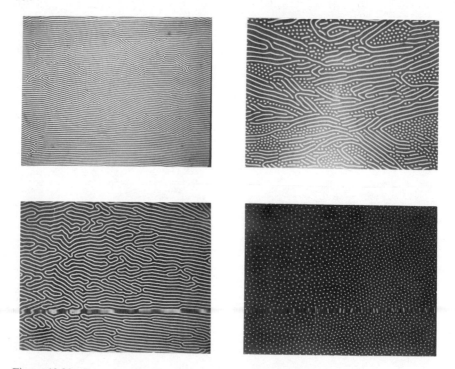

Figure 13.21 Domain patterns in a 2 mil thick $\{111\}$ platelet of $Ge_{0.94}Tb_{0.75}Er_{1.31}$-$Al_{0.5}Fe_{4.5}O_{12}$ garnet (upper left). Bias field is zero (lower left). Bias field of 50 Oe applied normal to the platelet surface (upper right). Same except strip domains have been 'cut' with a magnetic wire probe (lower right). Bias field is 150 Oe. The photographs represent a portion of the crystal 50×60 mils. The bubbles are $5\,\mu m$ in diameter.

found in these materials. Hagedorn *et al.* (1961) and others have reported mobilities greater than 10^3 for YIG. However, mobility measurements of YIG + Ga with sufficient, presumably stress induced, uniaxial anisotropy to support bubble domains gave a mobility of 385 cm/s Oe (Bobeck, unpublished). It can be reasoned that the mobility is lower in this case because the added domain wall energy has reduced the domain wall width.

Referring once again to the garnets of Table 13.4 the following observations can be made. Garnet S2 can be grown by the flux method with very uniform magnetic properties. Because of a compensation point just below room temperature its bubble parameters are very temperature sensitive. The other garnets are quite temperature insensitive; however, they contain either Ga or Al and grow with severe compositional gradients. Thus the bubble parameters are not consistent over the surface of a platelet.

The following model has proved useful for arriving at suitable crystal formulations. Anticipated uniaxial behaviour of garnet $A_xB_{3-x}Fe_5O_{12}$ for sectors beneath $\{211\}$ facets takes into consideration the sign and amplitude

Table 13.4

	Type cut	h (μm)	H_A (Oe)	$2r$ (μm)	πl (μm)	$4\pi M_s$ (gauss)	M_s	σ_w (ergs/cm^2)	K_u (ergs/cm^3)	H_K (Oe)	μ (cm/s/Oe)
S1 $Er_2Tb_1Al_{1\cdot1}Fe_{3\cdot9}O_{12}$	1	17	82	7.0	4.0	136	10.8	0.19	$2{\cdot}2 \times 10^4$	3800	55
S2 $Gd_{2\cdot34}Tb_{0\cdot66}Fe_5O_{12}$	2	15	75	7.5	4.8	137	10.9	0.23	$3{\cdot}3 \times 10^4$	6100	120
S3 $Gd_{0\cdot95}Tb_{0\cdot73}Er_{1\cdot3}Al_{0\cdot5}Fe_{4\cdot5}O_{12}$	1	11.5	140	3.0	1.1	181	14.4	0.083	$4{\cdot}4 \times 10^3$	620	60
S4a $Eu_2Er_1Ga_{0\cdot7}Fe_{4\cdot3}O_{12}$	1	18.0	182	5.0	2.0	247	19.7	0.31	$6{\cdot}2 \times 10^4$	6200	—
S4b $Eu_2Er_1Ga_{0\cdot7}Fe_{4\cdot3}O_{12}$	3	17.1	145	5.5	2.3	196	15.4	0.22	$3{\cdot}0 \times 10^4$	3800	165
S5 $Y_2Gd_1Al_{0\cdot8}Fe_{4\cdot2}O_{12}$	2	13.3	262	2.5	0.66	328	26	0.18	$2{\cdot}0 \times 10^4$	1500	180
S6 $Y_{1\cdot8}Eu_{0\cdot2}Gd_{0\cdot5}Tb_{0\cdot5}Al_{0\cdot6}Fe_{4\cdot4}O_{12}$	3	19.0	370	3.0	0.70	450	36	0.36	$7{\cdot}9 \times 10^4$	4400	186

of the magnetostriction constants λ_{111}-A and λ_{111}-B and the lattice constants $\AA_A$ and $\AA_B$ of the end member $A_3Fe_5O_{12}$ and $B_3Fe_5O_{12}$. A Type 1 garnet will result if $\AA_A > \AA_B$ and λ_{111}-A is positive and λ_{111}-B is negative. If $\AA_A > \AA_B$ and λ_{111}-A is negative and λ_{111}-B is positive, then the garnet will be of Type 2. No similar model has successfully predicted the characteristics of $\{110\}$ sectors.

13.6.3 Comparisons of uniaxial magnetic materials

The selection of materials for bubble devices must consider both the domain wall mobility and domain diameter (Bobeck, 1970). In Figure 13.22 we position

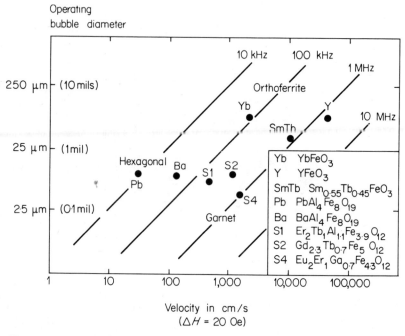

Figure 13.22 Uniaxial bubble materials are compared for data rate and storage density. For a fixed overdrive the orthoferrites attain the highest domain wall velocities but have the largest bubble diameters. The uniaxial garnets would seem to be the best overall material.

orthoferrite, garnet and hexagonal ferrite compositions on an operating bubble diameter versus domain wall velocity plot. Wall velocity rather than mobility is chosen as the abscissa so that we might conveniently include the hexagonal ferrites. Bubble domains respond to gradient fields and for Figure 13.22 a field difference $\Delta H = 20$ Oe is chosen since such a field is representative of that achieved in bubble propagation circuits.

Minimum usable separation between bubbles in, for example a bubble stream shift register, is limited by interaction. Experience has shown that a centre to centre separation of four bubble diameters is a practical compromise. Since this fixes the distance a bubble must travel we can also add lines of constant data rate. It is apparent from this figure that the orthoferrites and garnets allow comparable data rates and that devices constructed with hexagonal ferrites would be considerably slower. The garnets, with storage densities in excess of 10^5 bits/cm^2 seem to be the best overall bubble material.

13.7 BUBBLE DOMAIN PROPAGATION

Bubble domain propagation in orthoferrites has been reported by Bobeck (1970a), Perneski (1969), Bonyhard *et al.* (1970) and Copeland *et al.* (1971) and in garnets by Bobeck *et al.* (1971) and Danylchuk (1971). The magnetic field generated by current passed through a conductor loop positioned on the surface of an orthoferrite platelet near a bubble domain will produce a force on the domain (see Figure 13.23). With the current polarity chosen so that the

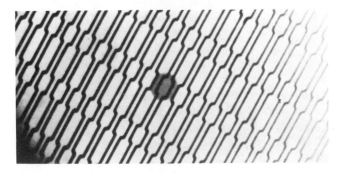

Figure 13.23 Bubbles in YFeO$_3$ can be moved in a shift register mode by a photolithographically printed circuit conductor array. Bubbles, 125 µm in diameter, are stepped at a 3 megahertz data rate by 0·5 ampere current pulses.

bias field within the loop is reduced the bubble will move, more or less retaining its original shape and size, under the conductor loop. A one-dimensional shift register is built by arranging a succession of such loops and energizing them in turn. Motion is extended on two dimensions by adding conductor loops in a second direction. In YFeO$_3$ bubble domains 0·1 mm in diameter have been moved at the rate of 10^7 steps per second. These data have been analysed to yield a domain wall mobility for YFeO$_3$ of 5000 cm/s/Oe, in agreement with other techniques.

In a second method of propagation the diameter of a bubble domain is enlarged and reduced by increasing or decreasing the bias field. Motion is achieved by manoeuvring this pulsating cylindrical domain in and out of asymmetrical magnetostatic energy traps. The circuit is illustrated in Figure 13.24. In so doing the expanding domain 'pushes' from its trailing edge while a

Figure 13.24 Pulsating domain in $LuFeO_3$ interacts with permalloy wedges and rings to produce unidirectional inchworm motion (counterclockwise). Permalloy is 0·75 μm thick and each wedge has a 100 μm wide base.

contracting domain 'pulls' on its leading edge. Motion at the rate of 2×10^4 steps per second in $Sm_{0.55}Tb_{0.45}FeO_3$ orthoferrite has been measured (W. Strauss and F. J. Ciak, private communication).

Motion of a bubble domain around a permalloy disc was described in Section 13.4.2. When a structured permalloy pattern is substituted for the disc, a field rotating in plane and acting on the structured pattern generates travelling positive and negative magnetic poles to selectively repel and attract, and thereby control the motion of a bubble. The structured pattern can take the form of the T–Bar shown in Figure 13.25. Bubble domains in $Sm_{0.55}Tb_{0.45}FeO_3$ ortho-ferrite have been moved at the rate of 5×10^5 periods per second. Danylchuk has reported 10^3 step Y–Bar shift registers using $Gd_{2.34}Tb_{0.66}Fe_5O_{12}$ garnet (see Table 13.4).

Next we consider some details of domain propagation in permalloy circuits of the T–Bar and Y–Bar type. The role of the permalloy patterns is to distort rotating in-plane magnetic fields thereby producing field components normal to the platelet surface which are effective on the bubble domain. An in-plane

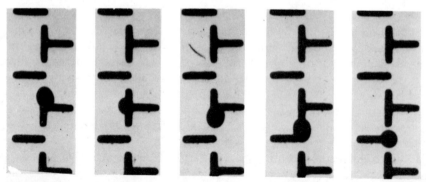

Figure 13.25 Cylindrical domains attracted to poles generated in a permalloy pattern by an in-plane rotating field. Bubble in $Gd_{2.34}Tb_{0.66}Fe_5O_{12}$ garnet moves one period of the T and Bar shaped pattern (20 μm) for each clockwise rotation of the field.

field will have a negligible direct effect on a domain if the anisotropy field of the uniaxial materials is much greater than the in-plane field, as is usually the case.

Consider Figure 13.26 where a permalloy 'T' is illustrated in contact with a uniaxial magnetic platelet of thickness h. An in-plane field, directed as shown, substantially saturates the upper part of the T. A bubble domain assumed magnetized downward experiences a force driving it to the right because of an interaction between the permalloy dipole of magnetic moment $2wtM_s'l_{12}$ and bubble domain dipole of magnetic moment $2\pi r_0^2 M_s h$ where w and t are the width and thickness of the permalloy strip and r_0 is the radius of the domain. A factor of two enters into the expression for the bubble domain dipole since we are considering a reversal volume of magnetic charge in an otherwise uniformly saturated platelet.

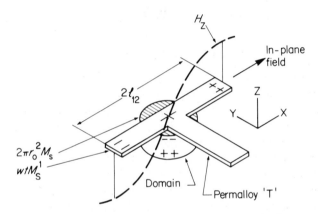

Figure 13.26 A bubble domain of diameter $2r_0$ is shown centred beneath a permalloy 'T'. The domain moves along the x axis in response to ΔH_z, the field differential across the domain.

In particular we desire the field differential across the domain $\Delta H_z = 2r_0 \partial H_z/\partial x$ since, with the relations $H_c = 8\,\Delta H/\pi$ and $v = \mu\,\Delta H/2$, we can predict details of the domain motion. Equation (13.5) which is based on interaction of a pair of cylindrical domains can, in fact, be reworked to yield

$$\Delta H_z = \frac{6wtM_s'hr_0}{l_{12}^4} \tag{13.7}$$

It is interesting to note that ΔH_z and therefore the bubble velocity $v = \mu[\Delta H_z - 8H_c/\pi]/2$ remains constant if the configuration of Figure 13.26 is scaled in dimension. However, the data rate ($\sim v/r_0$) increases if the size is decreased. Thus we see the double benefit of working at the smaller circuit sizes—increased storage density and data rate. Likewise the demagnetizing factor and therefore the in-plane field needed to establish the poles in the permalloy also remain constant.

We next calculate the data rate in a 25 μm period T–Bar circuit under the assumption that the bubble velocity calculated by (13.7) will be maintained throughout the circuit. Referring to Figures 13.25 and 13.26 the permalloy strip widths 'w' and the interpermalloy spacings are usually $0.4l_{12}$, thus the T–Bar period is $2l_{12} + 0.4l_{12} + 0.4l_{12} + 0.4l_{12} = 3.2l_{12}$. The permalloy thickness $t = w/2$ and the permalloy magnetic moment $M'_s = 600$.

The size of the bubble domain must, of course, match the dimensions of the permalloy circuit. Typically $r_0 = l_{12}/3$ and the T–Bar period is therefore 4.8 domain diameters in 'length'. For optimum platelet thickness $h = \pi r_0/4$. Substitution of all these factors in (13.7) gives $\Delta H_z = 25.1$ Oe. For $\mu = 100$ cm/s/ Oe, typical for the garnets, the data rate $\mu \Delta H_z/3.2l_{12} = 502$ kHz. This simplified analysis of T–Bar behaviour gives rather good agreement with data rates determined by experiment.

13.8 EPITAXIAL BUBBLE MATERIALS

Epitaxial garnet films (Mee *et al.*, 1971; Giess *et al.*, 1971) are currently the most favoured materials for bubble devices. Knowledge gained from the study of bulk garnet crystals has led to the preparation by liquid phase epitaxy (LPE) of epitaxial films with growth induced uniaxial anisotropy (see Chapter 2). Epitaxial garnet films with stress induced uniaxial anisotropy are also grown by LPE and chemical vapour deposition (CVD).

A tentative list of specifications for bubble materials is given in Table 13.5 (Bobeck *et al.*, 1971). Specifically this list applies to the characterization of epitaxial garnet films for use in bubble circuits with a periodicity of 25 μm that operate at a 10^6 bits/s data rate.

Table 13.5

Bubble material specifications—field access		
D	6 μm	Bubble diameter
h	6 μm $\pm$ 1%	Thickness
$4\pi M_s$	150 Gauss $\pm$ 1%	Magnetization
H_k	$> 1.5 . (4\pi M_s)$	Anisotropy field
μ	> 500 cm/s/Oe	Mobility
H_c	< 0.3 Oe	Coercivity
xs	$< 5/\text{cm}^2$	Defect density

The specification on the bubble diameter D is, of course, dictated by the circuit periodicity λ. Requirements of speed and detection on one hand and bubble–bubble interaction on the other hand lead to the compromise choice of $D = 6$ μm. This represents a departure from the so called ideal where the thickness is selected to minimize the bubble diameter. In garnets the bubble diameter ($D \sim M_s^{-2}$) is readily controlled and a specific diameter such as 6 μm easily achieved. The thickness is increased to increase bubble fringing fields and thereby enhance bubble detector output signals (Strauss and Smith, 1970).

The tolerance imposed on M_s is as important as the absolute value of M_s. With M_s held to $\pm 1\%$ the bias field range of $\pm 17\%$ for a Thiele ideal bubble is reduced to $\pm 15\%$. The absolute value of the magnetization not only relates to D and h but to the uniaxial anisotropy as well. DeBonte (1971) has shown that the range of bubble stability is severely reduced if the anisotropy field H_k is less than $6\pi M_s$. Another consideration in the H_k to M_s ratio is the desirability that the bubble parameters be reasonably insensitive to in-plane magnetic fields.

Energy losses in an operating bubble circuit are minimum if the domain wall coercivity is small and the mobility is large. It is doubtful whether a data rate of 10^6 bits/s can be achieved with a mobility less than 500 cm/s/Oe.

Epitaxial films with defect densities of 5/cm² and lower have been reported (Geusic et al., 1971). This implies that the yield of defect-free diced garnet chips, 4 mm on a side, will be approximately 50%. Y–Bar shift registers of 10,000 bit lengths have been operated on $Er_2Eu_1Ga_{0.7}Fe_{4.3}O_{12}$ garnet films. Details of these films are reported by Shick et al. (1971), Levinstein et al. (1971) and Geusic et al. (1971).

Table 13.6 lists the parameters of epitaxial films deposited on (111) $Gd_3Ga_5O_{12}$ substrates by LPE (Nielsen, 1971). These films encompass both growth and stress induced uniaxial anisotropies.

13.9 MASS MEMORY BUBBLE OPERATIONS

Bubble device development has been directed toward implementation of a mass memory (bubble file). An organization being pursued in the design of a bubble file is a major–minor loop configuration (Bonyhard et al., 1970). The basic bubble operations required to implement this design are bubble propagation, data input, detection and bubble transfer.

T–Bar, Y–Bar and chevron are under consideration as propagation elements. The chevron propagation channel consists of an array of closely spaced Λs as illustrated in Figure 13.27. The pattern of Λs operate much as T–Bar and Y–Bar circuits. Clockwise rotation of an in-plane field causes left to right movement of bubbles as they respond to the travelling poles generated in the permalloy pattern by the in-plane field. This pattern differs from the T–Bar and Y–Bar in that successive rows of elements can be very closely spaced as in Figure 13.28. Chevron circuits with multiple elements per column propagate island strip domains as well as bubbles. In Figure 13.29, 500 kbit/s and 1·0 Mbit/s data are presented for a two element chevron circuit etched from 4000 Å permalloy used in conjunction with the YGdYb epitaxial garnet of Table 13.6.

For insertion of binary data into bubble devices two alternatives exist. Either bubble domains are selectively generated (Bobeck, 1971) or continuously generated and diverted into a propagation channel as required. In either case an island domain, maintained on a large disc or rectangle of permalloy, is

Table 13.6

	$4\pi M_s$ (gauss)	h (μm)	l (μm)	σ_w (erg/cm^2)	Strip width zero-bias (μm)	H_{col} (Oe)	H_{S-B} (Oe)	μ (cm/s/Oe)
(1) $Er_2Eu_1Ga_{0.7}Fe_{4.3}O_{12}$	295	5.6	0.5	0.35	5.0	170	130	45
(2) $Er_{1.99}Gd_{1.01}Ga_{0.22}Fe_{4.78}O_{12}$	148	5.0	0.9	0.173	7.0	60	45	89
(3) $Y_{0.94}Gd_{1.07}Yb_{0.57}La_{0.42}Al_{0.7}Fe_{4.3}O_{12}$	240	7.3	0.46	0.21	5.1	130	110	275
(4) $Y_{1.03}Gd_{1.29}Yb_{0.68}Al_{0.7}Fe_{4.3}O_{12}$	175	2.1	0.51	0.125	4.0	70	45	2000

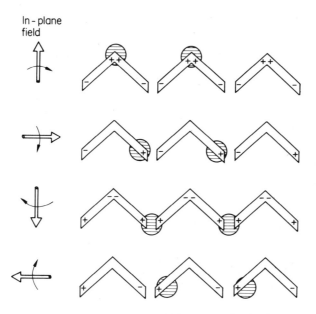

Figure 13.27 Step by step bubble propagation in chevron circuit caused by clockwise rotation of in-plane field.

subdivided to supply a train of bubbles. The process of subdivision can, of course, be continued indefinitely.

The stray field of bubble domains has been detected by means of Hall detectors (Strauss and Smith, 1970), magnetoresistance detectors (Almasi *et al.*, 1971) and induced voltage pickup. Optical detectors utilizing Faraday rotation have also been proposed. Magnetoresistance detectors seem the most promising

Figure 13.28 Closely packed chevron elements provide propagation in horizontal direction with lateral flexibility for multiple bubble or domain strip propagation.

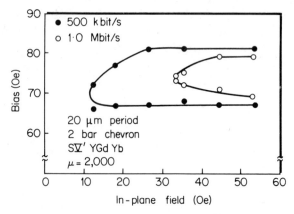

Figure 13.29   Operating ranges at 500 kbits/s and 1·0 Mbits/s using $Y_{1.03}Gd_{1.29}Yb_{0.60}Al_{0.7}Fe_{4.3}O_{12}$ garnet with a 2-Bar chevron circuit. The 11 Oe threshold at 500 kbits/s is only several oersteds higher than the quasistatic value.

as they can provide signals of millivolt level in a structure whose processing is completely compatible with other portions of bubble devices.

Magnetoresistance detectors use shaped permalloy structures which change in resistance in response to stray bubble fields. A change in resistance of 3% can be expected when the internal magnetization is rotated from parallel to transverse the path of a bias current. Energy from a detector scales with the volume of the domain being sensed thus improved detection results when detectors are located near domain stretchers. In Figure 13.30 a magnetoresistance detector design compatible with a 'bubble stretching' 3-Bar section is illustrated. W. Strauss, P. W. Shumate, Jr. and F. J. Ciak (Strauss and Smith, 1970) reported

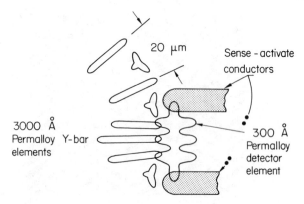

Figure 13.30   A 'Chinese letter' magnetoresistance permalloy detector located on a 3-Bar bubble stretcher. Combination has been inserted into a 20 μm period Y-Bar shift register.

quasistatic and 100 kHz detection of nominally 6 μm diameter bubbles using this circuit. They reported 350 μV signals when the detector was biased by 1 mA direct current.

Bubble transfer pertains to the transfer of bubble domains between normally independent propagation paths (Bobeck and Scovil, 1971). Transfer at a 100 kbit/s rate using current pulses to activate a conductor-chevron configuration has been reported by Bobeck (1972). A transfer coupling junction that operates by a temporary change in the direction of rotation of the in-plane field has been proposed by Bonyhard et al. (1970).

REFERENCES

Almasi, G. S., G. E. Keefe, Y. S. Lin and D. A. Thompson, 1971, *J. Appl. Phys.*, **42**, 1268.

Asti, G., M. Colombo, M. Guidici and A. Levialdi, 1965, *J. Appl. Phys.*, **36**, 3581.

Barns, R. L., 1971, *J. Appl. Phys.*, **42**, 1623.

Blank, S., L. K. Shick and J. W. Nielsen, 1971, *J. Appl. Phys.*, **42**, 1556.

Bobeck, A. H., 1967, *Bell. Sys. Tech. J.*, **46**, 1901.

Bobeck, A. H., 1970, *I.E.E.E. Trans. Mag.*, **MAG-6**, no. 3, 445.

Bobeck, A. H., unpublished work.

Bobeck, A. H. and H. E. D. Scovil, 1971, *Scientific American*, **78**.

Bobeck, A. H., R. F. Fischer, A. J. Perneski, J. P. Remeika and L. G. van Uitert, 1969, *I.E.E.E. Trans. Mag.*, **MAG-5**, 544.

Bobeck, A. H., E. G. Spencer, L. G. van Uitert, S. C. Abrahams, R. L. Barns, W. H. Grodkiewicz, R. C. Sherwood, P. H. Schmidt, D. H. Smith and E. M. Walters, 1970, *Appl. Phys. Letters*, **17**, 131.

Bobeck, A. H., I. Danylchuk, J. P. Remeika, L. G. van Uitert and E. M. Walters, 1971a, *Proc. Int. Conf. on Ferrites, July 1970*, Japan, 361.

Bobeck, A. H., R. F. Fischer and J. L. Smith, 1971b, *AIP Conf. Proc.*, no. 5, 45.

Bonyhard, P. I., I. Danylchuk, D. E. Kish and J. L. Smith, 1970, *I.E.E.E. Trans. Mag.*, **MAG-6**, no. 3, 447.

Charap, S. H. and J. M. Nemchik, 1969, *I.E.E.E. Trans. Mag.*, **MAG-5**, 566.

Chikazumi, S., 1964, *Physics of Magnetism* (Wiley: New York), p. 359.

Copeland, J. A., J. P. Elward, W. A. Johnson and J. G. Ruch, 1971, *J. Appl. Phys.*, **42**, 1266.

Craik, D. J. and P. V. Cooper, 1972, *Physics Letters*, **41A**, no. 3, 255.

Danylchuk, I., 1971, *Appl. Phys.*, **42**, 1358.

DeBonte, W. J., 1971, *AIP Conf. Proc.*, no. 5, 140.

Gianola, U. F., D. H. Smith, A. A. Thiele and L. G. van Uitert, 1969, *I.E.E.E. Trans. Mag.*, **MAG-5**, 558.

Giess, E. A., B. A. Calhoun, E. Clockholm, T. R. McGuire and L. L. Rosier, 1971, *Mat. Res. Bull.*, **6**, 317.

Geusic, J. E., H. J. Levinstein, S. J. Licht, L. K. Shick and C. D. Brandle, 1971, *Appl. Phys. Letters*, **19**, 93.

Hagedorn, F. B. and E. M. Gyorgy, 1961, *J. Appl. Phys. Suppl.*, **32**, 2825.

Iida, S., 1967, *J. Phys. Soc. Japan*, **22**, 1201.

Kozlowski, G. and W. Zietek, 1966, *A.C.T.A. Physica Polonica*, **29**, Fasc. 3.

Kooy, C. and U. Enz, 1960, *Philips Res. Rep.*, **15**, 7.

Kurtzig, A. J. and H. J. Guggenheim, 1970, *Appl. Phys. Letters*, **16**, 43.

LeCraw, R. C., R. Wolfe, A. H. Bobeck, R. D. Pierce and L. G. van Uitert, 1971, *J. Appl. Phys.*, **42**, 1641.

Levinstein, H. J., S. J. Licht, R. W. Landorf and S. L. Blank, 1971, *Appl. Phys. Letters*, **19**, 486.

Liu, T. W., A. H. Bobeck, E. A. Nesbitt, R. C. Sherwood and D. D. Bacon, 1971, *J. Appl. Phys.*, **42**, 1360.

Mee, J. E., G. R. Pulliam, D. M. Heinz, J. J. Owens and P. J. Besser, 1971, *Appl. Phys. Letters*, **18**, 60.

Nielsen, J. W., 1971, *AIP Conf. Proc.*, no. 5, 56.

Okada, T., K. Matsumi and H. Makimo, 1971, *Proc. Int. Conf. on Ferrites, July 1970*, Japan, 372.

Perneski, A. J., 1969, *I.E.E.E. Trans. Mag.*, **MAG-5**, 554.

Rado, G. T. and H. Suhl, 1963, *Magnetism*, Vol. III (Academic Press: New York), p. 106.

Remeika, J. P., 1956, *J. Am. Chem. Soc.*, **78**, 4259.

Rosencwaig, A. and W. J. Tabor, 1971, *J. Appl. Phys.*, **42**, 1643.

Rossol, F. C., 1969, *J. Appl. Phys.*, **40**, 1082.

Rossol, F. C., 1970, *Phys. Rev. Letters*, **24**, 1201.

Rossol, F. C. and A. A. Thiele, 1970, *J. Appl. Phys.*, **41**, 1163.

Sherwood, R. C., L. G. van Uitert, R. Wolfe and R. C. LeCraw, 1967, *Phys. Letters*, **25A**, 297.

Shick, L. K., J. W. Nielsen, A. H. Bobeck, A. J. Kurtzig, P. C. Michaelis and J. P. Reekstin, 1971, *Appl. Phys. Letters*, **18**, 90.

Strauss, W. and G. E. Smith, 1970, *J. Appl. Phys.*, **41**, 1169.

Thiele, A. A., 1969, *Bell Sys. Tech. J.*, **48**, 3287.

Treves, D., 1962, *Phys. Rev.*, **125**, 1843.

Umebayashi, H. and Y. Ishikawa, 1965, *J. Phys. Soc. Japan*, **20**, 2193.

Van Uitert, L. G., W. A. Bonner, W. H. Grodkiewicz, L. Peitrocki and G. J. Zydzik, 1970, *Mat. Res. Bull.*, **5**, 825.

Von Aulock, W. H., 1965, *Handbook of Microwave Ferrite Materials* (Academic Press: New York).

Williams, H. J., R. C. Sherwood, F. G. Foster and E. M. Kelly, 1957, *J. Appl. Phys.*, **28**, 1181.

Wolfe, R., A. J. Kurtzig and R. C. LeCraw, 1970, *J. Appl. Phys.*, **41**, 1218.

Author Index

782

784

786

788

Subject Index